P9-DTU-482

The Nature of Chaos

The Nature of Chaos

Edited by

TOM MULLIN
Department of Atmospheric, Oceanic and Planetary Physics
University of Oxford
and
Fellow of Linacre College

CLARENDON PRESS · OXFORD
1993

Oxford University Press, Walton Street, Oxford OX2 6DP
Oxford New York Toronto
Delhi Bombay Calcutta Madras Karachi
Kuala Lumpur Singapore Hong Kong Tokyo
Nairobi Dar es Salaam Cape Town
Melbourne Auckland Madrid
and associated companies in
Berlin Ibadan

Oxford is a trade mark of Oxford University Press

Published in the United States
by Oxford University Press Inc., New York

A catalogue record for this book is available from the British Library

Library of Congress Cataloging in Publication Data
The Nature of chaos / edited by Tom Mullin.
1. Chaotic behavior in systems. I. Mullin, Tom.
Q172.5.C45N388 1993 003'.7–dc20 92–41598
ISBN 0 19 853990 8 (hbk)
ISBN 0 19 853954 1 (pbk)

Typeset by Colset Pte Ltd, Singapore

Printed in Great Britain by
Bookcraft (Bath) Ltd
Midsomer Norton, Avon

Preface

This book arose out of a highly successful interdisciplinary graduate lecture course entitled 'Chaos and Related Nonlinear Phenomena' held in Oxford University in Hilary term 1990. It is not merely an edited version of the lecture notes as each contributor has added extensively to the material presented in the lectures. The remit given to each of the lecturers was to explain their research to an audience of first-year graduate students from all of the disciplines of science, engineering, and mathematics and this level has been maintained in the book. The response to the series was tremendous with lecture theatres filled to capacity and favourable comments from many who attended. I certainly consider myself privileged to have been part of it and I hope we have managed to capture some of the enthusiasm in this volume.

The main emphasis is on the application of modern mathematical ideas concerning nonlinear systems to a wide range of scientific problems in biology, chemistry, engineering, and physics. The book will therefore appeal to any science graduate who is interested in practical issues whether they are about to embark on a career in academic or industrial research. It would also be appropriate as a supplementary text to mathematics undergraduates who are taking one of the now very popular 'nonlinear mathematics' courses.

The book begins with a brief introduction to nonlinear phenomena written by myself. This is then followed by my four chapters on the application of the ideas to physical systems. These begin with an outline of some fundamental ideas using the example of a simple pendulum. Next there is a chapter on the important practical issue of the geometric representation of experimental time series in terms of 'attractors'. Organizing centres for chaos are then explained in the next chapter and this is followed by a description of the application of the above ideas towards an understanding of disordered fluid flows.

The next four chapters were written by David Holton from the notes and other material supplied by Bob May. The first of these is an introduction to the universal features of chaos in one-dimensional maps. The ideas are then extended and applied to particular population models in the next chapter. The topic of identifying deterministic chaos in time series data from biological systems is explained. Finally, the ideas developed in the preceding chapters are applied to models of the spread of disease.

The four individual chapters which complete the book begin with a discussion by Michael Thompson of the application of the ideas of low-dimensional dynamical systems to practical engineering problems and focus on a model of ship capsize. Next, Peter Read gives an informative account of the rich variety of nonlinear phenomena found in the atmosphere and discusses the potential role of chaos in their interpretation. Mark Child presents a theoretical approach to aspects of nonlinearity in three different classical chemical systems, including bifurcation phenomena and the origins of chaos in Hamiltonian systems. Finally, Jon Keating presents a lively account of the interplay between classical and quantum mechanics and addresses the question of whether information about classical chaos can be found using a simple quantum measure.

There are many people who made significant contributions to this volume whose names do not appear on the official list of contributors. Dr George Smith of the Materials Science Department planned and organized the lecture course. The following members of the Nonlinear Systems Group in the Clarendon either read manuscripts, made demonstrations, or produced figures and I am grateful to them all: Oksana Chubykalo, Alan Darbyshire, Jonathan Healey, Vivienne Heinrich, Jonathan Kobine, Keith Long, Francis Madden, Tim Price, and Anne Skeldon. Gideon Tearle and John Ward of the Clarendon Laboratory assisted in the presentation and recording of the lectures. John Hogan gave helpful comments on my contribution. Finally, Zoe typed my manuscript and Sylvia and Graham kept me sane.

Oxford T. M.
April 1993

Contents

Editor's introduction

One of the beliefs of modern science is that if an accurate mathematical description of a physical system can be found then both the potential for a deep understanding of the system's properties exists and predictions of its evolution can be made. These assertions have been shown to be true for a wide variety of phenomena which can range from the motions of planetary bodies to the fundamental constituents of matter. However, it is increasingly becoming realized that these notions are generally not true when dealing with nonlinear phenomena. This is something of a surprise to many. After all, it is common experience that although the fine details of the complete solutions of the full governing equations of any given system are not known, approximate versions of the laws may be used to make robust predictions about its behaviour. Furthermore, these predictions are very often confirmed by experimental observation.

An illustration of this 'simplified' scientific approach is given by the following. If we use the laws of planetary motion as formulated by Newton, then it is possible to predict accurately the orbits of the moon around the earth, for example, when the influence of other planets is ignored. Further, these predictions have been tested over centuries and are found to be robust. They are based on the sound principles which Newton established in his *Principia* over 300 years ago by considering the gravitational interaction of two planetary bodies [1]. In this case there is an analytical solution to the mathematical formulation and the results are borne out by observation.

Now, if a third smaller planet is introduced into Newton's mathematical description of the gravitational interaction of two massive bodies this gives rise to his famous intractable 'three-body' problem. Newton solved various restricted versions of the complete problem but he was unable to find a general solution to it. Two centuries later, Poincaré hinted that the motion of the third smaller planet orbiting in the gravitational field of two massive planets would, in general, be highly complicated [2]. Nowadays, we can readily obtain numerical solutions to the three-body problem on a small computer and it can be shown that the orbit of the third planet is indeed unpredictable in practice.

The above example illustrates that Newton's equations of motion, even whilst describing this relatively simple situation, can give rise to both predictable behaviour and complicated unpredictable motions. The appearance of the complications is not due to a breakdown in the validity of the equations but rather it is just one of their properties. Indeed, as we will see later in the first chapter of this book, we can take the conceptually much simpler system of an excited pendulum and show a systematic progression from one type of behaviour to the other. Thus in order to understand the appearance of the disordered behaviour we do not necessarily need to invent new physical laws. Rather we have to recognize that the presence of nonlinear terms in the governing equations can give rise to qualitatively different types of behaviour, which are not present in the solutions of the linearized versions of the equations. In addition, it is questionable if the intuitive insights learned from a study of the linear counterparts help in understanding these new types of behaviour.

Another example of the difficulties encountered in nonlinear problems is to be found in the field of aerodynamics. Powered controlled flight was first achieved by the Wright brothers nearly 90 years ago and the equations of motion for fluid flow have been known for even longer than that. They are nonlinear equations and we need to use a computer to obtain solutions to them for most realistic flows. Modern computers are at least three orders of magnitude more powerful than the first-generation valve machines and today can handle very complex problems. However, aircraft manufacturers still have to rely on empirical wind tunnel testing of new aircraft designs. In fact, despite the current rapid rate of progress in computer power, it would appear that the computer will not replace expensive wind tunnel testing in the foreseeable future, but will instead be used to give useful supplementary information.

It should not be forgotten that there have been many great scientific breakthroughs achieved using approximate, linear methods for a wide range of problems, and this has perhaps led us to think that all observable phenomena will ultimately yield to this type of approach in the course of time. However, many natural processes across the whole spectrum of science are inherently strongly nonlinear and it is becoming clear that simple adaptation of known methods may not be sufficient to resolve important issues, such as prediction of the weather or climate, for example. Therefore, we perhaps need to develop a new way of dealing with nonlinear processes and this is the subject of much current research. Many of these new research ideas are of course reinventions of older concepts but equally there are fresh outlooks on some classical problems and serious attempts to tackle difficult issues which have been brushed aside in the past.

I.1 Deterministic chaos

Perhaps the most familiar of these new ideas is that simple nonlinear deterministic equations can self-generate irregular outputs. Over the past decade this subject has become known through the technical phrase of 'deterministic chaos' which at first sight seems to be a strange mixture of concepts. If a system is governed by deterministic laws and we have a full description of its current state then future states are completely predictable in principle. Nowadays we can investigate the behaviour of a nonlinear deterministic system directly by representing its equations and the required input data on a computer. Of course this can only be done to some finite precision, but nevertheless calculations can be performed which can be compared with accurate experimental data. Typically one finds good absolute agreement between the two when the behaviour is linear or weakly nonlinear. However, as the nonlinearity increases it is often found that although the computer outputs may be smooth on short time-scales they appear random and unpredictable over longer periods. This is the phenomenon that has become known as deterministic chaos.

Although the ideas had been considered in abstract since the time of Poincaré, the first practical demonstration of the phenomenon is to be found in in the celebrated work of Lorenz [3]. He carried out a numerical study of an extremely crude model of atmospheric convection and found that when the integrations of the equations were started with two slightly different initial conditions very different outcomes were realized. This observation is an example of 'sensitivity to initial conditions' in that a change in the least significant digit of the starting conditions for the calculation will eventually lead to completely different outcomes despite the fact that the computations involve the representation of a deterministic law on a similarily deterministic calculating machine. In other words, if two calculations were started with the same initial conditions to within the accuracy of the machine then there would be no divergence of the outputs, i.e. the unpredictability is not an effect of the accumulation of rounding errors. However, if the same two calculations were initiated with two slightly different inputs then the numerical solutions would diverge from each other. Sensitivity to initial conditions is a fundamental feature of deterministic chaos.

I.2 Attractors

The above discovery is the first recorded observation of the appearance of a 'strange attractor' in the solution space of the equations, but before we consider this we must first discuss the basic concepts of attractors. A simple physical analogy of an attractor can be formed using a ball-bearing inside a car inner tube. First of all we will consider a two-dimensional cross-section of the tube. If we displace the ball to any point up the bottom half of the tube it will always roll back to a uniquely defined point at the bottom when it is released. If we now imagine that the motion of the ball is projected onto a sheet of paper below the tube then we can see that all trajectories of the ball will lead to a single point on the paper which we call the fixed point of the system. What is more, here it is called an attracting fixed point since any trajectory within the section of the tube will lead to this point. It is therefore the attractor for the system in this stationary state.

Now suppose we lay our tube flat on a turntable which is rotating at a constant speed. The ball will now run around the tube at some distance up its side which will depend on the selected speed of the turntable. The projection down onto the paper will now be a circle and we call this new attractor a limit cycle. If we again displace the ball from its attracting orbit then it will quickly return to it. In fact, if the ball is started at any point within the tube then it will eventually end up rotating on the uniquely defined attracting cycle for this system. We can then say that the basin of attraction for the cyclic attractor is the whole of the inner surface of the tube in this case.

If we now add to the turntable a large up and down sinusoidal motion then the ball will orbit around the inside surface of the tube. If the frequency of the up and down motion is not commensurate with that of the rotation then the trajectory of the ball will eventually cover the whole of the inside of the tube. This new attractor is called a torus. Unfortunately there is not a simple way of demonstrating that it is an attractor in the present example but when such objects are found in the solution space of nonlinear equations, points which are given initial conditions from within the basin of attraction of the torus are attracted on to it in a transient phase. The trajectories then continue to wind around the attractor as in the case of the ball in the tube.

We now return to a discussion of the concept of a 'strange attractor' where there are two competing effects present. In one, points which are near each other at some instant in time diverge exponentially fast due to the sensitivity to initial conditions, as we discussed in relation to the Lorenz equations. In the other, the trajectories must remain in a finite region of the solution space if the concept of an attractor is to have any meaning. As will be shown by example in the first chapter of the book, this means that strange attractor behaviour cannot exist in two dimensions or less for differential equations. If we tried to construct a deterministic model whose solutions were confined to a plane then trajectories could diverge from each other but they could never remain in a finite region, for if they tried to they would necessarily intersect each other. Thus a choice of direction would arise at the point of intersection of trajectories and the system would no longer be governed by deterministic rules. Thus solutions of a two-dimensional system cannot both occupy a localized region of the solution space of the equations and exhibit sensitivity to initial conditions. Finally, we remark that chaotic behaviour can be represented by lower-dimensional discrete maps but strange attractor behaviour in equations requires a minimum of three dimensions.

I.3 The geometry of time series

The previous section leads us naturally to consider representing the time series produced by dynamical systems in a geometrical way. This consideration of the geometrical structure of the solutions of the equations of motion of dynamical systems was first suggested by Poincaré but it was not really considered in a practical sense until the appearance of small powerful computers in the last decade or so. Technical details of the methods employed are discussed in Chapter 3 where an extensive list of references to the original work is also given. The present discussion will therefore focus on the generalities of the approach.

Here we demonstrate the value of this geometrical approach to time series analysis by considering the example shown in Fig. I.1a. There we show a time series which was obtained by adding together the outputs of two square wave laboratory oscillators and sampling the signal by computer. Next, we show in Fig. I.1b the power spectrum obtained using the standard laboratory technique of calculating the Fourier spectrum which may be considered as a plot of the amount of power contained in each frequency component of the signal. Thus, if we had added together linearly two sine waves then there would be two peaks in the spectrum, one at each frequency component. The reason that we see so many confusing peaks in the spectrum in the present example and not just the two fundamental ones is that the Fourier representation is a linear process where it is assumed that the time series can be decomposed into a sequence of sines and cosines. Thus, here we are attempting to represent the sharply changing square waves by a series of smooth functions (sines and cosines) and so in principle would need an infinite number of them.

Next, we show in Fig. I.1c a geometrical representation of the 'attractor' which has been reconstructed from the signal and we can see immediately that it has the distinctive geometrical form of a torus, albeit distorted. We can be reassured that it is a torus by inspection of Fig. I.1d which is a planar slice through the attractor. It can be seen that the points, which are the intersections of the individual trajectories with the plane, form a simple connected loop. Given that our signal is from two oscillators then this is what we should expect, in the case of the two frequencies being incommensurate. If we recall the example of the ball-bearing in the tube, then such a planar intersection would just be the circular cross-section of the tube. In principle, the section ought to be square since the object has been formed from the time series of two square waves but due to distortions introduced in the reconstruction a certain amount of rounding of the edges has occurred. The point that we wish to emphasize here is that traditional linear methods give an unnecessarily complicated interpretation of the contents of this relatively straightforward signal

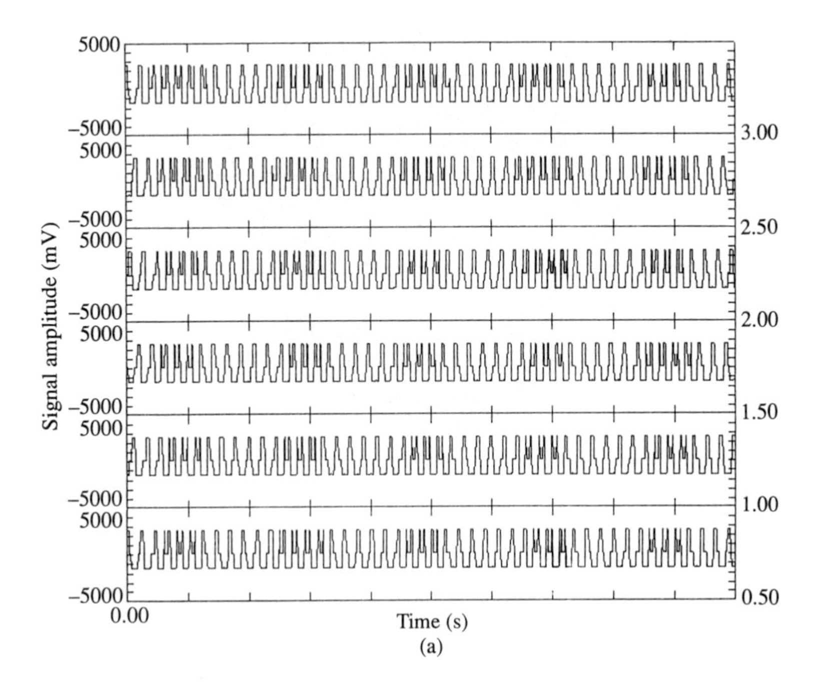
5000
−5000
3.00
2.50
2.00
1.50
1.00
0.50
Signal amplitude (mV)
0.00
Time (s)
(a)

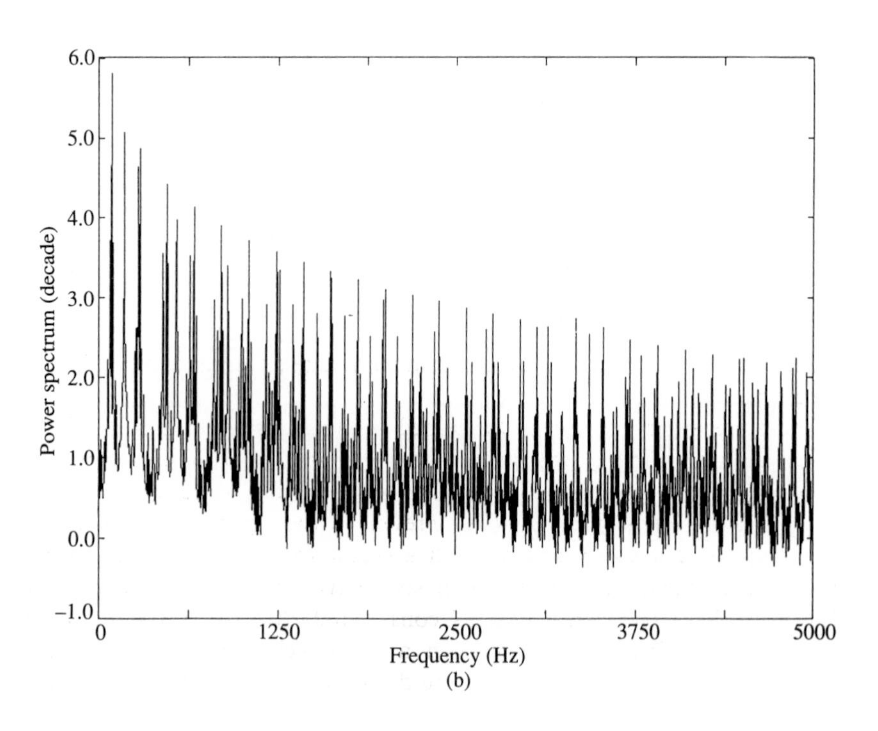
6.0
5.0
4.0
3.0
2.0
1.0
0.0
−1.0
Power spectrum (decade)
0
1250
2500
3750
5000
Frequency (Hz)
(b)

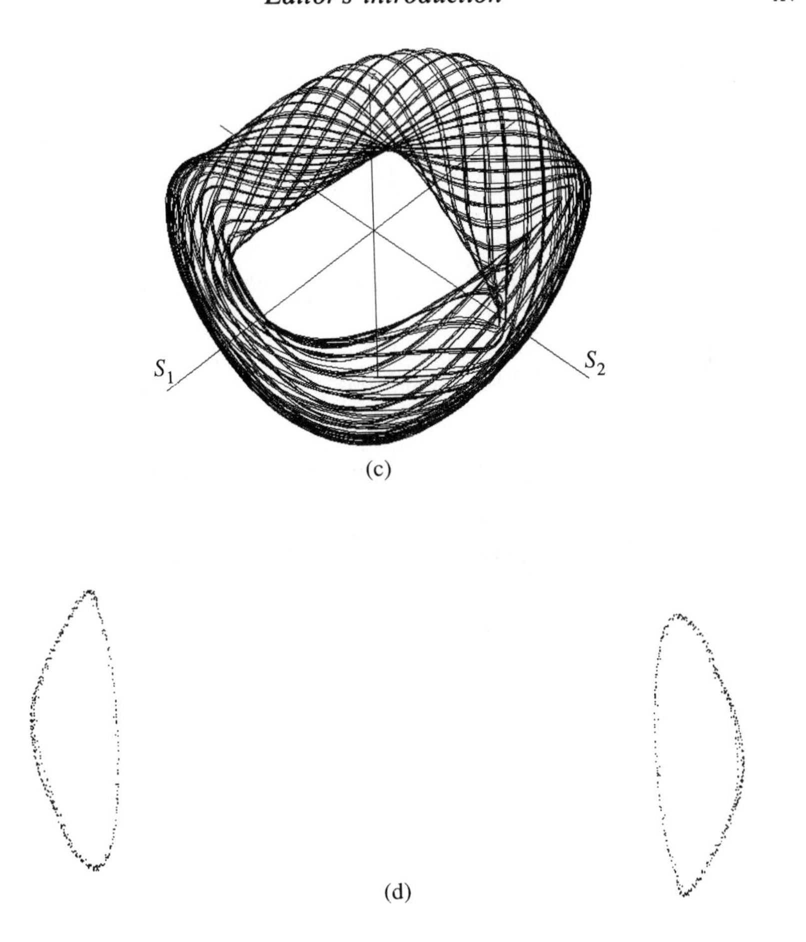

Fig. I (a) Time series formed by the addition of two square wave oscillator outputs; (b) frequency power spectrum; (c) reconstructed attractor; (d) Poincaré section taken through (c).

whereas the geometrical representation immediately gives a more informative description.

There are other benefits to be gained in taking this approach to time-series analysis. Firstly, any geometrical structure present in the reconstructed attractor from an apparently irregular time series implies there is some low-dimensional behaviour present in the signal. This can be of enormous benefit when dealing with a signal from an unknown system whose governing equations are either speculative or are known to be high dimensional. In particular, if the reconstructed attractor has a geometry or topology which is also found in a more well-understood problem, then predictions about changes in the dynamics with variation in control parameters of the unknown system can be made with some degree of confidence.

Next, both qualitative and quantitative descriptions of the underlying solution structure which gave rise to the observed behaviour can be made. For example, unstable fixed points of a system can be identified and then quantitative estimates of dynamical information near these fixed points can be used to make predictions about, as yet, unobserved behaviour. This feature is of immense value when investigating a system governed by partial differential equations and thus where the solution space is in principle infinite dimensional. The origins of observed highly complicated motions in one parameter range may be associated with the low-dimensional dynamical behaviour observed in another. Thus, identifying critical points about which theoretical analysis can be performed is often crucial in determining the important terms of the full equations which must be included in any modelling. In addition, quantitative estimates of the nature of the attractor can be made to try and establish the validity of finite-dimensional representation, the strength of the observed chaos and an upper bound on the number of modes which would be required in a model representation of the behaviour.

Last, but not least, it is has been recognized in recent years that the limits of prediction can be increased in deterministic dynamical systems when they are in a chaotic regime. Such systems are of course predictable in principle since they are deterministic, but because of sensitivity to initial conditions practical prediction is difficult. However, we know that the reconstructed attractor has some geometrical structure which we can hope to exploit. Thus the trajectories in the reconstructed phase space must follow an approximately defined track and their behaviour is not completely random. This feature can set both local and global bounds on the divergence of the trajectories and hence extend the range of prediction.

I.4 Multiplicity of states

It is worth remarking that chaos is just one feature of nonlinear systems, albeit an important one. One other crucial property is that different states can exist on the same boundary conditions in the nonlinear regime. These multiple states can have qualitatively different dynamics so that at some prescribed value of the parameters which govern a system there is the potential for several steady states, various types of periodic motion, chaotic states and nonperiodic motion. The particular state which is realized in practice will depend on the 'history' of its creation, e.g. the speed of change of parameter to the preselected point or the variation of more than one parameter.

Knowledge of multiple steady states is particularily important if we are to gain further understanding of the dynamical behaviour. This is because there is a robust mathematical theory of nonlinear steady state bifurcation phenomena which is called 'Singularity Theory'. It has been demonstrated to apply to a wide variety of physical systems which are governed by either ordinary or partial differential equations. As yet, a similar mathematical theory does not exist for dynamical phenomena and so knowledge must be built up on a case-to-case basis. However, it is clear that investigations of the interactions of various steady states give insight into the more complicated phenomena which are often found to be organized by the underlying steady solution structure. It should also be noted that the dynamical solutions can exhibit certain symmetry properties which are found in the steady regime such as the coexistence of pairs of states which break a geometrical symmetry. Finally, highly complicated motion can sometimes usefully be regarded as the manifestation of the presence of multiple states where the system does not settle into any particular attractor but instead wanders through a range of them.

I.5 Concluding remarks

It is undoubtedly true that chaos has become a very popular subject in recent years judging by the media attention and production of scientific papers. One of the penalties of a scientific subject gaining such popularity is that it is seen by some as the answer to everything. However, chaos seems to have survived the fashionable phase and perhaps one reason is that the natural world is inherently nonlinear. Therefore, one should expect to find chaos rather than order and perhaps we now have some tools for furthering our understanding of what was previously thought of as random noise. There are many deep mathematical ideas behind the dynamical systems approach to the study of nonlinear phenomena which are aimed at understanding the origins and structures of complicated behaviour, not merely describing it.

Naturally, there has been a tendency to extend some of these ideas into fields where there is no rigorous justification for doing so. These are often resolute efforts to tackle very difficult problems with new scientific ideas. The very least that one can say is that nonlinearity should play a key role in natural phenomena and therefore some of these modern concepts may well give a new insight into some unresolved problems. On the other hand it is also worth noting that an irregular time series formed from ice core samples or the monitoring of a bodily function for example need not necessarily be describable in terms of low-dimensional chaos. Therefore one must remain cautious about such studies for it is quite easy to misrepresent the above ideas by an imprecise application of techniques which have thus far only been successfully tested in well-controlled laboratory situations. However, if new insights into difficult areas are obtained using this approach, which amount to more than putting common sense into fancy mathematical language, then a great deal has been achieved.

References

1. Newton, I. (1686). *Principia*, Vols. 1 and 11. Motte's translation, revised by Cajori (1962), University of California Press, Berkeley.
2. Poincaré, J. H. (1892). *Les methodes nouvelles de la mecanique celeste*. English translation of Vols 1, 2, 3 (1967) as NASA TT F-450, TT F-451, TT F-452. NASA, Washington, DC.
3. Lorenz, E. N. (1963). Deterministic nonperiodic flow. *J. Atmos. Sci.*, **20**, 130–41.

Section headings

Contributors

M. S. Child: *Department of Theoretical Chemistry, University of Oxford, South Parks Road, Oxford OX1 3UB, UK.*

David Holton: *Department of Hydrogeology, Harwell Laboratory, Didcot, Oxon OX11 ORA, UK.*

Jonathan Keating: *Department of Mathematics, The University, Manchester M13 9PL, UK.*

Robert M. May: *Department of Zoology, University of Oxford, South Parks Road, Oxford OX1 3PS, UK.*

Tom Mullin: *Department of Atmospheric, Oceanic and Planetary Physics, Clarendon Laboratory, University of Oxford, Parks Road, Oxford OX1 3PS, UK.*

Peter L. Read: *Department of Atmospheric, Oceanic and Planetary Physics, Clarendon Laboratory, University of Oxford, Parks Road, Oxford OX1 3PS, UK.*

J. M. T. Thompson: *Centre for Nonlinear Dynamics and its Applications, Civil Engineering Building, University College London, Gower Street, London WC1E 6BT, UK.*

1
Chaos and its application to physical systems

Tom Mullin

The main issue to be addressed here is whether the ideas embodied in nonlinear dynamical systems and chaos are of any value in aiding our understanding of complicated physical phenomena. The views expressed will necessarily be selective and will reflect those features of modern thinking in dynamical systems theory which the author has found to be most useful in interpreting the outcome of experiments. The field may be viewed in the following way. On the one hand there are many splendid abstract mathematical ideas in discussions of dynamical systems and chaos, and on the other there are so-called 'real world' phenomena. As is usually the case in physics, the interpretation of experimental results in terms of a set of theoretical ideas is not always a straightforward matter and things are no different here. This is despite some claims in popular accounts that 'chaos theory' is the 'answer' to a wide variety of problems. An example of this oversimplification is to be found in the phenomenon of turbulence in fluid flows. Over the past decade the words 'chaos' and 'turbulence' have become synonymous in many popular accounts and yet the connection between the two is far from obvious even today. In the author's opinion the ideas of chaos have thus far added very little to the understanding of the phenomenon of turbulence. Yet many research groups, including the author's, continue to study this topic because a significant breakthrough in this very difficult subject seems tantalizingly close.

One attractive feature of research into nonlinear phenomena in classical physics is that the models and experimental systems investigated can be simple in principle. It is the interpretation of the calculated and or observed results which can prove to be difficult. The simple pendulum has therefore been chosen to illustrate a systematic experimental and theoretical study of one route to chaos. It highlights a combined experimental and theoretical approach which has been successfully used by the author to investigate more complicated systems.

The concepts of phase plane analysis are first introduced using the simple planar undamped pendulum. An external excitation is then added so that the system becomes suitable for a practical laboratory study. Next the concepts of bifurcation are introduced which naturally leads to a discussion of multiple states. Chaos is found to arise through the universal route of period doubling and the chapter ends with a discussion of the important idea of sensitivity to initial conditions.

1.1 The simple pendulum

First we will consider the very simple physical problem of an undamped pendulum. We will then develop some of the ideas which underlie dynamical systems theory, illustrating them with examples which involve pendulums. We use these simple physical models not to study them for their own sake but rather to see whether we can learn anything from them which will help interpret the more complicated problems we will meet in the later chapters.

Consider the simple frictionless pendulum shown in Fig. 1.1, consisting of a thin rigid massless rod of length l with a bob of mass m, suspended from point O. The forces are as shown and a balance of these gives us Newton's equation of motion $F = ma$. We can rewrite it for the present case in the form

$$F = ml\frac{\mathrm{d}^2\theta}{\mathrm{d}t^2} = -mg\sin\theta,$$

which can be rearranged as

$$\ddot{\theta} + \frac{g}{l}\sin\theta = 0, \tag{1.1}$$

where $\sin\theta$ is the nonlinearity in the equation and $\ddot{\theta}$ is the shorthand

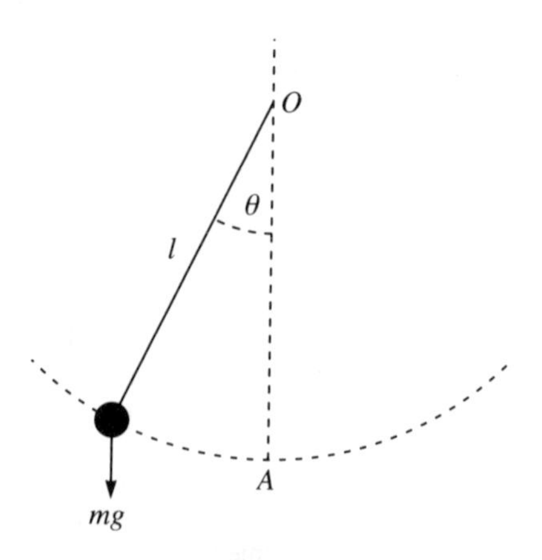

Fig. 1.1. Schematic diagram of simple pendulum consisting of a bob of mass m suspended on a massless rod of length l.

notation for $d^2\theta/dt^2$. Equation (1.1) is an example of a nonlinear second-order ordinary differential equation. We can obtain explicit solutions to it in terms of elliptic integrals. However, these are not of immediate value, and so in order to make analytical progress we usually linearize the equation. This is achieved by expanding the sine term in powers of θ and by truncating the expansion at first order in θ; we end up with the familiar linear ordinary differential equation

$$\ddot{\theta} + \frac{g}{l}\theta = 0. \tag{1.2}$$

This equation is valid for small angles of swing and it can be integrated directly to give us a solution of the form

$$\theta = \theta_0 \cos(\omega_0 t + \phi), \tag{1.3}$$

where $\omega_0 = (g/l)^{1/2}$ is the angular frequency from initial condition θ_0 and ϕ is the phase. Thus we would get a cosinusoidal output if the pendulum were to be released from rest at an initial angle θ_0 and it would swing from side to side with frequency given by ω_0. This solution is valid for small amplitudes of oscillation and of course it is a very important result in physics. For example it forms the basis of clocks and can also be used to determine g.

The linear pendulum or simple harmonic oscillator is a very familiar concept in physics, and the time series that one would obtain from such a system is a cosine wave as shown in Fig. 1.2. If we had given the pendulum slightly different starting conditions, then we would have obtained another cosine wave with the same period but with a small difference in the amplitude and phase from the one which is presented in Fig. 1.2. In addition, any two solutions to the equation can be added to give a third solution, which is one of the basic properties of linear systems.

An alternative way to view the motion of the pendulum is in terms of

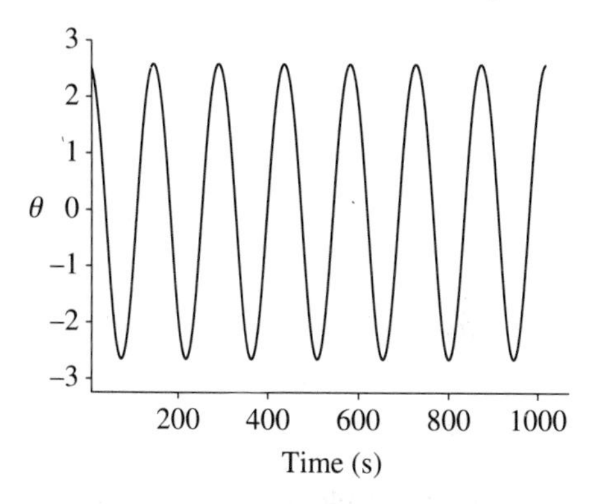

Fig. 1.2. Velocity time series for simple undamped pendulum.

the phase space representation. A thorough discussion of these ideas is given in the book by Jordan and Smith [1] and we will only outline the essential points here. In order to facilitate this, we are going to specify the system by two quantities: $\theta(t)$, the position or the angle which is a function of time, and the angular velocity $\dot{\theta}(t)$. So θ and $\dot{\theta}$ will be the coordinates of the phase space and a curve in it which describes the time evolution of the system will be called a trajectory. A set of such curves is called a phase portrait.

The construction of the phase space for the pendulum proceeds as follows. The pendulum equation (1.2) determines the orientation of the tangent to the trajectory in phase space. So, in principle, by successive extrapolation we can construct the phase trajectory. Given some initial position and velocity $(\theta(t_0), \dot{\theta}(t_0))$ at time $t = t_0$, we wish to find a new point $(\theta(t_0 + \delta t), \dot{\theta}(t_0 + \delta t))$ at a time δt later. For the linear pendulum equation (1.2), the construction may be carried out explicitly if we substitute $x = \theta$ and $y = \dot{\theta}$ in (1.2) and thus we have the relationship that $\dot{x} = y$. From (1.2) we get

$$\dot{y} = -\omega^2 x.$$

The equilibrium position for the pendulum is the position $(x, y) = (0, 0)$. Time does not appear explicitly in the phase space representation, and so to construct our phase portrait we eliminate t and the phase paths are the solutions of

$$\frac{\dot{y}}{\dot{x}} = \frac{\mathrm{d}y}{\mathrm{d}x} = -\frac{\omega^2 x}{y}, \tag{1.4}$$

which is separable and may be solved directly for this particular problem to give

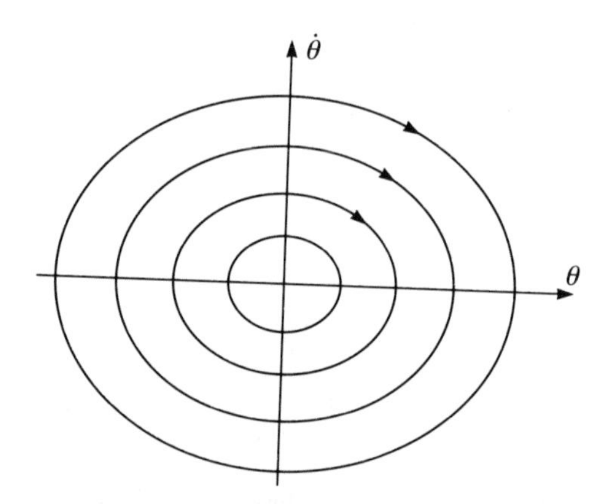

Fig. 1.3. Phase plane diagram for linear undamped pendulum.

$$y^2 + \omega^2 x^2 = C, \tag{1.5}$$

where C is the constant of integration. This is the equation for an ellipse and thus the phase portrait will consist of a series of ellipses centred around the origin, one for each amplitude of swing. An example of such a phase portrait is shown in Fig. 1.3.

The origin is a stable equilibrium point or fixed point solution in that trajectories with initial conditions in its neighbourhood will remain close to it for all time. Each of the ellipses corresponds to an orbit with a different amplitude but the same frequency, and the arrows indicate the direction of the evolution in time. Thus the simple side-to-side motion can be represented straightforwardly in this way. For the moment it may seem that we have expended a great deal of effort in obtaining this alternative representation of the motion without any apparent gain. However, if we now consider the nonlinear pendulum, the advantages of this approach will become apparent.

First, we show in Fig. 1.4 the time series for the nonlinear pendulum obtained by numerically integrating equation (1.1). We see that the frequency obtained from two different starting conditions, labelled 1 and 2, is now dependent upon the amplitude and the two time series appear to drift in phase relative to one another. The phase portrait can be constructed by the straightforward application of the method outlined above to equation (1.1). Now the equation for the phase paths is given by

$$\tfrac{1}{2}y^2 - \omega^2 \cos x = C. \tag{1.6}$$

The phase portrait is presented in Fig. 1.5 and again consists of simple closed curves; the qualitative nature of the motion is immediately apparent.

In the central region around point A, which corresponds to side-to-side motion, the trajectories are ellipses. In addition, the continuous curves on

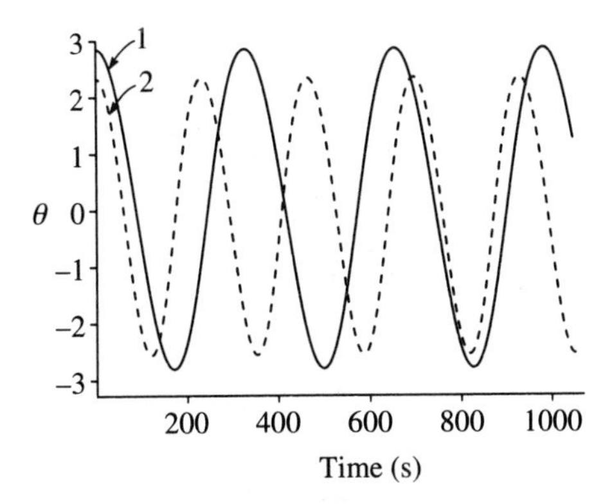

Fig. 1.4. Velocity time series for the nonlinear undamped pendulum showing the effects of changing the initial conditions.

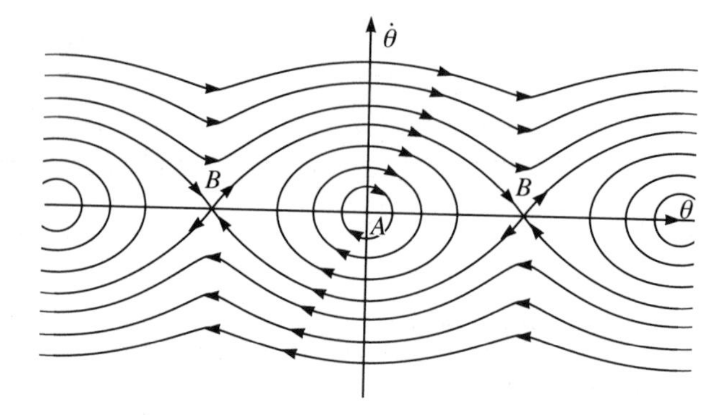

Fig. 1.5. Phase plane diagram for the nonlinear undamped pendulum. The separatrices passing through the points B are the boundaries between side-to-side motion (ellipses) and rotation (continuous curves).

the outside of the ellipses correspond to the pendulum going round and round in rotary motion in one direction at the top and the other direction at the bottom. The boundary between these two distinctly different types of motion is called a separatrix. Two equilibrium points are marked A and B, where A is the $(0, 0)$ position and is locally stable. As mentioned above, this means that if the system is given a small kick from A it will continue to oscillate nearby. However, if the pendulum is started from position B and given a small disturbance, then it will fall off to some completely different motion, i.e. it will either go round and round on a rotary solution or it will oscillate from side to side with large amplitude. Thus B is called an unstable equilibrium point and corresponds to the pendulum being exactly inverted. In this situation, the energy available in the system precisely matches that required to achieve the upside-down position. Thus the pendulum will take an infinite time to reach this point and will, therefore, only asymptote to it. We call the limiting trajectories which pass through such points heteroclinic cycles in the plane. These are the separatrices shown in Fig. 1.5.

Thus far we have described an idealization of the simple pendulum as it is considered to be frictionless. In any physical realization there will inevitably be some damping present in the system. We will model this by adding a term which is linearly proportional to the velocity. Nonlinear damping effects such as aerodynamic drag are therefore ignored. Nevertheless, this is found to be a good approximation in practice. Thus equation (1.3) will become

$$\ddot{\theta} + \gamma\dot{\theta} + \omega^2\theta = 0. \tag{1.7}$$

We can calculate the phase trajectories in the usual way and we obtain

$$\frac{\mathrm{d}y}{\mathrm{d}x} = -\gamma - \omega^2\frac{x}{y}, \tag{1.8}$$

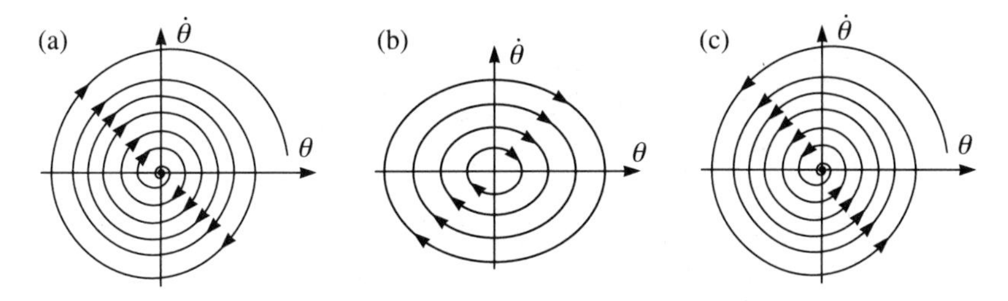

Fig. 1.6. Phase plane diagrams showing the effects of damping: (a) negative damping (the origin is a repeller); (b) zero damping (the origin is an equilibrium point); (c) positive damping (the origin is an attractor).

where $-\gamma$ is the new term (cf. equation (1.4)) which arises because of the presence of damping. Variation of the sign of the damping gives rise to the two qualitatively different cases separated by the undamped situation. We illustrate this point schematically in Fig. 1.6.

The first case is when γ is less than zero and we have negative damping. This is clearly a nonphysical case, but it shows the origin has now become an unstable repeller. In other words, any trajectories started near the origin will spiral away from it. In the case when the damping is zero we return to the linear case we had above, and the origin is again an equilibrium point. Finally, when we have positive damping, as in the normal physical situation, then the origin becomes an attracting fixed point. Therefore the origin is the attractor for the positively damped system. Thus trajectories started anywhere in the plane, except for exactly on the inverted solution, will decay into the origin. The basin of attraction of an attractor is defined to be the set of initial conditions of trajectories which will converge onto the given attractor. Thus the basin of attraction of the positively damped simple pendulum is the whole plane except the points which define the inverted solution.

In the nonlinear case, the phase trajectories for the damped problem can

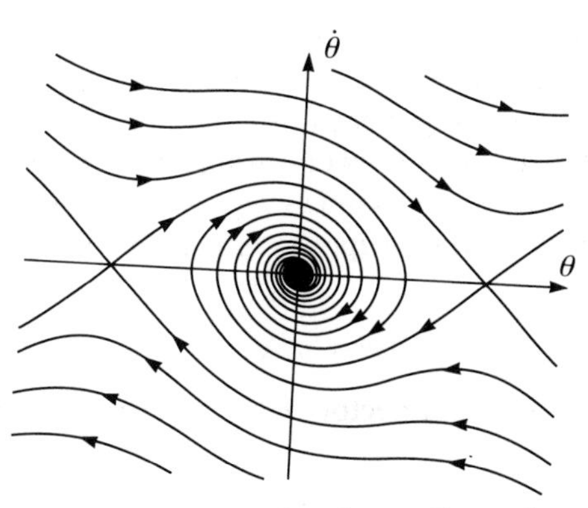

Fig. 1.7. Phase plane diagram for the nonlinear damped pendulum.

be calculated on a computer. These were done for the example shown in Fig. 1.7, where we can see that the oscillations persist for a long time near the origin and are less elliptical far from it. Thus the only attracting fixed point of a damped unforced pendulum is the fixed point at the origin.

1.2 The parametrically excited pendulum

The pendulum systems we have discussed up to this point can all be represented on a two-dimensional phase space. We must also recall that trajectories are not allowed to cross in these deterministic systems. If the trajectories ever did cross, then the system would have a choice at the point of intersection and it would be a probabilistic rather than a deterministic system. Therefore, as discussed in the previous section, we will only find equilibrium points and periodic motion in the undamped case and fixed points when it is damped. In order to produce more interesting dynamical behaviour which might include finite-dimensional chaos, we must make the phase space three dimensional. This increase in dimension of the phase space will permit more complicated nonintersecting trajectories as will be described below. One way of increasing the dimension of the phase space is to add an external driving force to the system.

There are two obvious ways of doing this in practice in the laboratory. Either a sideways periodic forcing can be added to the pendulum pivot point or it can be moved up and down in a sinusoidal manner. The latter method is called parametric excitation of the pendulum and it may be modelled mathematically by making g a function of time. We will choose the second option as in this case there is a trivial solution of zero swinging to the equation of motion and the appearance of swinging motion can be represented as a bifurcation from this state. In addition, the equation of motion can be investigated quite straightforwardly on many modern small computers and a physical demonstration provides a convincing and spectacular illustration of the relevance of chaos to simple physical systems. A schematic diagram of the set-up is shown in Fig. 1.8.

The equation for the nonlinear undamped parametrically excited pendulum is

$$\ddot{\theta} + \omega_0^2(1 + h\cos\omega t)\sin\theta = 0, \tag{1.9}$$

where

$$h = g_1/g_0, \qquad \omega_0 = (g_0/l)^{1/2}$$

and g is our 'parameter' which we have set to be a function of time:

$$g(t) = g_0 + g_1\cos\omega t.$$

Thus ω_0 is the natural frequency of the pendulum and ω is the applied

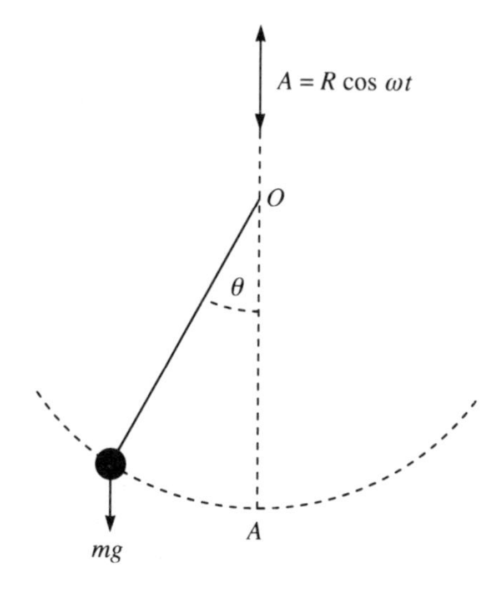

Fig. 1.8. Schematic diagram of the parametrically excited pendulum geometry.

frequency of the driving motion. First, we will linearize the equation in θ and make the substitution $\tau = \omega t$ to get

$$\ddot{\theta} + (\alpha + \beta\cos\tau)\theta = 0, \tag{1.10}$$

where $\alpha = \omega_0^2/\omega^2$ and $\beta = g_1/l\omega^2$.

This is called Mathieu's equation. It was first derived by Mathieu [2] in connection with the vibration of elastic plates and shells. The solution to this equation can be found in terms of what are called Mathieu functions, which can be evaluated numerically. However, for small values of the parameter β, explicit solutions can be found. Specifically, there are resonant tongues on the line $\beta = 0$ where the trivial solution loses stability to swinging motion of a specific frequency when $\alpha = \frac{1}{4}n^2$, where n is an integer. We show the situation schematically for the first few resonances in Fig. 1.9 and an interpretation of their physical consequences is given in the next paragraph. A good introduction to the Mathieu equation is given by Jordan and Smith [1] and an extensive treatment together with a discussion of its physical applications is given by McLachlan [3].

The physical interpretation of the resonant responses will now be made using the notation of equation (1.10) and with reference to Fig. 1.9. This is a chart of the response of the system plotted as a function of the two main control parameters, the frequency and amplitude of the excitation. In the unshaded region, outside of the resonant tongues, the simple up-and-down motion of the pendulum is stable and there is no swinging from side

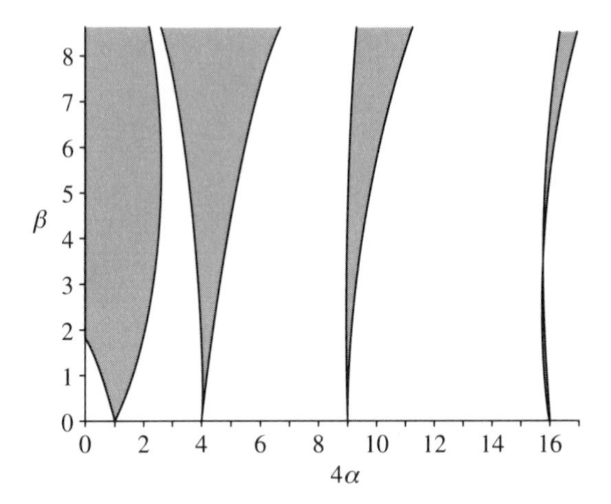

Fig. 1.9. First four resonances of the Mathieu equation. Up-and-down motion is stable in the unshaded regions and becomes unstable to swinging motion with frequencies giving $\alpha = \frac{1}{4}$, 1, $\frac{9}{4}$, and 4.

to side. We have already referred to this as the trivial solution. If a parameter, e.g. the frequency, is now changed so that a resonant region is entered, then the pendulum begins to swing from side to side. In other words the up-and-down state has become unstable and is replaced by side-to-side swinging motion. It is inside these resonant regions where the pendulum is said to be parametrically excited.

We will now concentrate on the region which starts at $\alpha = \frac{1}{4}$ on the x-axis of Fig. 1.9. This is called the primary region of instability, and because of its shape it is often referred to as a 'tongue'. Inside this resonant tongue, which covers the widest range of parameter space of any of the resonant bands, the pendulum oscillates at half the driving frequency. By analogy we can think of this as a child pushing down on a swing twice every cycle. This phenomenon is well known and was demonstrated by Lord Rayleigh in a lecture at the Royal Institution at the turn of the century using champagne corks attached to tuning forks by pieces of string.

We can introduce the effect of damping in the system by adding a term proportional to the velocity as previously. The only effect of the inclusion of damping is to move the resonant tongue off the $\beta = 0$ axis as shown in Fig. 1.10. Now we need a finite-size excitation amplitude proportional to the damping to promote the instability, as seems intuitively reasonable.

The Mathieu equation is linear and therefore it can simply describe the loss of stability to motion with a certain period when a resonance curve is crossed. At the crossing point there is an exponential growth of a new

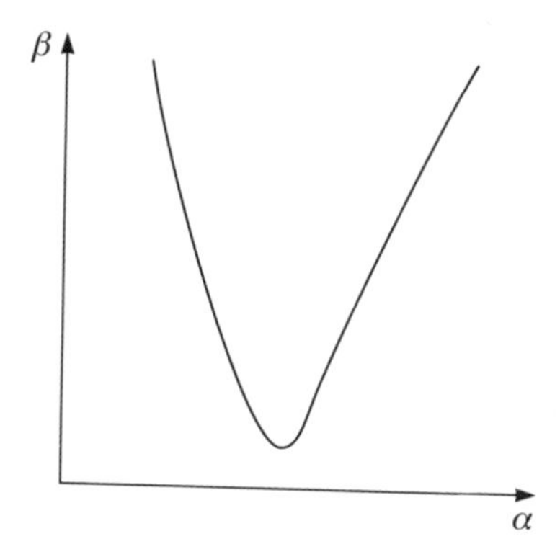

Fig. 1.10. Primary resonant tongue ($\alpha = \frac{1}{4}$) for the damped Mathieu equation.

solution in this mathematical formulation which is clearly nonphysical. Therefore, some form of nonlinearity is required to saturate the growth of this new solution. We will now consider the full nonlinear equation (1.9) to obtain a deeper understanding of the observed physical phenomena. In order to do this, we must invoke some modern numerical methods for the treatment of nonlinear ordinary differential equations to be outlined below. Thus we must jump from nineteenth- to twentieth-century mathematical techniques as there is very little one can say about the general solutions of nonlinear equations using standard analytical methods.

1.3 Principal bifurcation structure of the parametrically excited pendulum

The exchange between the trivial state of up-and-down motion and side-to-side swinging as a resonance curve is crossed signifies that a bifurcation has occurred in the solutions to the equations. For the specific resonant tongue being considered here, it is an example of a simple period-doubling bifurcation where the trivial solution exchanges stability with a pair of swinging solutions whose frequency is half that of the drive. As outlined above, we need to study the full nonlinear equation to investigate this change of state. This may be done using the numerical bifurcation package called AUTO developed by Doedel [4]. It can be used to calculate equilibrium solutions of first-order nonlinear ordinary differential equations which means our second-order equation (1.10) must be transformed into a first-order equation. The transformation is a straightforward substitution procedure.

The principle of the numerical method is to start with some known solution to the equation and then to change a control parameter of the equation by a small amount. Iteration is then performed towards the nearby solution at the new parameter value under the assumption that small changes in parameters will lead to similary small changes in the solutions to the equation. Thus in the present case we could start with the trivial zero swing state and then vary either the frequency or amplitude of the drive as these are the control parameters of the present system. By these means the solution can be extended to new values and its path followed by stepping along the branch as a parameter is varied.

The technique may also be used to locate bifurcation points and follow any stable or unstable solutions which arise there. This latter feature can be very useful as the unstable solutions, which are unobservable in any experiment, can nevertheless play an important role in organizing the dynamical behaviour, as we shall see later. Using these techniques we can thus form a chart or map of the bifurcation points which will help locate local multiple bifurcation points where two or more lines meet. As we will see in later chapters, these local phenomena can provide organizing centres for global dynamical phenomena.

We give an example of the results of an AUTO calculation for the parametric pendulum in Fig. 1.11. It is a plot of the maximum angle of swing versus amplitude of the drive. The amplitude of swing is chosen as the linear functional which allows us to distinguish between solutions of different type in a practical way. On the left-hand side of the figure we have only the trivial solution with no swing. As the amplitude of the

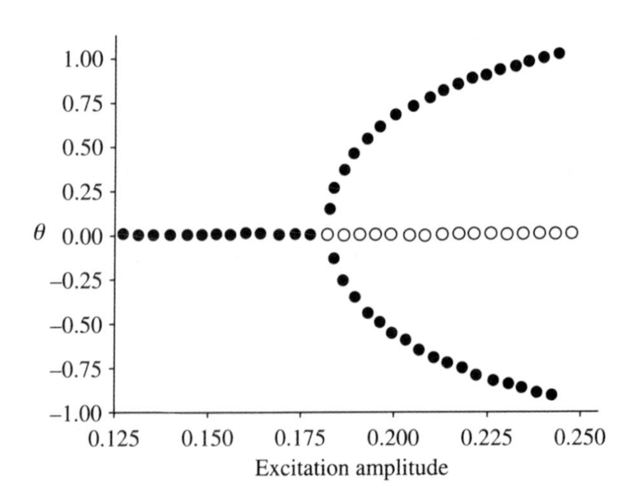

Fig. 1.11. AUTO calculation of the first bifurcation to swinging motion for the parametrically excited pendulum. Here, the bifurcation parameter is the amplitude of the applied excitation (β) and the functional which is used to discriminate between solution types is the amplitude of the response. (Note that this would correspond to crossing the curve shown in Fig. 1.10 to the right of the minimum).

excitation is increased, a bifurcation occurs and a pair of new solutions arise which correspond to the swinging motion at half the driving frequency. Each branch is just phase-shifted from the other by π and their amplitudes grow, as the square root of the parameter, local to the bifurcation point. The open circles indicate an unstable solution so that the up-and-down solution has now become unstable. Thus, if the initial conditions were set to the trivial solution in this regime, then a full time integration of the equations or a physical experiment would show the rapid departure from this initial state to a swinging motion at half the drive frequency. This is an example of a supercritical pitchfork bifurcation, which here represents the exchange of stability between the trivial state and side-to-side swinging. In the present special case we have the additional restriction that the swinging solutions correspond to oscillations at half the drive frequency and it can therefore be considered as an example of a simple period-doubling bifurcation.

A more practical way of demonstrating this bifurcation in an experimental arrangement is to vary the frequency of the drive at a fixed amplitude. This is shown schematically in Fig. 1.12(a) together with the appropriate portion of the parameter space diagram in Fig. 1.12(b). If we start at a value of the frequency parameter ω outside of the resonant band to the right of XY in Fig. 1.12(b) and reduce the frequency of excitation, then, as the locus of bifurcation points XY is crossed, side-to-side motion at half the drive frequency sets in. In the bifurcation diagram shown in Fig. 1.12(a) we again have a pitchfork bifurcation but now the control parameter is the frequency ω. Thus, if we were to give the pendulum initial conditions of zero swing at a value of ω to the left of XY, side-to-side oscillation at half the driving frequency would rapidly develop.

We can represent the resulting motion in phase space by a connected loop as shown in Fig. 1.13(a). It appears slightly distorted because it represents the full motion and thus includes a component at the drive frequency. The complete motion is represented in this three-dimensional phase space by the combination of two ellipses, one for the up-and-down and one for

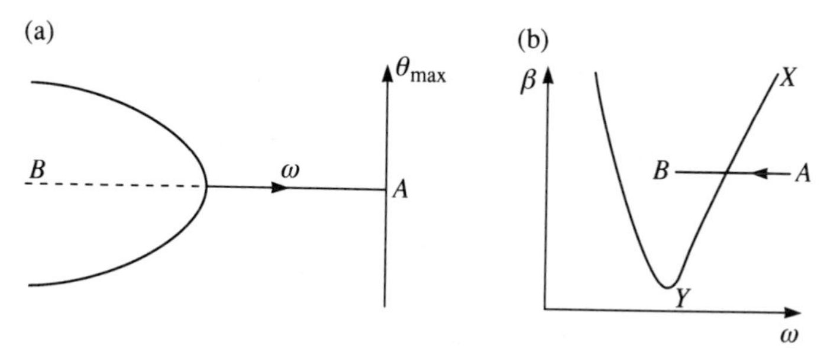

Fig. 1.12. (a) Schematic bifurcation diagram showing the first bifurcation found when the frequency is varid from A to B in (b).

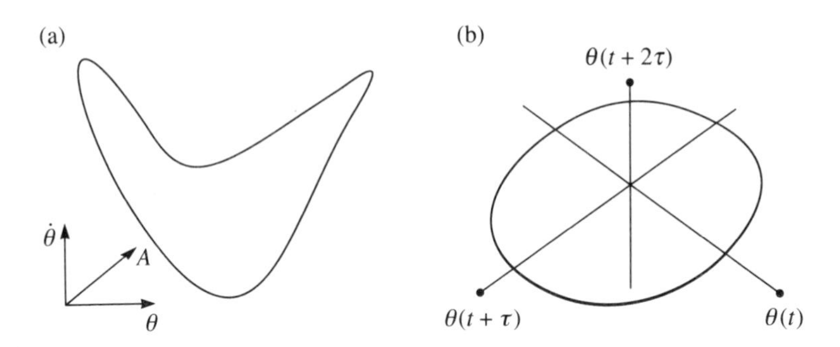

Fig. 1.13. Phase portraits for the swinging pendulum: (a) obtained from integrating the equation of motion; (b) reconstructed from an experimental recording of the angle of swing.

the swinging motion. This will have the topological form of a doughnut or torus in phase space with its two component frequencies related to each other by a factor of 2, as the swinging is at half the driving frequency. Thus, this is a highly degenerate torus since the two frequencies are integrally related. If the two frequencies were unrelated to each other, then we would call the motion quasi-periodic and the surface of the torus would be filled in after a suitable number of orbits.

A similar 'phase portrait' can be constructed from the experiment by the methods to be outlined in the next chapter from a recording of the angle of the pendulum and this is shown in Fig. 1.13(b). This recording does not contain any component due to the up-and-down motion and so is a simple ellipse which is slightly broadened by the presence of experimental and measurement noise. Thus the attractor for the system in this representation is a simple limit cycle.

So far we have concentrated on the right-hand side of the resonant tongue, where the behaviour is fairly straightforward. Next, the results of an investigation of the left-hand side of the resonant band are presented schematically in Fig. 1.14. We know from the linear Mathieu equation that there will be a loss of stability of the trivial solution as the locus YZ is crossed in Fig. 1.14(b). The nonlinear form of the bifurcation can then be obtained by the application of AUTO to the full equation. In Fig. 1.14(a) we see that the bifurcating solution branches initially bend backwards and are found to be unstable. The solution branches then bend forwards through the limit points C, so they then represent a pair of stable solutions which are again related to each other by a phase shift. This form of bifurcation is called subcritical and has hysteresis associated with it.

In an experiment, one would observe the sudden appearance of a large amplitude oscillation as the locus YZ was crossed and the swinging would

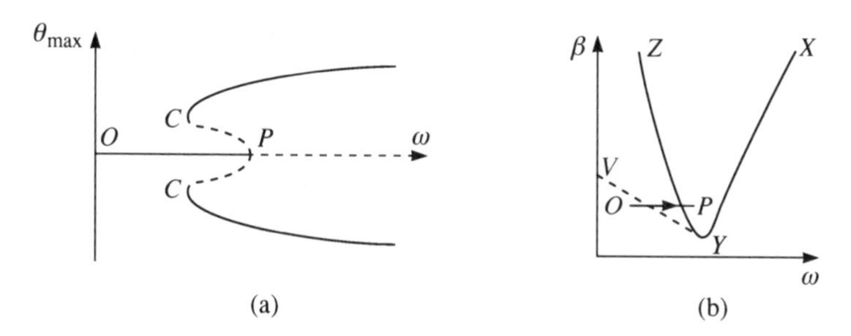

Fig. 1.14. (a) Schematic bifurcation diagram for the left-hand side of the Mathieu 'tongue'. This type of bifurcation is called subcritical and there is hysteresis involved in the onset of swinging. The swinging motion which appears as YZ in (b) is crossed at P with increase in α but remains until VY is crossed when α is reduced. The locus VY maps the path of the limit points C of (a) as α and β are varied.

persist on the reduction of ω until VY was crossed. Thus there would be hysteresis present on the left-hand side of the resonant band, which cannot be predicted from the linear analysis. This also means that between VY and ZY in Fig. 1.14(b) there are three stable solutions separated by two unstable solutions. The three stable solutions are the trivial state and the two swinging motions. Thus there is a solution multiplicity of five. Immediately to the left of XY the multiplicity is three and is thus also odd, i.e. new solutions appear in pairs. Multiplicity of solution states is a common feature of nonlinear systems, which means in practice that the realization of any particular equilibrium state of a system is dependent on the path taken through parameter space. In addition, the different equilibrium states can interact with each other as external control parameters are varied and this can lead to more complicated behaviour. This theme will be explored at length in the chapters on electronic oscillators and fluid dynamics.

1.4 Transition to chaos

Now that we have built up a picture of the basic bifurcation structure of the parametrically excited pendulum, we will discuss the more complicated dynamical behaviour which can arise, including chaos. If we numerically integrate the equations of motion for parameter values well inside the resonant band, then we can obtain a phase portrait typified by that shown in Fig. 1.15. Also, if we were to run an experiment at the same point, then we also may observe irregular motion with the pendulum rotating and oscillating in an apparently random fashion. However, one must be careful to remember that, because of multiplicity, other attractors are

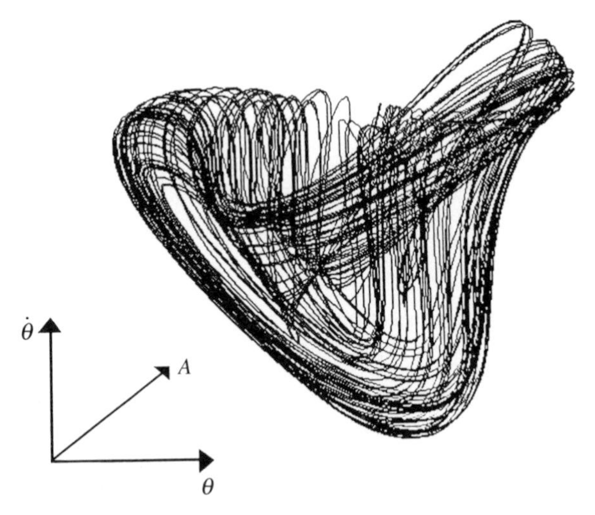

Fig. 1.15. Phase portrait obtained from the integration of the full equation with parameter values in the chaotic region.

available and there may not always be a one-to-one correspondence between numerics and experiments in this regime. In other words, some initial conditions may produce chaotically rotating and vibrating motions, while immediately adjacent ones may finally settle on the trivial solution after a transient phase. Thus the relative level accuracy in the control of the input conditions in the two realizations becomes important.

We now wish to address the issue of where the chaotic behaviour comes from and see whether it is possible to obtain a deeper understanding of its origins. The bifurcation sequence which leads to chaos is shown in Fig. 1.16 and was obtained using AUTO. We have only plotted one side of the bifurcation diagram to ease the presentation. If we start with a value of ω to the right of A and reduce it, then there will be the exchange of stability between the trivial branch and the side-to-side oscillation at half the driving frequency as before. Now, if we continue to decrease ω the solution will pass through a symmetry-breaking pitchfork bifurcation at D. The resulting motion will consist of a slightly higher swing to one side of the vertical than the other. These solutions exist in pairs because of the underlying symmetry of the system about the vertical axis. We have found the presence of spatial symmetry-breaking bifurcations of this type to be rather common in a variety of physical systems and we will encounter more in later chapters.

Yet further reduction in ω leads to the onset of a period-doubling cascade to chaos starting at point labelled E. Period-doubling cascades of this type are usually associated with Feigenbaum [5] and are discussed exten-

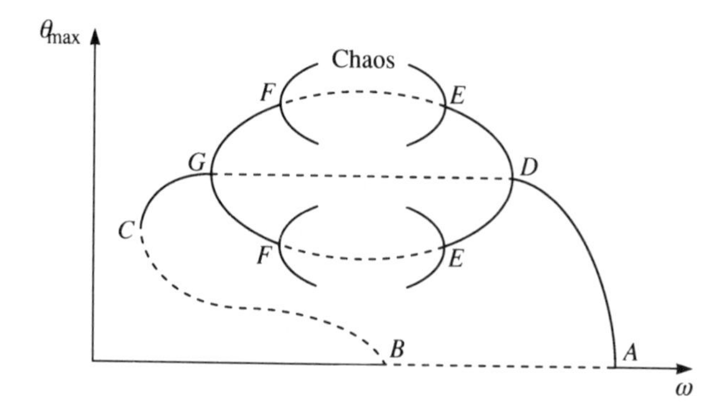

Fig. 1.16. Schematic bifurcation diagram showing the complete sequence on the route to chaos. A and B are the first bifurcations from the trivial nonswinging motion. D and G are symmetry-breaking bifurcations such that the swing is slightly higher to one side than the other. The points labelled E and F mark the beginning of the period-doubling cascades to chaos. (Note that we have only shown half of the bifurcation diagram as there will be a mirror image of this sequence because of the other solution branches arising at A and B.)

sively by Holton and May in this book with reference to one-dimensional maps. This route to chaos has been observed in a wide variety of physical systems from lasers to fluid flows as discussed by Berge *et al.* [6]. In the present case, the period-doubling sequence exists over an extraordinarily small parameter range, so that it can only be observed in the numerical results if a great deal of care is taken. In addition, it is not possible to observe it in the experiment and most of the 'chaos' seen there is probably transient behaviour. Nevertheless, the knowledge of the presence of the period-doubling route to chaos gives a deeper understanding of the observed phenomena. We will return to this point below.

Continued decrease in α leads to a reversal of the above process through an inverse cascade to the point labelled F, an inverse symmetry breaking at G, and finally the limit point C as discussed above. It should also be noted that for values of ω to the left of B there is an overlap of the stable trivial solution and the chaotic solution and so both can be realized in this parameter range. Thus chaotic motion and the simple nonswinging solution can coexist at the same point in parameter space.

The above sequence of events corresponds to a single sweep through the resonant band and shows one way that chaos arises in this system. We present in Fig. 1.17 the two-parameter diagram showing the region where the symmetry breaking and period doubling of the oscillating solutions occurs. The single pendulum has also been studied recently by Bryant and Miles [7] using an alternative numerical method and good

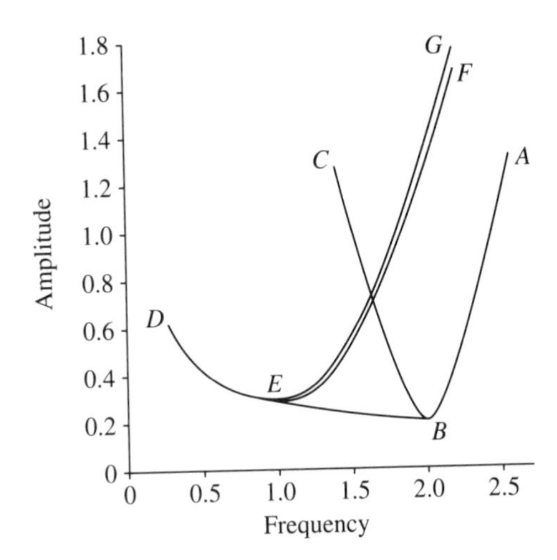

Fig. 1.17. Calculated bifurcation set for the parametrically excited pendulum. *ABC* is the locus of bifurcations from the trivial nonswinging solution, *BD* the locus of the limit points for the subcritical bifurcation, *EF* the locus of symmetry-breaking bifurcations, and *EG* that for the first period doubling. (Note that *EF* and *EG* do not meet the curve *BD*; they merely asymptote to it but we are unable to show this feature properly on the scale used here.)

agreement is found with the results presented here. An important point to note is that the symmetry-breaking and period-doubling curves do not intersect the limit point curve but merely asymptote to it. If they did meet then they would do so at a multiple bifurcation point which, as we shall see in later chapters, would provide an organizing centre for the chaotic phenomena. This does not happen in the single pendulum case which perhaps provides a reason why the chaos exists in a very narrow region of parameter space. The situation appears to be degenerate in the following sense. Chaos is generally found over wide parameter ranges in more complicated dynamical systems and is thus typical behaviour. Indeed this is also found to be the case for a parametrically excited double pendulum system studied by Skeldon [8]. The absence of a multiple bifurcation point in the parametrically excited simple pendulum is probably the reason for this special behaviour.

Period doubling of the vibrating solutions is not the only route to chaos for the pendulum, as the solutions associated with the rotary motion can also become chaotic via period doubling. The interested reader should consult the original reference of Leven *et al.* [9] for a full discussion of the rotary route to the onset of chaotic motion. Leven *et al.* also attempted to investigate the onset of chaos experimentally and showed some evidence for asymmetric states and period doubling. However, as seems to be typical for investigations of nonlinear phenomena and chaos, even with a very carefully controlled experiment it required considerable effort to obtain quantitative results.

1.5 Sensitivity to initial conditions

We will now use the pendulum equations to illustrate the very important feature of sensitivity to initial conditions of chaotic solutions. Presented in Fig. 1.18 are two time series which were obtained from integration of the pendulum equation in a parameter range where chaos is known to exist. The two were given slightly different initial conditions, differing by one part in 10^6. It can be seen that initially the two traces follow each other very closely, and then they diverge so that they are no longer well correlated. Small differences in initial conditions lead to very different outcomes with the progress of time. Thus, in the chaotic regime, there is extreme sensitivity to initial conditions and trajectories started close together will diverge from each other exponentially fast.

However, we can see from the chaotic phase portrait of Fig. 1.15 that the chaotic solution occupies a finite volume of phase space. i.e. the chaotic solution exists on an attractor in phase space. Thus, although trajectories must continue to diverge for all time, they must do so within a finite volume. This is achieved by the nonintuitive procedure of stretching and folding on the attractor and is perhaps best illustrated by considering the example of the Rössler attractor shown in Fig. 1.19.

The Rössler equations are

$$\dot{X} = -Y - Z, \qquad \dot{Y} = X + pY, \qquad \dot{Z} = q + XZ - rZ. \tag{1.11}$$

They are a set of coupled ordinary differential equations which model a chemical oscillator as discussed by Berge *et al.* [6]. They exhibit a period-

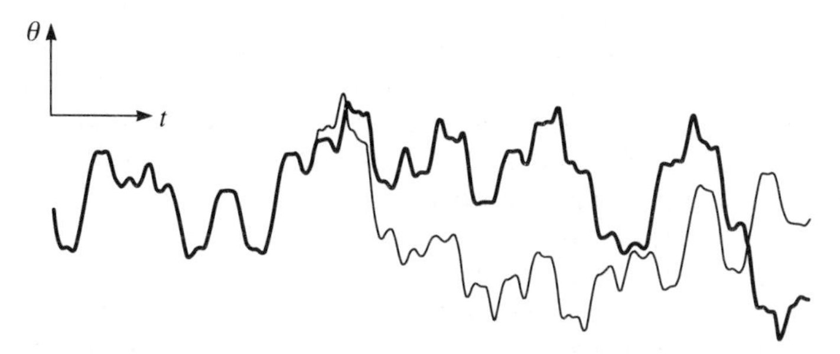

Fig. 1.18. Two velocity time series records obtained from integrations of the equation of motion of the parametrically excited pendulum in the chaotic regime. The initial conditions for the integration are set to be different by 1 part in 10^6.

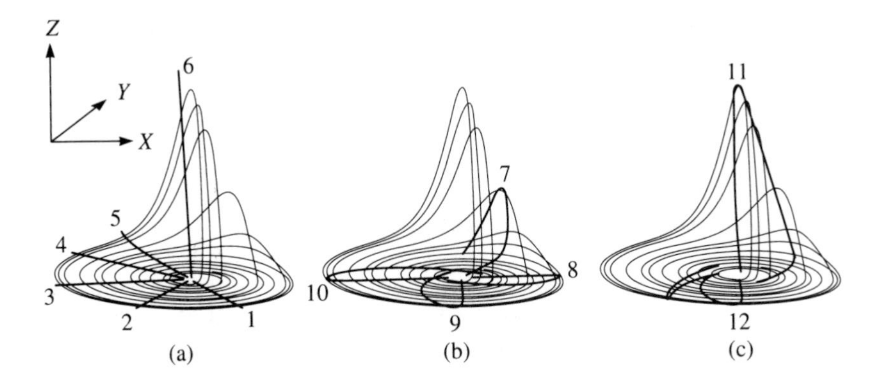

Fig. 1.19. A sequence of snapshots of the evolution of a line of 21 points as it is swept around a strange attractor found in the Rössler equations for parameter values $p = q = 0.2$ and $r = 0.5$. The figures are given sequentially (a) → (b) → (c) and the events are labelled numerically within each figure.

doubling bifurcation sequence to chaos as in the case of the pendulum discussed above. Their phase space is three dimensional and we show a phase portrait for the chaotic state with parameter values $p = q = 0.2$ and $r = 5.7$ in Fig. 1.19. The sequence of figures shows a series of time-lapsed images of the evolution of a line of 21 points superposed on the attractor. Thus the line represents a set of 21 different initial conditions which are all started at the same instant in time.

The line of points is initially stretched as it progresses around the attractor to position 5 in Fig. 1.19(a). Subsequently, some of the points leave the plane and are folded back in to form the horseshoe shape shown in position 7 of Fig. 1.19(b). The process of stretching and folding repeats with each sweep around the attractor. Thus chaotic mixing of the original line of points occurs and the initial stages of this can be seen in Fig. 1.19(c). This is an example of 'strange attractor' behaviour, where we have an attracting region of phase space inside of which the trajectories exhibit sensitivity to initial conditions.

Finally, we would like to make some remarks here concerning the dimension of the attractor. A limit cycle can be embedded in two dimensions or a plane but locally it can be considered as a line with dimension 1. Similarly, a torus can be embedded in three dimensions and it may locally be described by a surface of dimension 2. Therefore the dimension of the attractor is less than the dimension of the space in which it can be embedded. The strange attractor of the Rössler model described above can be embedded in three dimensions, but we also know that to avoid intersections of trajectories in this deterministic system the dimension of the attractor must be greater than 2. Thus the dimension of this attractor must lie somewhere between 2 and 3 and is therefore fractional. Therefore, strange attractors exhibit all the scaling properties associated with fractal objects.

1.6 Final remarks

We have shown that the simple deterministic physical system of a parametrically excited pendulum can exhibit complicated nonlinear behaviour including chaos. As mentioned previously, we could have alternately introduced external forcing into the system by moving the pivot point from side to side instead of up and down. Here the trivial state will be swinging motion at the driving frequency. There are resonant tongues as in the parametrically excited case and it can be shown that the two problems are related through the Mathieu equation. The main difference is that the edges of the tongues in the forced case are symmetry-breaking bifurcations. Inside the resonant tongues, period-doubling sequences and chaos are again found. A discussion of this problem is given in the book by Baker and Gollub [10] and the references therein.

An additional feature of parametric excitation is that it can also be used to stabilize an otherwise unstable equilibrium. In other words, the addition of an external driving force need not necessarily produce instability and chaos. In the case of the pendulum, the inverted state can be stabilized if the frequency of excitation is high. This solution was known to Mathieu and interestingly it has also been shown to become chaotic through symmetry breaking and period doubling by Skeldon and Mullin [11].

In summary, we have shown one route to chaos in an apparently simple physical system. The chaos is not a result of errors in the input of the numerical simulation on the computer or the physical realization in the laboratory. It is a fundamental property of the equations of motion. Moreover, the route to chaos appears to follow that found in simple one-dimensional maps and a wide variety of physical systems including lasers and fluids. However, as we shall see in later chapters, a feature of some more complicated physical systems is that the period-doubling sequences are only part of a global dynamical structure which is 'organized' by multiple bifurcation points in the underlying solution set. Such points do not exist for these very simple oscillators, and so their behaviour could be considered to be degenerate in some sense. Nevertheless, they are very instructive systems to study since they can be investigated very straightforwardly on any small computer and provide spectacular demonstrations in the laboratory.

References

1. Jordan, D. W. and Smith, P. (1987). *Nonlinear ordinary differential equations*. Oxford University Press.
2. Mathieu, E. (1868). Memoire sur le mouvement vibratoire d'une membrane de forme elliptique. *J. Math. Pures Appl.*, **13**, 137–203.
3. McLachlan, N. W. (1947). *Theory and application of Mathieu functions*. Oxford University Press.
4. Doedel, E. (1981). AUTO, a program for the automatic bifurcation analysis of autonomous systems. *Cong. Numer.*, **30**, 265–84.
5. Feigenbaum, M. J. (1979). The universal metric properties of nonlinear transformations. *J. Statist Phys.*, **21**, 669–706.
6. Berge, F., Pomeau, Y., and Vidal, C. (1984). *Order within chaos*. Wiley, New York.
7. Bryant, P. J. and Miles, J. W. (1990). On a periodically forced, weakly damped pendulum. Part 3: Vertical forcing. *J. Austral. Math. Soc.* Ser. B, **32**, 42–60.
8. Skeldon, A. C. (1990). Bifurcations and chaos in a parametrically excited double pendulum. D. Phil. thesis, Oxford.
9. Leven, R. W., Pompe, B., Wilke, C., and Koch, B. P. (1985). Experiments on periodic and chaotic motions of a parametrically forced pendulum. *Physica* D, **16**, 371–84.
10. Baker, G. L. and Gollub, J. P. (1990). *Chaotic dynamics: An introduction*. Cambridge University Press.
11. Skeldon, A. C. and Mullin, T. (1992). Mode interaction in a double pendulum. *Phys. Lett. A*, **166**, 224–9.

2

A dynamical systems approach to time series analysis

Tom Mullin

The aim of this chapter is to show how the information extracted from a physical experiment or numerical computation may be analysed using modern signal processing techniques and indicate how the results can be related to ideas from finite-dimensional dynamical systems. The ultimate objective is to see whether a direct connection can be made between the information contained in the time series obtained from a continuous infinite-degree-of-freedom system and the well-developed field of chaos in ordinary differential equations and maps. Before we proceed with the description of the modern nonlinear techniques, we will first of all briefly discuss the traditional linear methods of signal processing. Next we will introduce the concepts of phase portrait reconstruction using the straightforward 'method of delays'. The principles are extended by a discussion of the singular value decomposition technique which provides a systematic way of selecting a coordinate system in which the attractor can be constructed and which also deals with the universal problem of noise in experimental observations. A brief discussion of the construction of Poincaré maps is then presented as a useful technique for reducing the dimension of the reconstructed attractor. Finally, we discuss the merits and practical limitations of methods of obtaining quantitative estimates from these nonlinear signal processing techniques.

2.1 The time series

The time series may be obtained from a transducer which measures a state variable of an experimental arrangement as a function of time. We will assume that the measure is typical in the sense that it contains all the relevant dynamical information about the system. In practice, this is not usually a restrictive assumption because in the nonlinear regime there is normally strong coupling between the excited modes of a dynamical system. Obviously certain modes will make larger contributions than others depending on the location of the measuring point. However, experience shows that it is normally possible to obtain typical measures with sufficient care.

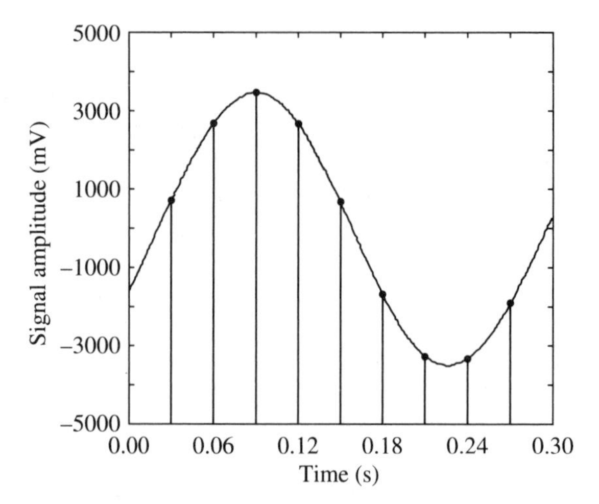

Fig. 2.1. A schematic diagram of the conversion of a sine wave into a set of discrete numbers using an analogue-to-digital (A/D) converter.

The continuous analogue signal from the transducer is fed into the analogue-to-digital (A/D) converter of a computer, which then samples the time series at regular intervals. It converts the voltage levels of the signal at the predetermined intervals into a set of binary numbers as shown schematically for a sine wave Fig. 2.1. The time between samples is referred to as the sample time τ_s. The continuous time series is thus converted into a sequence of numbers which are sampled to a finite precision. It may be represented in matrix form as

$$X = (\boldsymbol{x}_0, \boldsymbol{x}_1, \ldots, \boldsymbol{x}_{N-1}),$$

where N is the total number of samples.

In the numerical integration of a set of equations, a similar series of numbers could be generated and thus one would also obtain a discrete representation of a continuous signal. However, even if single precision arithmetic is used for the numerical integration the accuracy is still approximately 10^3 times better than that available through the A/D converter. This can have serious consequences when trying to implement without modification many of the modern signal processing techniques which have been derived from dynamical systems theory. The reason for this is that many of the methods have been developed and tested using data from maps or from the integration of ordinary differential equations. Thus the potentially serious practical problems associated with the finite precision sampling of noisy signals have not always been considered. This can lead to erroneous or misleading results if the 'model' techniques are applied without modification.

Finally, we end with a few comments on the sources of noise in experimental time series. These include mechanical vibration, parameter fluctuations, transducer noise, etc. In addition, the electronic circuits which may be used to convert the raw transducer output into a suitable time series form prior to A/D conversion are also sources of noise and chaos since they themselves are often nonlinear deterministic systems. Thus the time series on which we wish to perform the analysis is a sum of experimental information with its associated fluctuations together with the measurement noise. In general, it is very difficult to deconvolve this combination, and so it is imperative when making delicate experimental observations on dynamical processes to minimize *all* sources of noise.

2.2 Linear signal processing

The most common and very useful way of analysing a time series using linear signal processing techniques is to construct the power spectrum and its Fourier transform pair the autocorrelation function. The fundamental assumption of these techniques is that the stationary signal can be decomposed into a series of sines and cosines. The spectrum gives a measure of the amount of power in a given frequency band over a selected frequency range. If it consists of discrete lines, then the time series is periodic and the spectrum can be used to obtain estimates of the relative power in each frequency component of the time series. On the other hand, irregular, chaotic, and random time series all have a continuous part in their spectra indicating that the power is spread over a range of frequencies. Therefore spectral analysis can usefully be used to investigate the transition between the discrete and continuous cases as a control parameter in an experiment is varied.

The autocorrelation function for a periodic signal is itself periodic and can often give a less confusing representation of the data than the power spectrum. This is particularily so when there is a high harmonic content in the signal. It therefore provides a very useful complementary representation of the data. Irregularity in the time series gives rise to a decay in the autocorrelation function and the rate of decay gives a measure of the degree of irregularity. However, the extraction of quantitative estimates from the autocorrelation function can be problematical and it is usually used as a complement to power spectral analysis.

The discrete Fourier power spectrum is defined as

$$P(f) = 2|H_n|^2,$$

where

$$H_n = \sum_{k=0}^{N-1} x_k \mathrm{e}^{2\pi i k n/N}.$$

The connection between the discrete Fourier transform (DFT) and the familiar integral formulation is not trivial and the interested reader should consult the book by Hamming [1] for a lucid discussion of this point. In practice an efficient computer implementation of the DFT called the fast Fourier transform (FFT) is normally used. Algorithms for calculating the FFT are now widely available and a good discussion of them is given in the book by Press *et al.* [2].

There are two main practical points to be considered when using these signal processing techniques. First, the time series is usually assumed to be stationary, i.e. transient behaviour is not present. There are ways of dealing with transients using FFT techniques but they need to be used with care. Secondly, the highest frequency which can be accommodated is called the Nyquist critical frequency which is given by

$$f_c \equiv \frac{1}{2\tau_s}$$

where τ_s is the sample time. Thus the highest frequency component of interest in the time series must be sampled at least twice per period to obtain a proper representation of it in the power spectrum. If this condition is not met, then aliasing errors arise which result in a folding back of the undersampled higher-frequency components into the range of interest to give a distorted spectrum.

Since the power spectrum is the square of a complex function, then phase information is lost in this type of analysis. In addition, the spectrum is usually averaged over many blocks of the data in order to obtain a good statistical representation of the frequency content of the time series. Therefore, in order to obtain phase information, one must calculate and examine the DFT directly.

The autocorrelation function is defined as

$$R(n\tau_s) = \frac{1}{N-1} \sum_{i=0}^{N-1} x_i x_{i+n\tau s}$$

and the spectrum and autocorrelation function form a Fourier transform pair. In fact some commercial spectrum analysers calculate the autocorrelation function first because it is a particularly robust procedure in the presence of noise.

We present a set of power spectra and their corresponding autocorrelation functions in Fig. 2.2. The spectrum for periodic motion in Fig. 2.2(a) shows that the power is concentrated in a single main spike at the frequency of the wave and the autocorrelation function is shown in Fig. 2.2(b) is a simple cosinusoid. A subharmonic period-8 time series was used to produce spectrum shown in Fig. 2.2(c). There we see a peak at the main frequency and a sequence at the subharmonics together with all the harmonics and sums and differences. Although it looks very complicated, the corresponding autocorrelation function shown in Fig. 2.2(d) has a simple form. It must be remembered that the spectrum and the autocorrelation function contain the same information, but in some cases one is easier to visually interpret than the other.

In a purely random process all frequencies are excited at the same energy level and so the power spectrum would be flat over the whole frequency range if the calculation is carried out for infinite time. However, in practice noise processes are bandwidth-limited or coloured so that all frequencies are not given equal weighting. An example of such a spectrum is given in Fig. 2.2(e) and the corresponding autocorrelation in Fig. 2.2(f). It can be seen that the presence of noise is indicated by a decay in the autocorrelation function. Chaos also gives continuous spectra over a limited range of frequencies as shown in Fig. 2.2(g) and so it is not always possible to distinguish it from coloured noise directly using this standard technique. In addition, the corresponding autocorrelation function for the chaotic signal given in Fig. 2.2(h) shows a decay. Thus it is clear that situations may be encountered when it is not possible to distinguish between coloured noise and chaos. Finally, we show a power spectrum autocorrelation pair in Figs. 2.2(i, j) constructed from a time series taken in a turbulent flow field. Again it is very difficult to distinguish this from noise immediately, although there are techniques such as calculating higher-order moments which help do this. However, one obvious difference between this spectrum and the one formed from the chaotic signal is that the energy is spread over a much wider bandwidth.

We do not intend to give the impression that the above techniques are valueless in the study of finite-dimensional dynamics. On the contrary, they are extremely important and of immense practical value in the laboratory. However, the development of nonlinear signal processing techniques is highly desirable since they will not only permit us to distinguish between chaos and noise but also enable us to develop a closer link between theory and experiment. In addition, it is surely desirable to develop nonlinear signal processing techniques to analyse the time series obtained from nonlinear dynamical systems.

2.3 Phase portrait reconstruction

2.3.1 *Method of delays*

The technique which is central to modern nonlinear signal processing is phase portrait reconstruction, where the attractor is formed in a pseudo-phase space from a single time series measurement. The theoretical phase portraits introduced in the previous chapter on pendulums consisted of

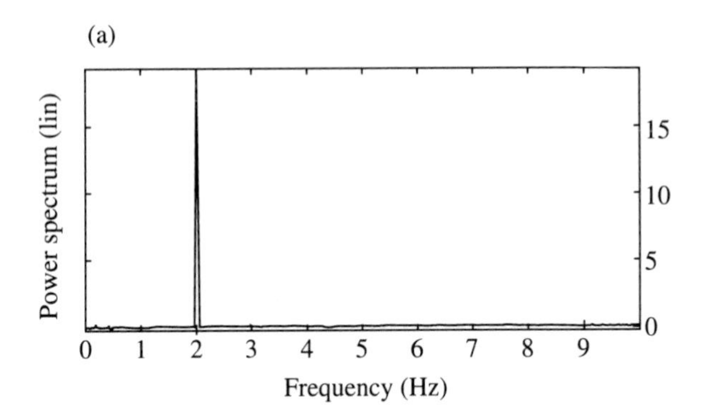
(a)
Power spectrum (lin)
15
10
5
0
0 1 2 3 4 5 6 7 8 9
Frequency (Hz)

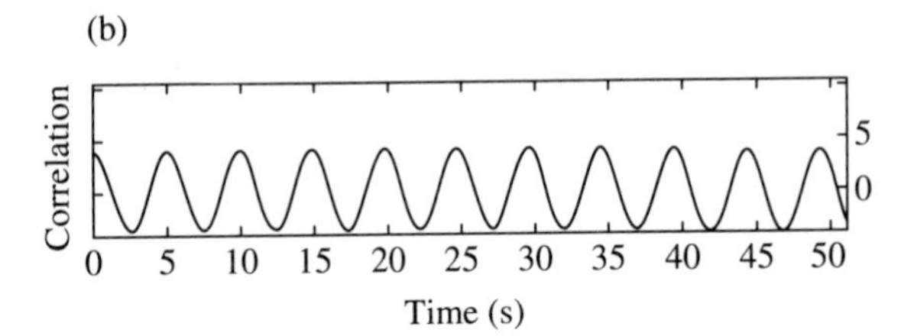
(b)
Correlation
5
0
0 5 10 15 20 25 30 35 40 45 50
Time (s)

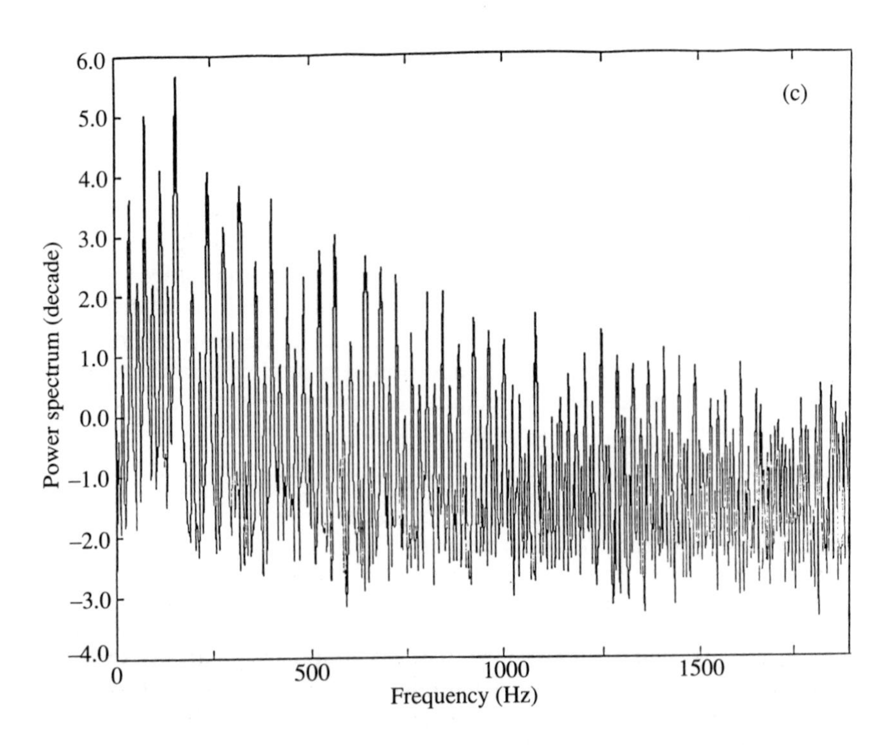
(c)
Power spectrum (decade)
6.0
5.0
4.0
3.0
2.0
1.0
0.0
−1.0
−2.0
−3.0
−4.0
0 500 1000 1500
Frequency (Hz)

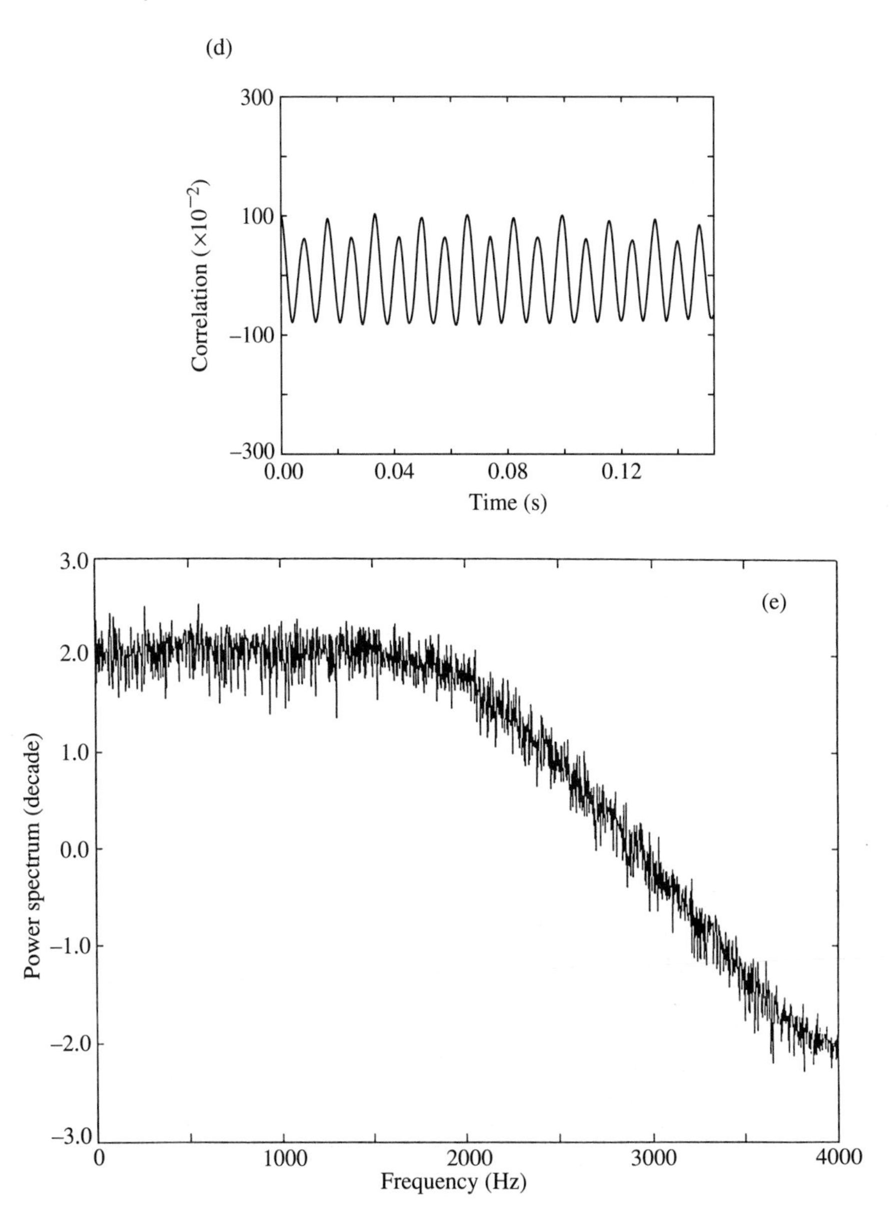

Fig. 2.2. Some commonly encountered time histories found in the study of dynamical systems and their representations obtained using standard linear signal processing techniques: (a, b) the power spectrum and autocorrelation function for a singly periodic signal; (c, d) the power spectrum and autocorrelation function for a subharmonic period-8 time series; (e, f) the power spectrum and autocorrelation function for sample random noise; (g, h) the power spectrum and autocorrelation function for a weakly chaotic time series; (i, j) the power spectrum and autocorrelation function for a signal from a turbulent flow field.

(f)

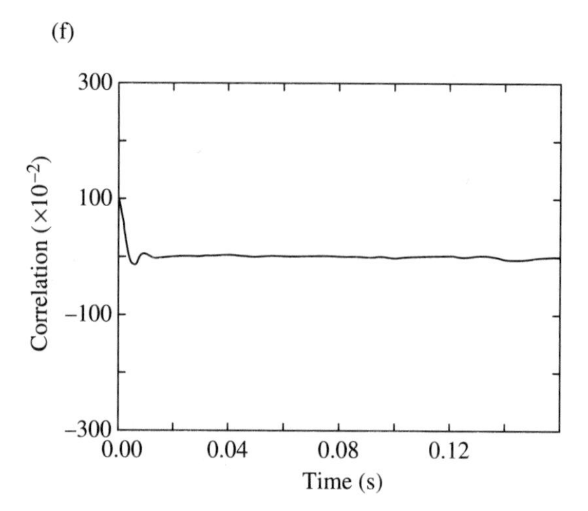

300
100
−100
−300
Correlation (×10⁻²)
0.00
0.04
0.08
0.12
Time (s)

(g)

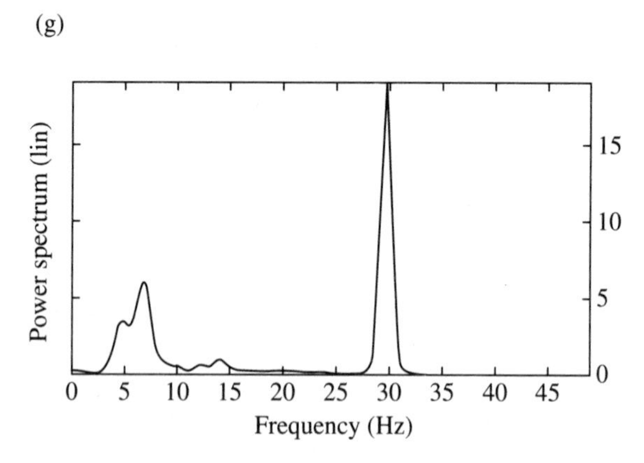

Power spectrum (lin)
15
10
5
0
0 5 10 15 20 25 30 35 40 45
Frequency (Hz)

(h)

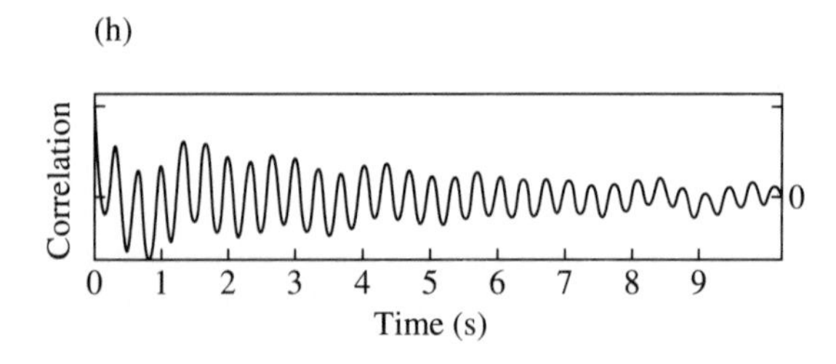

Correlation
0
0 1 2 3 4 5 6 7 8 9
Time (s)

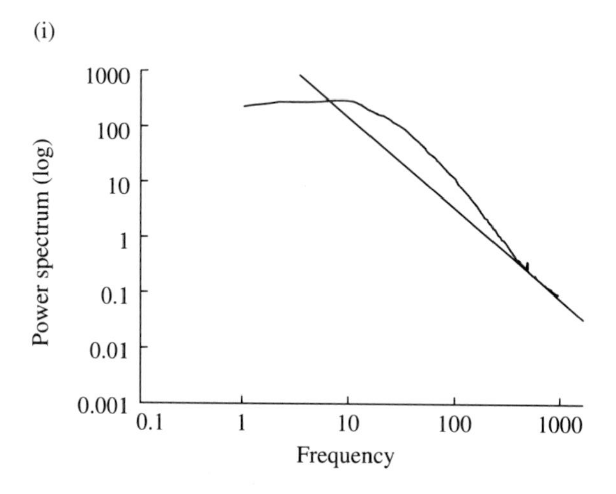
(i)
Power spectrum (log)
1000
100
10
1
0.1
0.01
0.001
0.1
1
10
100
1000
Frequency

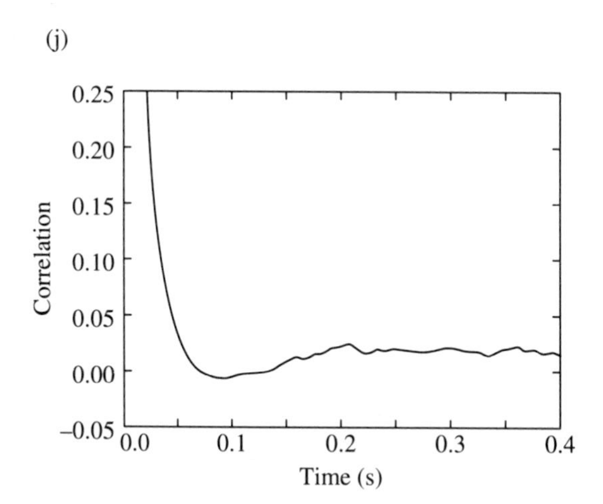
(j)
Correlation
0.25
0.20
0.15
0.10
0.05
0.00
–0.05
0.0
0.1
0.2
0.3
0.4
Time (s)

Fig. 2.2. *contd.*

plots of variables versus their derivatives. Although it is in principle possible to carry out this procedure with experimental data, problems will arise in practice due to the presence of noise on the finitely sampled data. This would give artificially high derivatives which will at best severely distort the reconstructed phase portrait. A practical solution to this difficulty would be to use filters, but this has obvious limitations if the time series covers a wide frequency range. Therefore we tend to use a simple practical alternative called the method of delays which was first proposed by Packard *et al.* [3]. The connection between the phase portraits reconstructed by this technique can be shown to be topologically equivalent to the actual attractors as shown by Takens [4]. One example of this equivalence will be given below for the Rössler attractor. In addition, although it is difficult to apply the same mathematical rigour to noisy experimental data sets, there is good reason to believe that the important dynamics are correctly reproduced by these methods.

In the primitive method of delays, the first sample is plotted against the second, the second against the third, and so on, i.e. $x(i)$ is plotted against $x(i+1)$ where i goes from 1 to N and N is the total number of samples. This method gives a two-dimensional projection of the phase portrait and we show a simple construction for a sine wave in Fig. 2.3. One immediate problem with this approach is that the neighbouring samples are highly correlated so that the ellipse in Fig. 2.3 is strongly aligned along the [1, 1] direction. This can be a major problem in practice as a complicated time series needs to be highly sampled to retain as much information as possible. Therefore, there will be a high correlation over a range of samples and the resulting portrait would be highly distorted using this naive approach.

In order to overcome this practical limitation, we must choose a delay time $n\tau_s$, where τ_s is the sample time, and the results for the sine wave

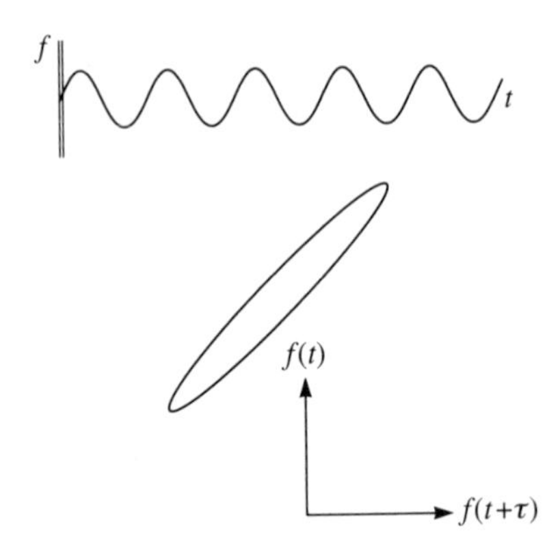

Fig. 2.3. Phase portrait of a sine wave reconstructed using sequential samples.

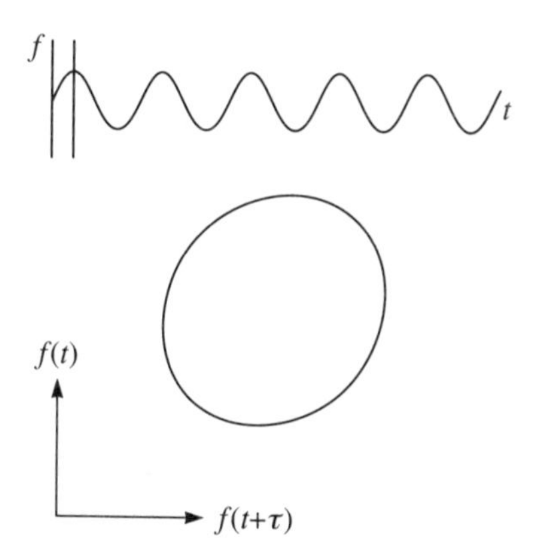

Fig. 2.4. Phase portrait of a sine wave reconstructed using an optimal delay time. Here it is one-quarter of the period of the oscillation.

are shown in Fig. 2.4. In practice, there are a number of suggested techniques for choosing n and these are reviewed in Buzug *et al.* [5]. In fact, the choice of n will not alter the topology of the attractor, but this is of little consolation when dealing with some complicated geometrical form as the compression and stretching of the structure by a poor choice may distort pertinent features to the extent that they are very difficult to analyse.

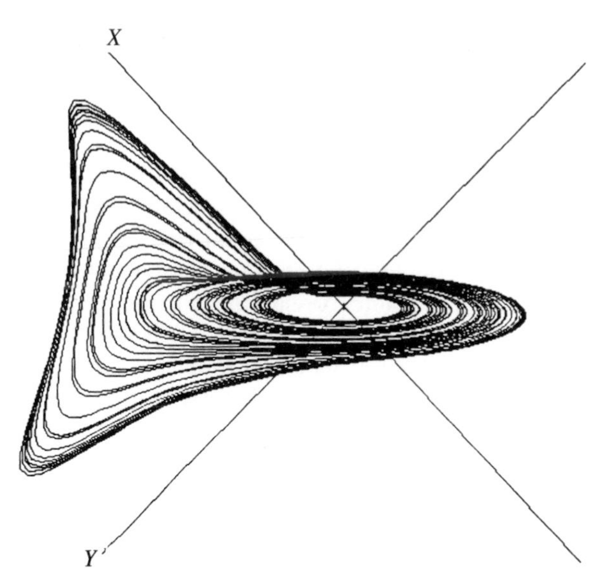

Fig. 2.5. A reconstructed phase portrait of the Rössler attractor obtained from a method of delays construction of the summed $(X + Y + Z)$ time series from an integration of the equations.

In addition, if noise is present and the object is highly compressed in one or more directions by the reconstruction process, then the topology could become completely obscured.

The above technique can be extended to three or higher dimensions by introducing second and third delayed coordinates, etc. In order to demonstrate that the technique works on something more complicated than a sine wave, we show the reconstructed Rössler attractor in Fig. 2.5. This has been calculated from a single time series obtained by the numerical integration of the equations and a simple summation of all the components. It should be compared with the full attractor shown in Fig. 1.19 and one may see that they are qualitatively the same. In fact one can show that they are also quantitatively the same by calculating their respective dimensions by the techniques which will be discussed in Section 2.5.2.

2.3.2 *Singular value decomposition*

The main practical problem with the standard method of delays is that the phase portrait reconstruction is carried out by projecting the time series onto an arbitrary basis. Thus no systematic account is taken of the information content of the signal in the reconstruction process, nor are the effects of finite sampling and experimental noise in the time series catered for. The technique of singular value decomposition (SVD) was proposed as a solution to these difficulties by Broomhead and King [6]. It is used to calculate an optimal basis for the projection of the attractor which is reconstructed from the time series. In addition, the technique was developed to deal with noise and is, therefore, ideally suited to experimental data.

The procedure is carried out as follows. First of all we form a trajectory matrix X whose rows contain the n-dimensional vectors used in the method of delays with the delay time set equal to τ_s. In practice we set n large enough to capture the lowest frequency component in the signal and n can be selected using the autocorrelation function. We refer to this as the window length of our trajectory matrix. Each successive row in the matrix is then given by sequential data windows.

The objective of the SVD technique is to find the unit vectors which are optimally aligned with the position vectors of the trajectory matrix X. These unit vectors will form the coordinate system onto which the time series will be projected in the phase portrait reconstruction. One may think of this technique as finding the principal moments of an 'object', which in this case is the attractor, i.e. it is a method of extracting the optimum projection of the phase portrait from the data. The process involves diagonalization of the covariance matrix X^TX to obtain the set of eigenvectors which are the orthogonal singular vectors. The square roots of the corresponding eigenvalues give the singular values. Thus the singular

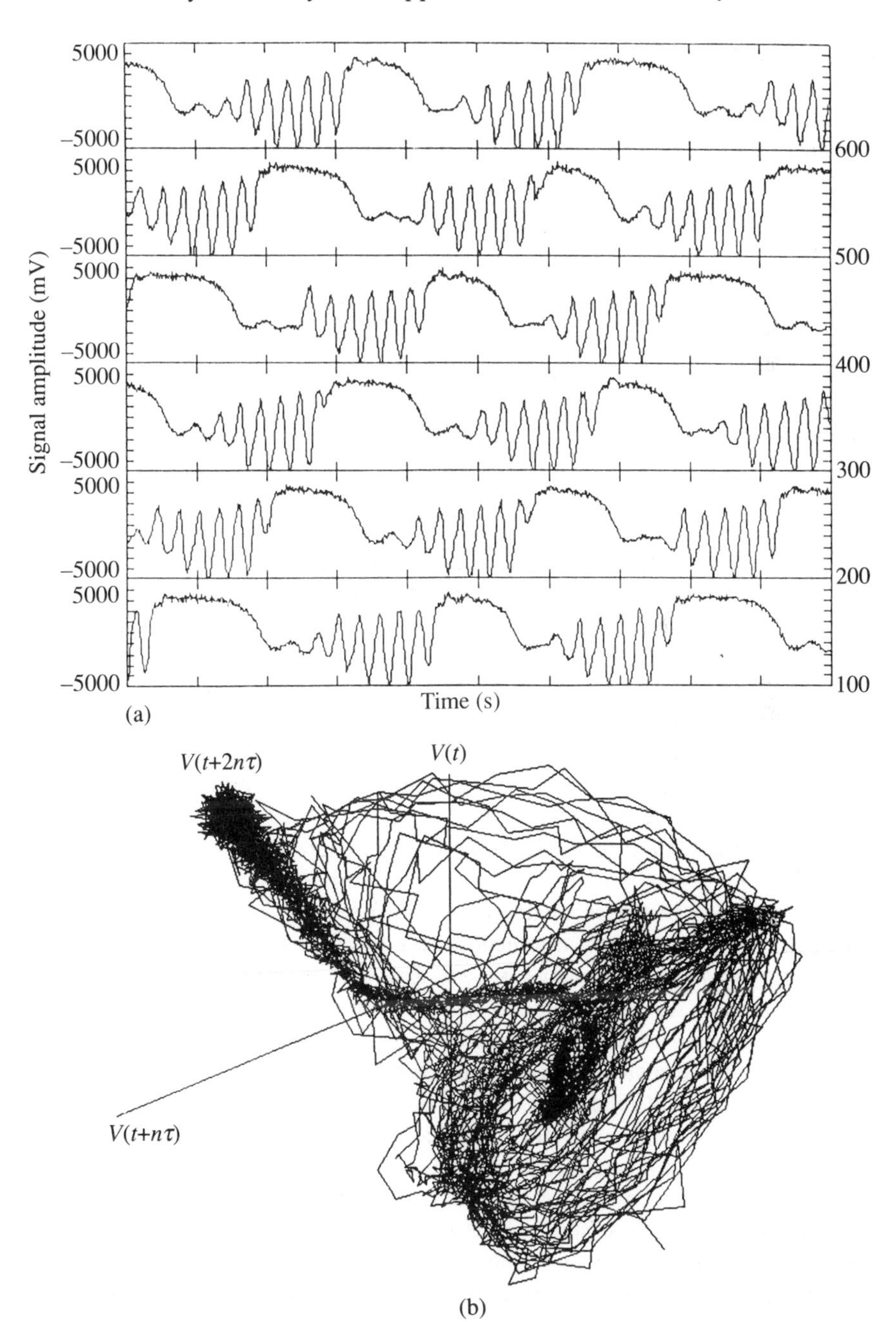

Fig. 2.6. (a) The velocity time series for a quasi-periodic fluid flow motion. (b) The reconstructed attractor obtained using the method of delays with the optimal delay time.

vectors give the directions of the coordinate axes and the corresponding singular values give the weightings for each. The phase portrait is constructed on this coordinate system and each point on the trajectory is a weighted version of the sampled point in the time series.

Noise will be present in all the singular values and will appear in the singular value spectrum as a flat noise floor at its upper end. The significant singular values will appear above the noise floor and their number gives an upper limit on the embedding dimension of the attractor. Thus, this is a way of distinguishing the deterministic part of the signal from the noise. It should be noted that this process does not give the dimension of the attractor directly. However, the scaling of the singular values can be used to give an estimate of the local dimension of an attractor (see [7] for details).

We now turn to a specific example to show some of the practical advantages of this technique when applied to experimental data. We show in Fig. 2.6(a) a time series obtained from a fluid flow experiment and below it in Fig. 2.6(b) the phase portrait reconstructed by the method of delays using the optimal delay time. We will not discuss here the details of the experimental system which provided this example as they will be fully explained in Chapter 4. It is sufficient to note for the present discussion that this is essentially quasi-periodic motion on a two-frequency torus. The same data set was used to construct the phase portrait shown in Fig. 2.7(b) using the SVD technique. Its singular spectrum is shown plotted on a logarithmic scale in Fig. 2.7(a), where it can be seen that the presence of three significant singular values suggest that it is justifiable to plot the phase portrait in three dimensions. The difference between Figs. 2.6(b) and 2.7(b) is evident and is due purely to the nonoptimal projection of the trajectories in the reconstruction shown in Fig. 2.6(b). In essence the trajectories are exploring dimensions which contain only noise, and their projection onto three dimensions can give an apparently poorer signal to noise content. The obvious improvement in the reconstruction process using SVD is not just cosmetic and may be of considerable importance when attempting to extract quantitative measures, such as local dimensions etc., from phase portraits which are reconstructed from experimental time series.

Finally, we should point out that phase portrait techniques do have their limitations. It is not possible to construct a meaningful phase portrait from a time series which contains no low-dimensional dynamical structure. While this may appear to be stating the obvious, we would like to emphasize that these techniques are not filtering processes which in some way reduce the data to a finite-dimensional form. To illustrate this point, we show a phase portrait reconstructed from a time series measured in a disordered fluid flow in Fig. 2.8 using the SVD technique. It may be clearly seen that there is no obvious form associated with this phase portrait and the motion explores a high-dimensional state space.

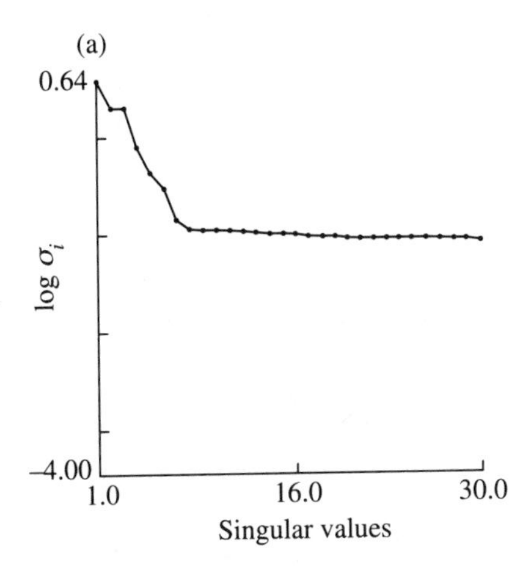

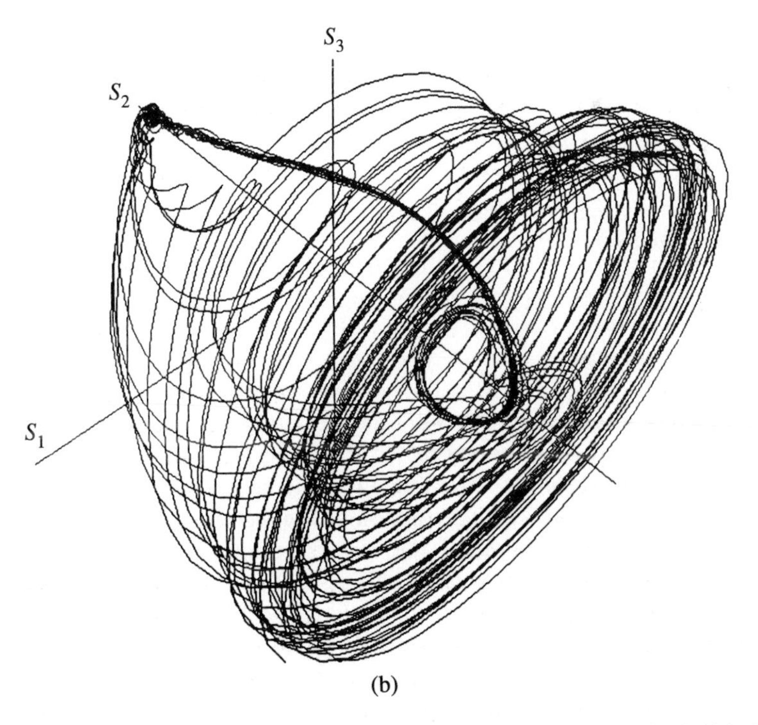

Fig. 2.7. (a) The singular value spectrum for the time series of Fig. 2.6(a). The y-scale is logarithmic and so there are three significant singular values and a definite noise floor. (b) The reconstructed attractor plotted on a coordinate system of the first three singular vectors.

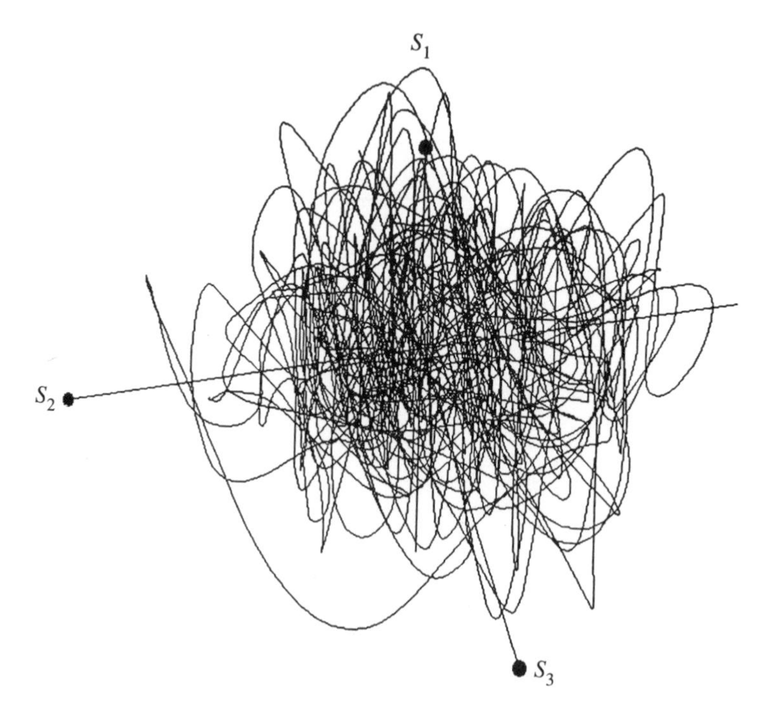

Fig. 2.8. A reconstructed phase portrait from the time series measured in a disordered fluid flow.

2.4 Poincaré maps

An extremely useful way of extracting the vital qualitative characteristics of a phase portrait is to construct its Poincaré map. This can be formed by either stroboscopically sampling the time series at a particular point in a cycle or by forming the intersection of the reconstructed phase portrait with a plane. The mapping of the discrete points by the action of the flow on the attractor then gives a very useful insight into the underlying dynamics. Perhaps the simplest example of such a map is given by the intersection of a limit cycle and an orthogonal plane. An ideal limit cycle would cut the plane at two points each corresponding to a 180° phase shift from the other. If we consider just one of the points, then this contains all the dynamical information about the signal as successive points are mapped onto each other at some fixed phase in the cycle.

Next we show in Fig. 2.9 the phase portrait and Poincaré section through a torus obtained from a fluid flow experiment. The trajectories would normally be represented by lines but here we have left them as points to try and clarify the attractor. A plane cuts across the torus and we have

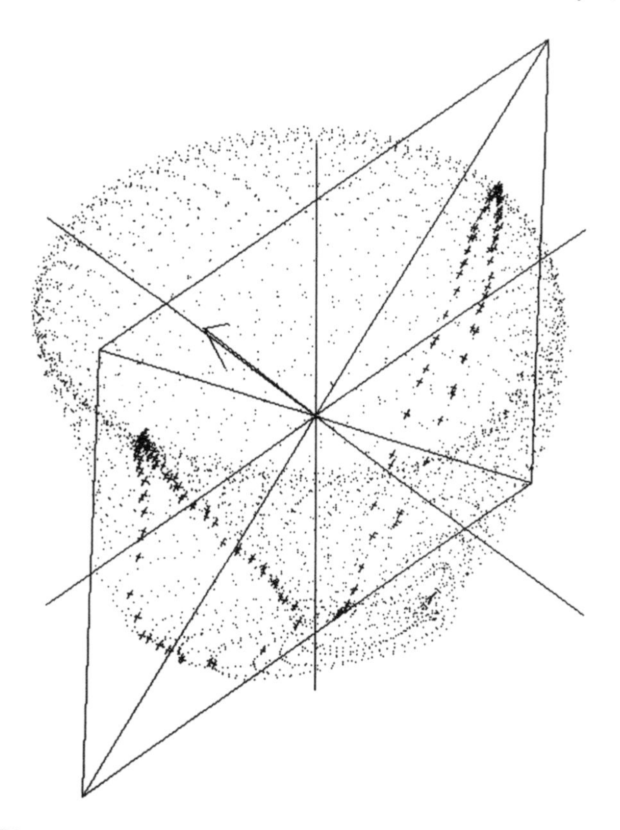

Fig. 2.9. The reconstructed attractor for a quasi-periodic flow. The individual sample points may be seen and the points at which the trajectories cut a selected plane are shown by crosses. We refer to this as Poincaré section since it shows how points are mapped around the plane under the action of the flow on the torus.

marked the points of intersection of the trajectories with the plane by crosses. The two frequencies which make up the torus are not commensurate with each other and the Poincaré section would form a closed loop after a large number of cycles. Points are thus successively mapped around the closed loop for the present case. On the other hand, if there was an integer relationship between the frequencies then the points would form clusters whose number would depend on the frequency ratio between the two oscillations.

In Fig. 2.10 we show the subsequent sequence of Poincaré sections obtained from the same experiment as the motion changes from regular to chaotic when the control parameter is varied. We see that the circle initially splits into two in Fig. 2.10(a), so that one of the frequencies on the torus has period-doubled. Thus the motion must now take place on a doubled

(a)

(b)

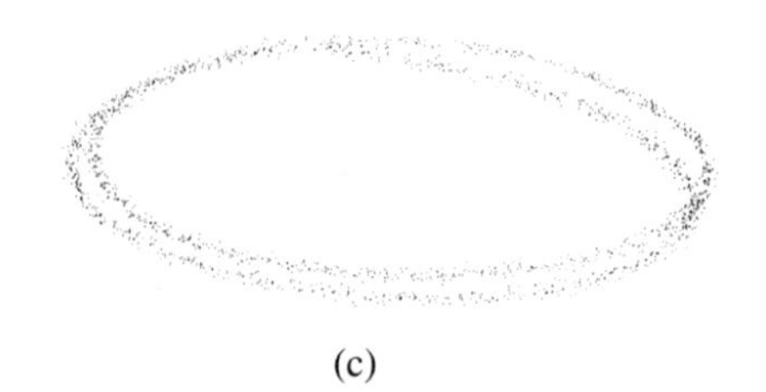

(c)

(d)

Fig. 2.10. A sequence of experimental Poincaré sections showing the appearance of period doubling on a torus: (a) quasi-periodic motion; (b) one of the winding frequencies has doubled, giving rise to a splitting of the section; (c) a three-dimensional Poincaré section formed by the intersection of the doubled torus with a hyperplane; (d) the Poincaré section obtained when the motion has become chaotic.

torus which exists in a minimum of four dimensions, but if we carefully select a section through the three-dimensional representation then the splitting of the torus is clear, as shown in Fig. 2.10(b). If we now take a three-dimensional section of this four-dimensional object, then we obtain the period-doubled loop shown in Fig. 2.10(c). Thus the Poincaré section may usefully be used to reduce the dimension of phase portraits in a meaningful way. Finally, we see in Fig. 2.10(d) that the motion has become irregular, so that the mapping is now not so evidently structured. At this point one might expect to see evidence of fractal structure, etc., in the Poincaré sections but these features are very difficult to discern in fluid mechanical data. We will return to a discussion of this point in Chapter 4.

The Poincaré map can alternatively be constructed without using the phase portrait by stroboscopically sampling the data, as mentioned above. This is particularly useful for periodically driven systems where the sample point can be set at a particular phase angle of the drive frequency. An example of such a system is provided by the parametrically excited pendulum where a fixed phase in the sinusoidal drive cycle can be used as the reference. The book by Moon [8] contains several practical suggestions on how to do this. Finally, care has to be used when interpreting any of these maps when the dynamical behaviour becomes complicated since they are, of course, only a local measure on the attractor.

2.5 Quantitative measures from attractors

A variety of techniques has been developed for obtaining quantitative measures from reconstructed attractors and here we will discuss two of the most extensively used ones. They are estimates of the Lyapunov exponents and calculations of the attractor dimension. In the next chapter we will discuss a novel technique which gives local information about particular points on an attractor. This latter approach can be very useful if there is an 'organizing centre' controlling the chaos.

2.5.1 *Lyapunov exponents*

One representative feature of strange attractor behaviour, which we have already discussed in connection with the Rössler model is sensitivity to initial conditions and the consequent exponential divergence of neighbouring trajectories. This exponential divergence may be characterized by an estimate of the Lyapunov exponent which for a one-dimensional map is defined as

$$d(t) = d_0 e^{\lambda t},$$

where d_0 is a measure of the initial distance apart. Thus if λ is negative this would correspond to a contraction mapping, whereas if it is positive then there will be exponential divergence of nearby initial conditions.

This concept can be extended to higher dimensions, but here we must consider a spectrum of Lyapunov exponents as we are now concerned with the evolution of an n-dimensional sphere of initial conditions in n-space. The sum of all the exponents will be negative if there is an attractor and there will be at least one positive exponent if the attractor is strange. For example, a three-dimensional system which is itself embedded in three dimensions will have three exponents associated with it. We illustrate the situation in Fig. 2.11, where we see in (a) that if all three exponents are negative then all trajectories would converge in towards a fixed point. In Fig. 2.11(b) the system has a limit cycle and thus there are two attracting directions. Nearby trajectories will be attracted onto the limit cycle which itself has a Lyapunov exponent of zero because points on the limit cycle neither converge nor diverge. In Fig. 2.11(c) we show a torus and there is one attracting direction onto the torus and two zero exponents associated with the two winding directions.

Finally we sketch a strange attractor in Fig. 2.11(d), where one of the Lyapunov exponents in our three-dimensional model is now positive and trajectories with neighbouring initial conditions will diverge from each other. In addition there is a negative and zero exponent because the system as a whole is an attractor with a preferred direction of winding. Thus, if

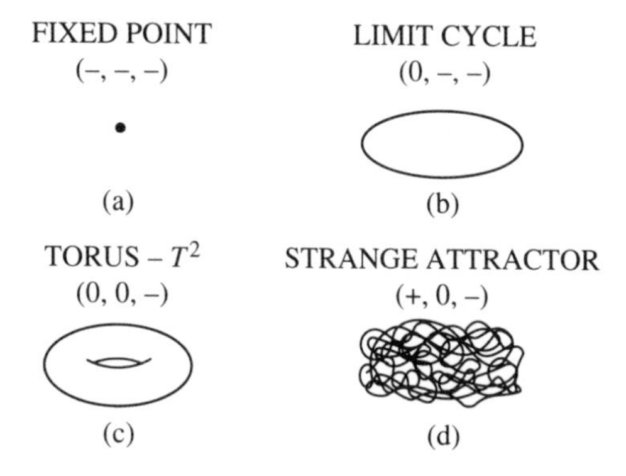

Fig. 2.11. A schematic representation of the function of Lyapunov exponents as measures of the state of a dynamical system. Here, we consider a three-dimensional system. (a) All three exponents are negative and there is a single attracting fixed point. (b) There is one zero exponent corresponding to motion around the limit cycle and two attracting directions. (c) There are now two winding directions on the torus and one attracting direction. (d) One of the Lyapunov exponents is positive, indicating sensitivity to initial conditions. The zero corresponds to the winding direction, and the negative exponent an attracting direction.

we consider a set of initial conditions on the attractor to be enclosed by a sphere and the attractor is embedded in n dimensions, then the sphere will be distorted into an ellipsoid as it moves along the trajectories with the progress of time. The spectrum of Lyapunov exponents is given by

$$\lambda_i = \lim_{t \to \infty} \frac{1}{t} \ln \frac{d_i(t)}{d_i(0)}$$

where the $d_i(t)$ are the lengths of the principal axes of the ellipsoid at time t. It is in principle possible to calculate the Lyapunov spectrum from a time series using the method proposed by Eckmann and Ruelle [9], but this is a very difficult task in practice unless the amplitude of the noise is less than the thickness of the 'skin' of the attractor. This condition is rarely met in practice but recent work by Darbyshire and Broomhead [10] using a technique based on SVD has been successfully used to calculate Lyapunov spectra from experimental data sets.

One other practical consideration is that exponential behaviour will be lost when $d(t)$ becomes too large, and so a new $d_0(t)$ must be chosen. The process can then be repeated and an average formed over the whole attractor. The first Lyapunov exponent can now be calculated from

$$\lambda = \frac{1}{t_N - t_0} \sum_{k=1}^{N} \log_2 \frac{d(t_k)}{d_0(t_{k-1})} .$$

If $\lambda > 0$ then this is taken as an 'indicator of chaos', or if $\lambda \leqslant 0$ then regular behaviour is found.

A practical method of calculating the largest Lyapunov exponent is to use the algorithm developed by Wolf *et al.* [11]. The procedure involves firstly selecting a point $A(0)$ on a trajectory of the phase portrait which is embedded in some n-dimensional phase space. The nearest point $P(0)$ on the attractor is then found, which is a distance away greater than some predetermined noise threshold. This is one practical difficulty with the method, as one does not always know what the noise threshold is and so some empirical experimentation is often required.

Assuming that this difficulty can be overcome, let the scalar distance between the two points be $L(0)$. If the trajectories are now followed for a short fixed time T from the two initial points and a new distance between them $L(T)$ is calculated, then the Lyapunov exponent is given by

$$\lambda_1 = \frac{1}{T} \log_2 \frac{L(T)}{L(0)},$$

which will be positive if the trajectories diverge. The process is then repeated for the point $A(T)$, where a new point $P_1(T)$ is found with the additional restriction that it must be oriented in same direction as $P(T)$. This may be thought of as simulating the distortion of the n-sphere into

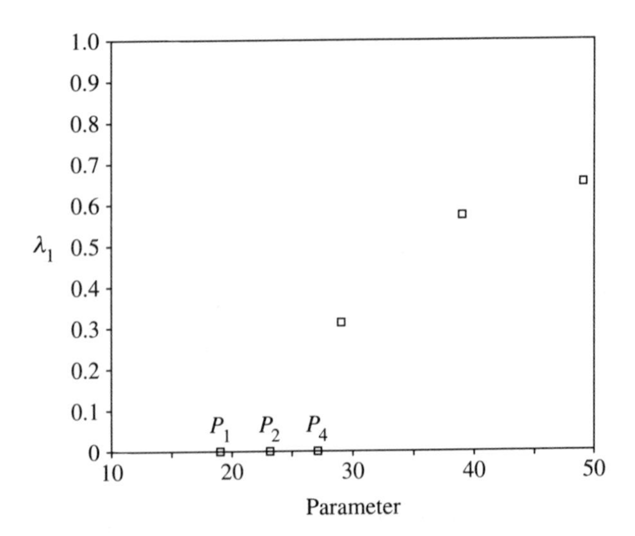

Fig. 2.12. The largest Lyapunov exponent calculated for an experimental phase portrait obtained from a nonlinear electronic oscillator. The first three points are for parameter values where there is a limit cycle, a period-2 orbit, and a period-4 orbit, respectively. Above the parameter value of 28, the output of the circuit is chaotic and the largest Lyapunov exponent is positive.

an ellipsoid as discussed above and preserving the orientation of the major axis of the ellipsoid. The process is then repeated and an average taken over the attractor.

An example of the application of this method is shown in Fig. 2.12 for the experimental phase portrait from an electronic oscillator. The time series for each of the calculations was 25 000 records long and the evolution time T was approximately $\frac{2}{3}$ of the period of the basic oscillation which corresponds to 60 records. We see that the largest exponent is zero to within experimental error for the limit cycle and period-doubled orbits. However, it is positive for the disordered state giving evidence for a chaotic attractor with sensitivity to initial conditions.

It is reasonable to expect this indicator of chaos to give a meaningful result in the present example since the system is governed by a set of ordinary differential equations and the data set provides a reasonably uniform covering of the attractor. The application of these techniques to experimental time series from systems governed by partial differential equations such as fluid flow problems must be treated with caution since the theoretical basis is not yet developed. Finally, phase portraits of the type shown in Fig. 2.7(b) have a nonhomogeneous distribution of data points on the attractor. This clustering is due to the presence of unstable fixed points which as we will see in the next two chapters are responsible for the organization of the dynamical behaviour. Near these points strong contraction will occur in certain directions so that the presence of noise will be more strongly felt, which can lead to erroneous results.

2.5.2 *Dimension calculations*

The simplest geometrical characterization of a reconstructed attractor is to count the minimum number of hypercubes of linear size ε which are required to cover the attractor. This gives an estimate of what is termed the capacity dimension. If the number of hypercubes is $N(\varepsilon)$, then, as ε is varied, $N(\varepsilon)$ varies as ε^{-D}, where D is the dimension. For example, if the trajectory is a line segment of length L, then $N(\varepsilon) = L\varepsilon^{-1}$ and therefore D is 1. In the limit as ε tends to zero, D can be defined as the capacity dimension:

$$D = \lim_{\varepsilon \to 0} \frac{\ln N(\varepsilon)}{\ln(1/\varepsilon)} .$$

The value of the capacity dimension need not be an integer. A set which has a noninteger dimension is said to be fractal and thus has the usual scaling properties associated with such objects. Unfortunately, this direct method of calculating the dimension is impractical to implement on most computers as the power of the number of required hypercubes scales in proportion to the dimension. Thus the combined requirements of small-scale hypercubes with minimum dimensions of 3 for chaos means that large amounts of computer memory are required; for example, an attractor embedded in three dimensions with a covering of $1000 \times 1000 \times 1000$ boxes could require 125 Mb. of computer memory. There are several suggested schemes for coping with this problem and a review of some practical methods is given by Glazier and Libchaber [12].

Another measure of the dimension of an attractor is given by the correlation dimension. Consider a hypersphere of radius r whose centre is on an object which is composed of discrete points. It can be shown that the number of points $N(r)$ contained in the hypersphere scale as r^ν, where ν is the dimension of the object. The correlation dimension is defined as

$$D_c = \lim_{r \to 0} \frac{\ln C(r)}{\ln r},$$

where

$$C(r) = \lim_{N \to 0} \frac{1}{N^2} [\text{number of pairs } (x_i, x_j) \text{ such that } |x_i - x_j| < r].$$

It can be shown that $C(r)$ is proportional to r^{D_c} when r, the radius of the hypersphere, is small compared with the size of the attractor.

The most widely used algorithm for calculating the correlation dimension from an experimental time series is due to Grassberger and Procaccia [13] and a recent review of its uses and limitations is given by Ruelle [14]. It is usually used to calculate $C(r)$ for a range of r over the attractor. In order to make the calculation more practical, the normal practice is to select random pairs of (i, j) rather than systematically evaluating every pair on the attractor. In this way, a plot of $\log C(r)$ versus $\log r$ is produced and the dimension is given by the linear part of the slope of the line. The nonlinear parts of the graphs are set by the resolution of the discrete sampling at one end and the finite size of the attractor at the other. The embedding dimension of the attractor is then increased and a new graph is calculated. The slope will become independent of embedding dimension when this is set to be large enough.

This procedure was followed for the results produced in Fig. 2.13 for the phase portrait obtained from an electronic oscillator in an apparently chaotic state. Here, 1000 centres were chosen at random from the total set of 25 000, and at each centre 1000 points were also chosen randomly to calculate $C(r)$. Three distinct regions can be seen in the graphs. At small values of r, and hence low $C(r)$, the fluctuations can be attributed to the discrete nature of the representation of the trajectories and statistical fluctuations. On the other hand, at large r there is saturation when r becomes comparable with the size of the attractor. The linear portion in between these two regimes is where the scaling can be seen and also the

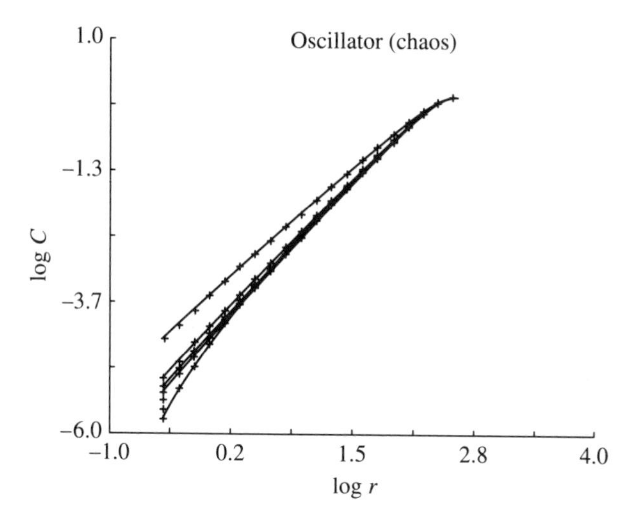

Fig. 2.13. Log–log plot of $C(r)$ versus r for a range of embedding dimensions for the time series obtained from the electronic oscillator in a chaotic regime.

lines are parallel for embedding dimensions greater than 2. This suggests that the chaotic attractor is finite-dimensional and the correlation dimension D_c is given by the slope of this linear portion. It is perhaps appropriate to comment here that the determination of this slope is not always as clear-cut as in the present case.

Now, if we repeat the process with a time series obtained from a random

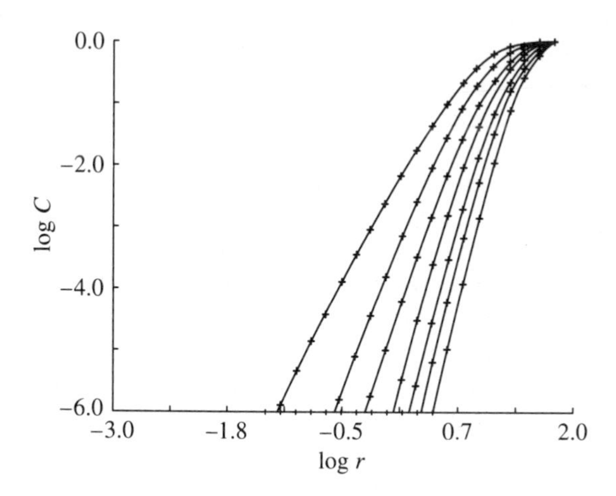

Fig. 2.14. Log–log plot of $C(r)$ versus r for a range of embedding dimensions using the time series obtained from a random noise generator.

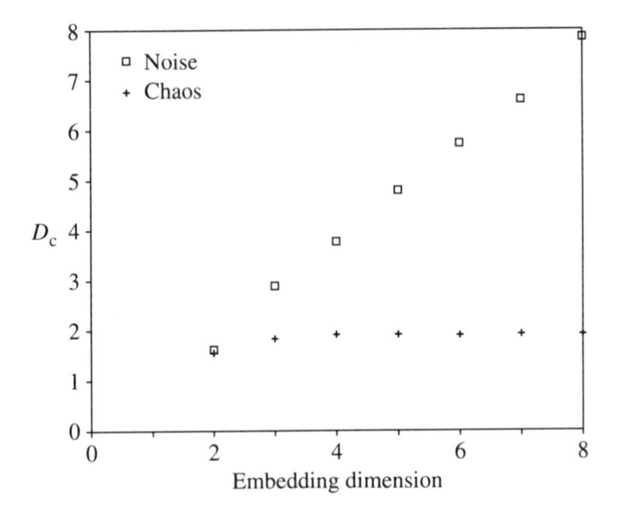

Fig. 2.15. Estimates of the dimension of the attractors for the chaotic system and random noise generator as a function of embedding dimension.

noise generator, we find the results presented in Fig. 2.14. It is clear that the dimension is increasing with embedding dimension as the lines do not become parallel and in fact become almost perpendicular to the horizontal axis. The difference between chaos and noise is very evident in Fig. 2.15 where we show the dimension estimates versus embedding dimension for the two cases. The correlation dimension for the chaotic oscillator clearly saturates at a value around 2, while the estimate for the noise continues to increase. Thus it would appear that this is a useful technique for distinguishing between noise and finite-dimensional chaos.

Before concluding this section, we should note that there can be severe practical difficulties with using this method if the dimension is large. As pointed out by Ruelle, the number of sample points that are required in these calculations is proportional to $10^{d/2}$ where d is the dimension of the attractor. Thus it becomes impractical to carry out such calculations for values of $d > 7$. To emphasize the point, Ruelle made the following amusing observation. If we had a time series sampled once per second since the universe began, approximately 10^{21} seconds ago, then the maximum detectable dimension would be 42, a number that has obtained notoriety in fiction.

2.6 Summary

We have shown how to reconstruct a phase portrait from a time series obtained from an experiment or the numerical integration of a set of equations. These powerful techniques are a significant advance over traditional Fourier techniques which can mask the important dynamical information contained in the time series from a nonlinear system. The straightforward method of delays has deficiencies when dealing with experimental data where noise is inevitably present. This type of data is best analysed using singular value decomposition to obtain the optimum coordinate system for projection.

Once the phase portrait has been reconstructed the aim is to obtain some quantitative estimates from it. The first of these involve calculating the Lyapunov exponents which monitor the sensitivity to initial conditions associated with strange attractor behaviour. These techniques can be shown to work to some extent in relatively simple systems, but the extension of these ideas to situations governed by partial differential equations is still an open question. Finally, some of the techniques for calculating the dimension of the attractor were discussed and the practical limitations of the techniques were explained. The aim of this part of the process is first of all to establish whether the dimension of the attractor is small and secondly to see if it is noninteger thus giving evidence for fractal structure. However, it should be pointed out that it is very difficult to support such scaling arguments using experimental data sampled through a standard 12-bit A/D converter.

The above techniques are descriptive in nature and thus far do not give any indication of how one might relate the results of an experimental time series to allied theoretical work. We will outline in the next chapter some new ideas which show how to establish a direct connection between theory and experiment using phase portrait reconstruction techniques and subsequent quantitative analysis of particular points on the attractor. This not only permits a good description of the dynamics but may also be used to predict parameter values where qualitatively different types of dynamical behaviour can be found.

References

1. Hamming, R. W. (1989). *Digital filters*. Prentice-Hall, New Jersey.
2. Press, W. H., Flannery, B. P., Teukolsky, S. A., and Vetterling, W. T. (1988). *Numerical recipes in C*. Cambridge University Press.
3. Packard, N. H., Crutchfield, J. P., Farmer, J. D., and Shaw, R. S. (1980). Geometry from a time series. *Phys. Rev. Lett.*, **45**, 712–15.
4. Takens, F. (1981). Detecting strange attractors in turbulence. In: *Dynamical systems and turbulence* (ed. D. A. Rand and L. S. Young). Springer, Berlin.
5. Buzug, T., Reimers, T., and Pfister, G. (1990). Optimal reconstruction of strange attractors from purely geometrical arguments. *Euro. Phys. Lett.*, **13**, 605–10.

6. Broomhead, D. S. and King, G. P. (1986). Extracting qualitative dynamics from experimental data. *Physica* D, **20**, 217–39.
7. Broomhead, D. S. and Jones, R. (1990). Time series analysis. *Proc. R. Soc. Lond.* A, **423**, 103–21.
8. Moon, F. C. (1987). *Chaotic vibrations*. Wiley, New York.
9. Eckmann, J. P. and Ruelle, D. (1985). Ergodic theory of chaos. *Rev. Mod. Phys.*, **45**, 617–56.
10. Darbyshire, A. C. and Broomhead, D. S. (1992). The calculation of Lyapunov exponents from experimental data. (In preparation.)
11. Wolf, A., Swift, J. B., Swinney, H. L., and Vastano, J. A. (1985). Determining Lyapunov exponents from a time series. *Physica* D, **16**, 285–317.
12. Glazier, J. A. and Libchaber A. (1988). Quasiperiodicity and dynamical systems: An experimentalist's view. *IEEE Trans. Circuits Syst.*, **35**, 790–809.
13. Grassberger, P. and Procaccia, I. (1984). Characterization of strange attractors. *Phys. Rev. Lett.*, **52**, 2241–4.
14. Ruelle, D. (1990). Deterministic chaos: The science and the fiction. *Proc. R. Soc. Lond.* A, **427**, 241–8.

3
A multiple bifurcation point as an organizing centre for chaos

Tom Mullin

In this chapter we will show how qualitatively different types of bifurcations in a nonlinear system can interact with each other to produce an organizing centre for more complicated phenomena such as chaos. First, we will introduce static and time-dependent bifurcations and describe their practical features in terms of physical examples. Then we will discuss the details of an interaction between the two types of bifurcation and show how it produces a multiple bifurcation point in the specific example of a nonlinear electronic oscillator. Finally, we discuss how the 'global dynamics' of the system are controlled by this local bifurcation structure.

3.1 The symmetry-breaking pitchfork bifurcation

Perhaps the most widely discussed example of a static symmetry-breaking pitchfork bifurcation is that of an elastic strut under a compressive load commonly referred to as the Euler strut problem. A sketch of the set up is shown in Fig. 3.1, where we show an elastic strut under a compressive end loading P. The physical problem may be modelled mathematically as follows. The equilibrium equation for a strut of length L is given by

$$B \frac{\mathrm{d}^2\theta}{\mathrm{d}x^2} + P \sin \theta = 0, \tag{3.1}$$

where B is the elastic bending stiffness and P is the applied load. This equation describes a balance between the applied load and the bending stiffness multiplied by the curvature. If the equation is linearized by replacing $\sin\theta$ by θ, which is valid for small angles of curvature, then explicit solutions to this equation can be found by standard methods. One solution is the zero deflection state and we refer to this as the trivial solution. It can be shown that this becomes unstable above a critical load P_c, so that an exchange of stability takes place with a pair of buckled states which have the form of arches deflected to the left or right. The situation may be realized crudely in practice by the end-loading of a plastic ruler using your hands. For small loads the ruler is straight, but as the

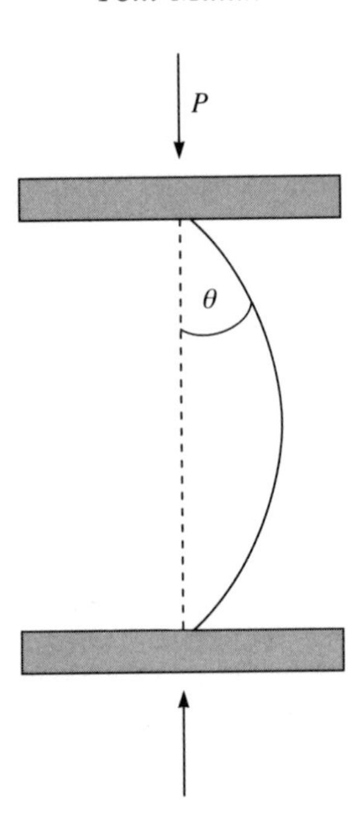

Fig. 3.1. Schematic diagram of the buckling of an elastic strut due to end compression by an applied load P.

load is increased then the ruler will deflect to the left or right.

The mathematical situation is shown schematically in Fig. 3.2, where we can see that the solution structure has the form of a pitchfork bifurcation. A change in the number of solutions from one to three occurs at P_c with increase of load. Above P_c there are two stable buckled states separated by the unstable straight solution. The buckled states have lost the side-to-side symmetry of the trivial state and so we refer to this as a symmetry-breaking bifurcation. It is equivalent to the symmetry-breaking bifurcation we discussed previously in connection with the parametrically excited pendulum, where in that case the amplitude of the swing was higher to one side than the other.

Algebraic forms for bifurcations of this type may be reduced from equations such as (3.1) by use of a technique called Lyapunov–Schmidt reduction. The form for the present case is

$$G(x, \lambda) = x^3 - \lambda x, \tag{3.2}$$

where $\lambda = P/B$. If we set $G(x, \lambda) = 0$, then we obtain the equilibrium solution which shows that the amplitude of the buckled states grows quadratically with increase in load.

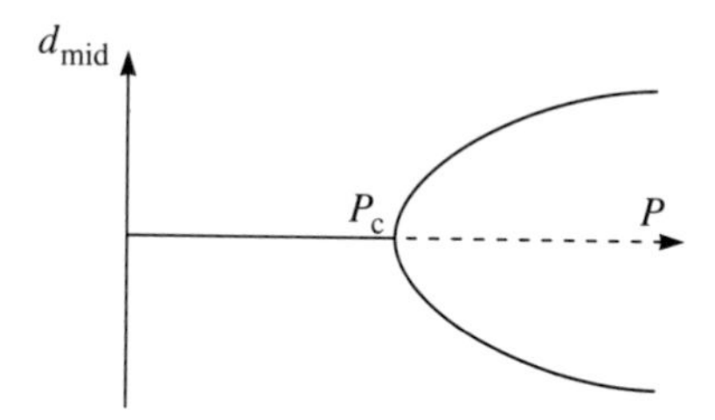

Fig. 3.2. Bifurcation diagram for the buckling of an elastic strut. The bifurcation parameter is the end load P, and the functional which discriminates between solutions is the midlength displacement d. Below P_c the zero deflection state is the unique solution, while above P_c there are three solutions. Two stable buckled states separated by an unstable straight solution.

One clear criticism of the description thus far is the assumed perfect symmetry of the situation. It is obvious in a demonstration of the buckling of a plastic ruler that there will always be a preferred direction of bending of the ruler and that it is never convincingly straight at small loads. The perfect symmetry assumed in the mathematical model can never be realized in practice, even if we use an elaborate experimental arrangement. In other words, perfect symmetry is a non-robust mathematical abstraction, which can only be approximated to in the real world. However, the physical situation can be modelled by adding an imperfection term to the algebraic form as follows:

$$G(x, \lambda) = x^3 - \lambda x + I, \tag{3.3}$$

where I is a constant and is known as the imperfection term. This has the effect of disconnecting the pitchfork bifurcation, as shown in Fig. 3.3. Thus there will now be the smooth evolution of one state with increase of load and a disconnected mode will exist a critical load. The disconnected state can only be reached by a discontinuous jump in load and it will snap back to the preferred state below a critical value. This disconnected form

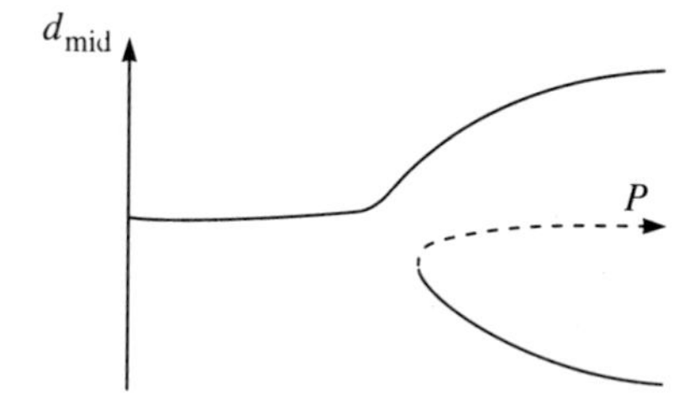

Fig. 3.3. Schematic bifurcation diagram showing the effects of including symmetry-breaking imperfections which will be present in any physical realization of the problem.

of the pitchfork is a good model of the physical system and a lucid modern account of the Euler strut problem is given by Zeeman [1].

3.2 The van der Pol oscillator

A schematic circuit diagram for the nonlinear LCR van der Pol oscillator is shown in Fig. 3.4 and a complete discussion of the current investigation is given in Healey *et al.* [2]. In addition to the normal LCR resonant circuit, we have included two extra elements a and b in the feedback loop. These are nonlinear components whose magnitude may be varied using external high-precision decade potentiometers. We will investigate the bifurcation structure of this nonlinear circuit by variation of the control parameters μ_1 and μ_2 which are proportional to the sizes of a and b.

The simplest bifurcation we find is illustrated in Fig. 3.5, which is a plot of voltage versus the parameter μ_1. At small values of μ_1, the zero volts state is the stable attracting fixed point solution. If μ_1 is now increased, the solution passes through a symmetry-breaking pitchfork bifurcation. Here, a pair of asymmetric steady states are produced which have two non-zero voltages, positive and negative in the sense of our measurements. In addition, as in the case of the Euler strut, the bifurcation is slightly disconnected by the presence of inevitable small asymmetries in the circuit, e.g. no two resistors will ever be perfectly matched. Thus the fixed point solution will not be at exactly zero volts. In addition, only one of the pair of asymmetric voltage states will be reached by small increments of the control parameter. However, the other state may be obtained by suddenly switching on the circuit above a critical value of μ_1. There will then be a 50–50 chance of obtaining either solution. Once the

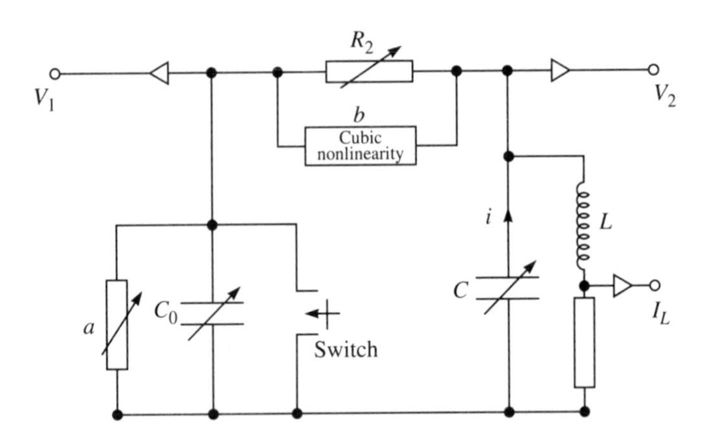

Fig. 3.4. Schematic circuit diagram of the nonlinear van der Pol oscillator circuit.

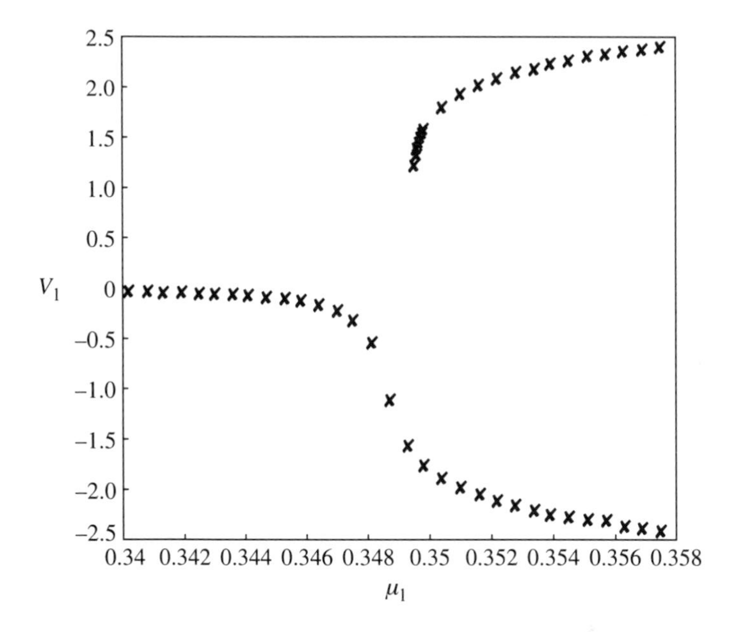

Fig. 3.5. An experimentally determined pitchfork bifurcation for the steady-state voltages in the van der Pol oscillator.

disconnected state has been reached it will disappear catastrophically at a critical value of μ_1 as this parameter is reduced.

There is another 'simple' bifurcation in the oscillator which is called a Hopf bifurcation. The Hopf bifurcation gives rise to a simply periodic time-dependent state in the oscillator from a stationary solution, as illustrated in Fig. 3.6(a). At small values of our parameter μ_1 the solution will initially be in a stable steady state, and thus if we give the system a small perturbation it will just return to this steady state. As we increase μ_1, the steady state will become unstable and it will be replaced by simple periodic motion which grows in amplitude as $(\mu_1 - \mu_c)^{1/2}$ from the bifurcation point at μ_c. This may be represented in phase space as the exchange of stability of an attracting fixed point with a limit cycle, as shown in Fig. 3.6(b). The simplest way of thinking of this construction is to consider it as end-on view of the bifurcation diagram shown in Fig. 3.6(a).

If we continue to increase the parameter μ_1, then the periodic orbit produced through the Hopf bifurcation will initially grow in amplitude but then will typically pass through a period-doubling cascade to chaos. We show a typical sequence of reconstructed experimental phase portraits in Fig. 3.7, which were obtained from the time-series output of the van der Pol oscillator circuit. The period-doubling cascade, through periods 1, 2, and 4 leading to chaos in Fig. 3.7(d) is evident. We note that this circuit

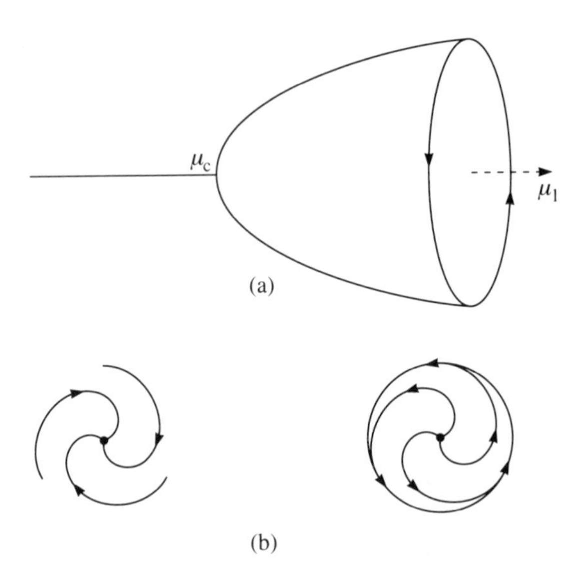

Fig. 3.6. (a) A schematic diagram of a Hopf bifurcation to simply periodic motion; (b) a phase space representation of a Hopf bifurcation.

is not an analogue model of a one-dimensional map, where the output is fed back in the iterative sense. The oscillator is governed by its own nonlinear ordinary differential equations, which can be derived using Kirchhoff's laws. Thus the description of this sequence of period doublings in terms of iterative maps gives an incomplete understanding of the process, and as we shall see below the period-doubling cascade is part of a much deeper solution structure.

3.3 Codimension of a bifurcation

First, we must define some terminology which is essential to the understanding of the bifurcation structure of the problem. The codimension of a bifurcation may be defined to be the smallest dimension of parameter space which contains the bifurcation in a persistent way. (Note that this is not the only definition of codimension in common usage; there is another in singularity theory and the interested reader should consult the book by Golubitsky and Schaeffer [3] for a clear exposition of these ideas.) In the nonlinear oscillator we have two parameters μ_1 and μ_2 and hence a two-dimensional parameter space. Thus, we would expect typically to encounter codimension-1 bifurcations if we restrict our study to varying a single para-

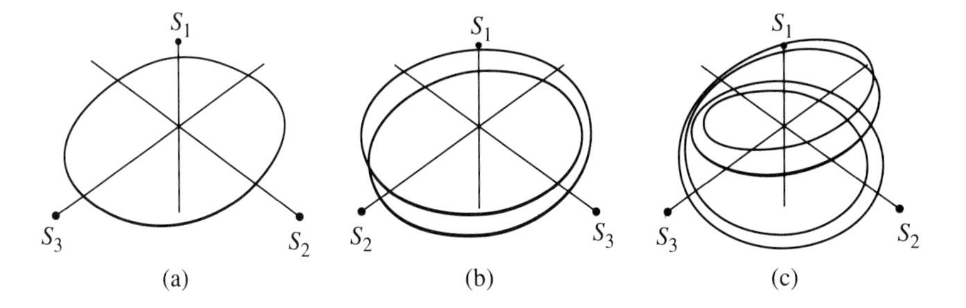

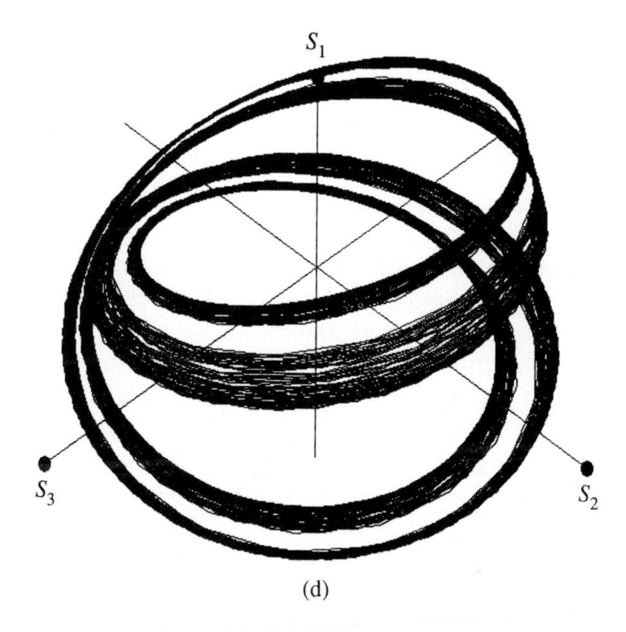

Fig. 3.7. A sequence of reconstructed attractors showing a period-doubling route to chaos: (a) period 1; (b) period 2; (c) period 4; (d) chaos.

meter at a time. In the present problem there are two codimension-1 bifurcations, viz. the pitchfork and the Hopf bifurcations. Thus, if we take one-dimensional slices through our two-dimensional control parameter plane, we would typicallv expect to find Hopf or pitchfork bifurcations as the most common types. (Note that pitchfork bifurcations are only codimension-1 under the restriction of perfect symmetry, which we will assume for the present to simplify the discussion.) Golubitsky and Schaeffer discuss this point at length in their book.

Next, consider the situation shown in Fig. 3.8, where the two codimension-1 bifurcations occur sequentially when μ_1 is varied. This

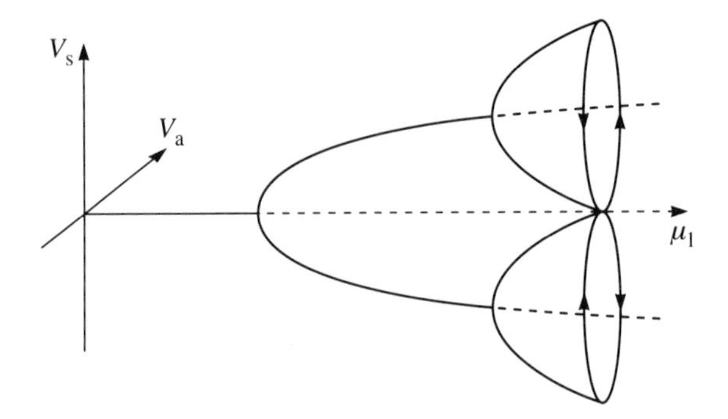

Fig. 3.8. Schematic bifurcation diagram showing the sequence of static and Hopf bifurcations. V_s and V_a are the symmetric and antisymmetric voltage components respectively.

figure has been drawn with the assumption of perfect symmetry in the circuit to ease presentation. The effects of disconnection due to imperfections will be discussed later. The first bifurcation is the pitchfork, which gives rise to the pair of asymmetric states, and if μ_1 is increased further a pair of codimension-1 Hopf bifurcations will arise, one on each asymmetric branch. Further, the critical values of μ_1 at which the bifurcations occur are also dependent on the second parameter μ_2. Thus it is possible for the two bifurcations to occur at the same critical value of μ_2 for a specific value of μ_1. In other words, the loci or lines of bifurcation points in the (μ_1, μ_2) plane could intersect at a special point, which we call the codimension-2 point.

The codimension-2 point can be located in the experiment by investigating the dependence of the two bifurcation points on the control parameters μ_1 and μ_2. In practice we do not vary both parameters simultaneously but instead take a series of one-dimensional slices through the parameter space by holding one of them fixed and varying the other. This technique has been used to obtain the results shown in Fig. 3.9, which presents a considerably simplified version of the full bifurcation set in order to focus attention on the most significant features. There, we can see that the two curves meet at the codimension-2 point labelled *B*.

3.4 The bifurcation structure of the oscillator

The locus *AB* in Fig. 3.9 is that for the pitchfork bifurcations to asymmetric states. Thus each crossing of this line from left to right by increasing μ_1 would give rise to a pair of asymmetric states. As noted above the pitchfork will of course be slightly disconnected in the experiment owing to the

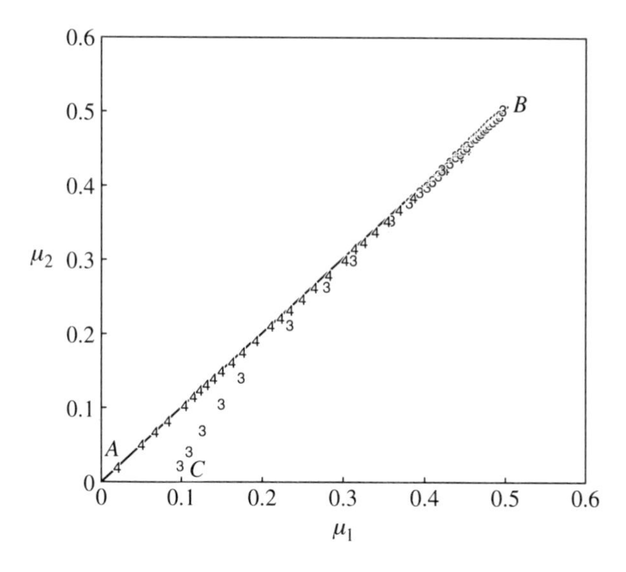

Fig. 3.9. Experimentally determined bifurcation set for the static and Hopf bifurcations plotted as functions of the control parameters. *AB* is the locus of estimates of the pitchfork bifurcation shown compared with theoretical values (solid line) and *BC* is that for the Hopf bifurcations.

presence of an inevitable small bias in the circuit. Thus our estimate of the bifurcation point is taken to be the limit point of the disconnected portion. It is found by jumping to the disconnected branch by suddenly switching on the circuit and then reducing μ_1 slowly until a catastrophic collapse of the asymmetric state is found.

Similarily, *BC* is the line of Hopf bifurcations to a singly periodic oscillation. Note we have only plotted one of a possible pair of lines as there are actually two Hopf bifurcations, one on each branch of the pitchfork. For consistency with the above estimates of the pitchfork bifurcation, we have chosen to measure the Hopf bifurcation on the disconnected branch. As may be seen, they are distinct lines of codimension-1 bifurcations in parameter space but they do meet at the codimension-2 point *B*. It is known that near such points, more complicated dynamical behaviour often occurs (See, for example, the discussion of codimension-2 points in the book by Wiggins [4]).

Now we will concentrate on the bifurcation structure near the codimension-2 point *B*. A sketch of the bifurcation diagram and its related phase portrait is shown in Fig. 3.10, which is again drawn assuming perfect symmetry. It can be seen that the limit cycles on the asymmetric branches grow in amplitude until the orbits meet at a saddle connection point as the control parameters are varied. Then the orbits continue to grow and

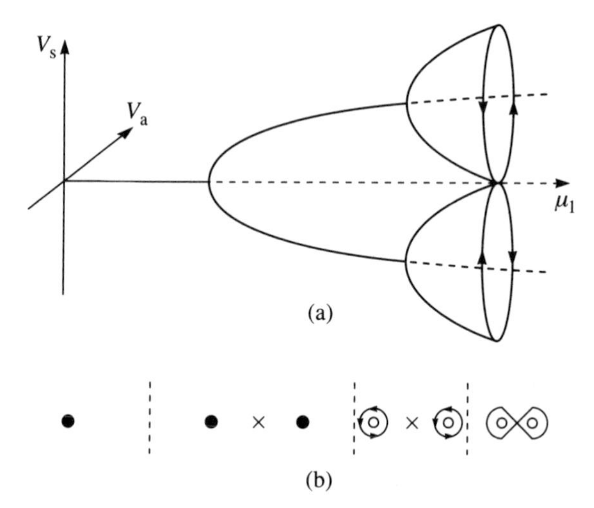

Fig. 3.10. (a) Schematic bifurcation diagram showing how the periodic orbits which arise on each of the asymmetric branches connect through the fixed point corresponding to the unstable trivial solution. Axes labelled as in fig. 3.8 (b) End on view.

surround all three fixed points with yet further increase of the parameters.

In the extreme left of the figure there is one attracting fixed point and that is replaced by two fixed points separated by an unstable equilibrium after the pitchfork. Further increase in μ_1 leads to the pair of Hopf bifurcations and the motion can be represented by two limit cycles in phase space. These in turn grow in amplitude as μ_1 is further increased and connect through the unstable equilibrium point in a heteroclinic orbit. Just as in the case of the separatrix in the pendulum problem discussed previously, the period of this limiting orbit will be infinite. The connected single large orbit formed with further increase in μ_1 surrounds all three unstable fixed points and continues to grow with yet further increase in the parameter.

3.5 The Silnikov mechanism

Thus far all the dynamical motion we have discussed has taken place in the plane. Therefore, in order to obtain an explanation for the more complicated dynamical behaviour including chaos, the dimension of the phase space will have to be increased by at least 1. We now turn to the experiment to see what actually happens for parameter values very close to the codimension-2 point. A typical phase portrtait is shown in Fig. 3.11, where we can see that it is just a simple connected loop or limit cycle which is

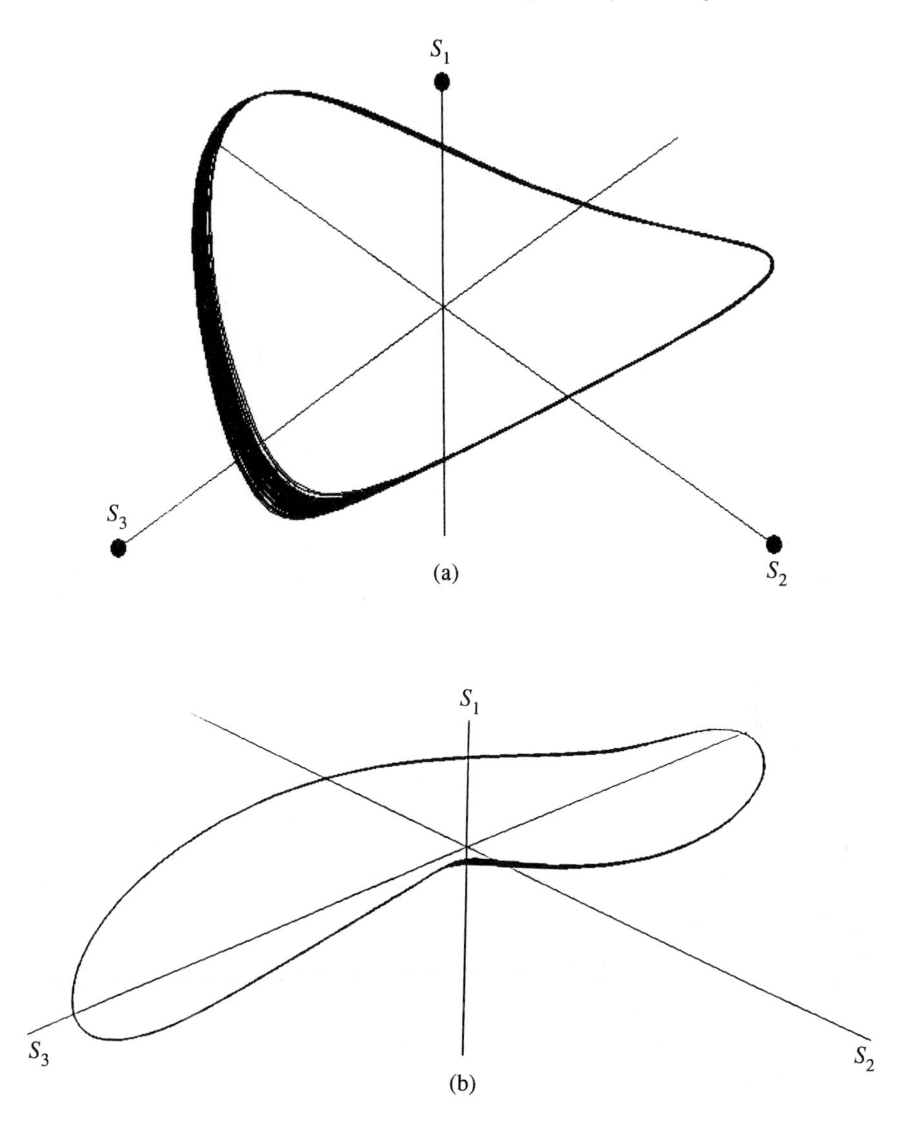

Fig. 3.11. (a) Limit cycle reconstructed from experimental data taken near the codimension-2 point. Note how the noise is spread in the plane as the trajectory slows down near the unstable fixed point. (b) Experimental limit cycle taken just after connection has taken place.

close to homoclinicity. There is some noise present in the experiment, which has the effect of spreading the trajectories into a band near the unstable fixed point. The spreading arises because the motion on each trajectory slows down as it is attracted in towards the saddle point, and thus the noise has time to grow so that the trajectories are then smeared into a band. The important feature to notice is that the noise spreads the trajectories in the plane, showing that the motion is very close to being two-dimensional.

If we continue to increase μ_1, then we get the connected planar cycle shown in Fig. 3.11(b). Note that the critical cycle, where the connection takes place, cannot be located precisely as it would take an infinite time to complete such a cycle and in any case small experimental imperfections will necessarily disconnect it. The large cycle encompasses all three unstable fixed points but approaches the saddle point in the centre where slowing of the trajectories occurs. This is the complete sequence very close to the codimension-2 point. One observes the growth of the period and amplitude of the limit cycle, which then suddenly flips through to a large cycle and the period then decreases. As stated above this two-dimensional system cannot give rise to chaos and, of course, the amplification of the small-scale noise is not the origin of chaos.

Next we consider the reconstructed phase portraits shown in Fig. 3.12, which were measured a small distance away in parameter space from the codimension-2 point. It is now evident that three-dimensional effects have become important as spiralling motion out of the plane can now be seen. The spiralling of the trajectories is perhaps easier to see in the connected portrait of Fig. 3.12(b), where the spirals have opposite senses on either side of the saddle point. There is also evidence for the spreading of the trajectories due to the presence of noise, but it is not as apparent as in the planar case. In any event, the motion clearly now takes place in three dimensions and thus there is the potential for more complicated motion including chaos. This assertion of three-dimensionality can be justified quantitatively by investigating the scaling properties of the local singular values near the unstable fixed point. This is done by centring a series of concentric spheres on the trajectories and investigating the scaling properties of the data with sphere radius. The linear scaling of the significant singular values of the data sets then gives an estimate of the local dimension of the attractor. Healey *et al.* [2] did this for the present case and clearly demonstrated the change in dimension from 2 to 3 as the distance in parameter space from the codimension-2 point was increased.

One mechanism which gives rise to the chaos and which is relevant to the present case was first studied in abstract by Silnikov [5] and later extended to a specific model equation by Glendinning and Sparrow [6]. Their model contained homoclinic orbits which were of the so-called saddle focus type. We will now discuss the implications of their work in relation to our specific practical example.

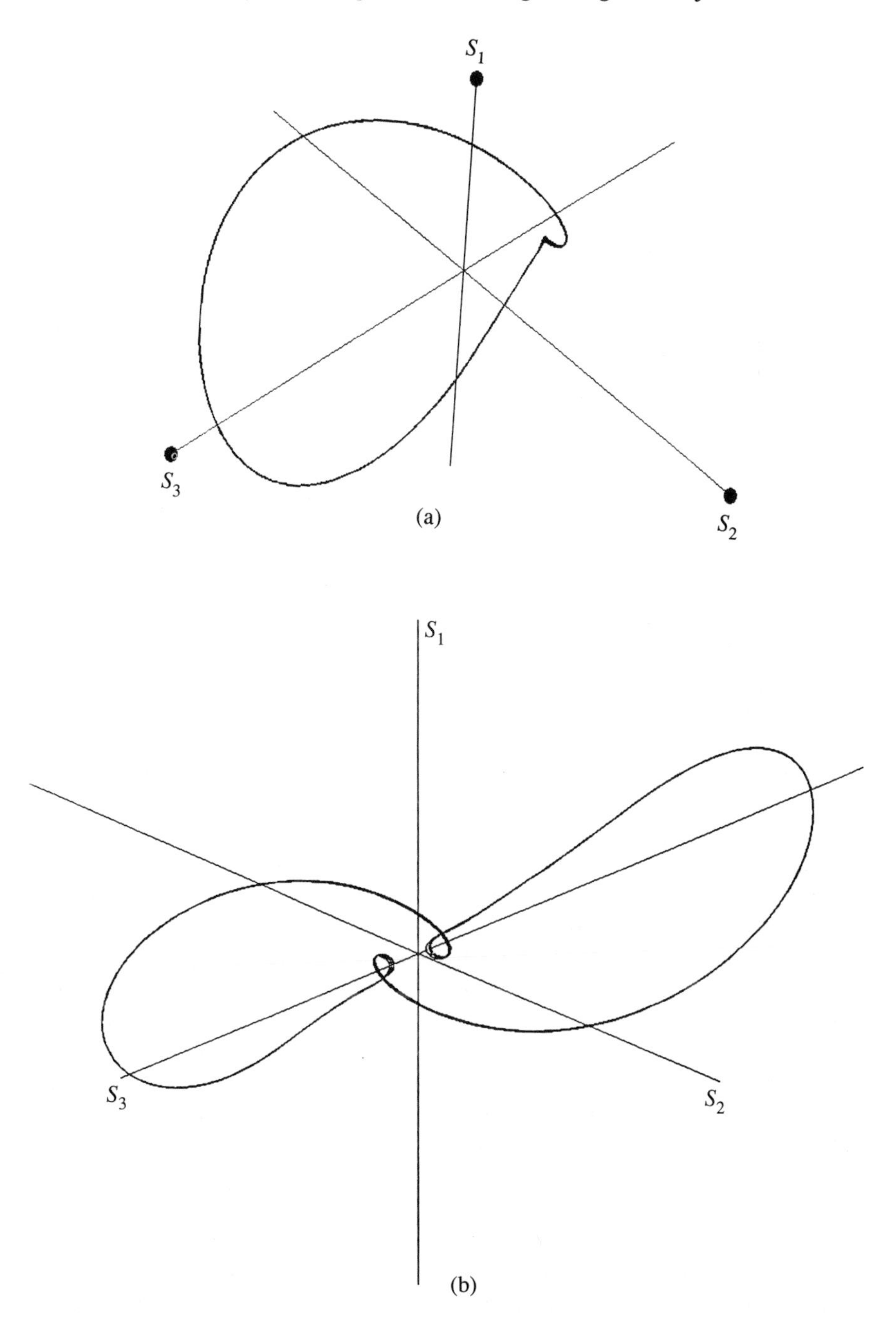

Fig. 3.12. An illustration of the connection sequence further from the codimension-2 point so that three-dimensional spiralling is now more evident: (a) the single-sided cycle; (b) the connected cycle.

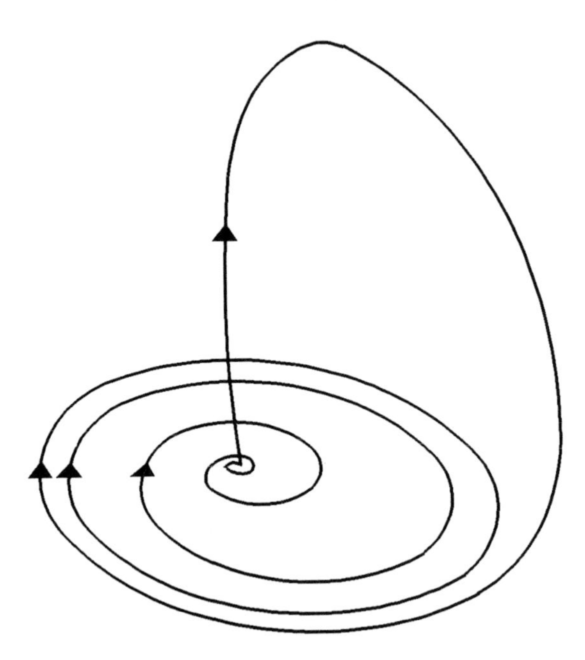

Fig. 3.13. A sketch of a limit cycle close to a stationary point of saddle focus type.

We show in Fig. 3.13 a sketch of a phase trajectory on an orbit close to a stationary point of saddle focus type. The trajectory spirals around in the plane towards the fixed point but then is ejected out of the plane along the unstable direction. There is a pair of these orbits in our system and we have already seen one of them in Fig. 3.12(a). The reason that there is a pair is that the original pitchfork, which gave rise to the two asymmetric states, is slightly disconnected. Thus the heteroclinic connection of our idealized symmetric circuit will also be disconnected to form a pair of homoclinic connections which become of saddle focus type as the control parameters are varied away from the planar region.

Glendinning and Sparrow [6] showed that the type of observed dynamical behaviour near such orbits depends critically on a parameter δ. This may be loosely defined as

$$\delta = \frac{\text{Speed of rotation}}{\text{Speed of ejection}}.$$

It is actually the ratio of the real parts of the eigenvalues associated with the stable and unstable directions of the fixed point, but we choose to give it this less sophisticated physical interpretation. In the Glendinning and Sparrow model, when $\delta > 1$ there will be monotonic approach to homoclinicity (inifinite period) with variation of a control parameter. In

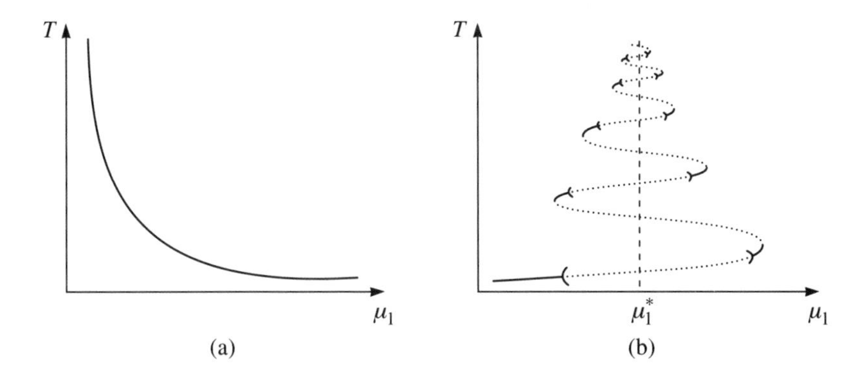

Fig. 3.14. Plots of the variation of period with parameter in the two limiting cases of $\delta > 1$ or $\delta < 1$: (a) $\delta > 1$ and the increase of period is monotonic as homoclinicity is approached; (b) $\delta < 1$ and a wiggle develops in the period versus parameter curve so that multiple stable states now exist.

the second situation, when $\delta > 1$ a more complicated route to infinite period is found, which includes forward and reverse period-doubling cascades. We now discuss these two routes in more detail.

To reiterate, when $\delta > 1$ and we vary a parameter in the experiment (say μ_1) so that the orbit approaches homoclinicity, the period of the orbit will approach infinity in a monotonic way, as shown in Fig. 3.14(a). Alternatively, if $\delta > 1$ then a wiggle will develop in the period versus

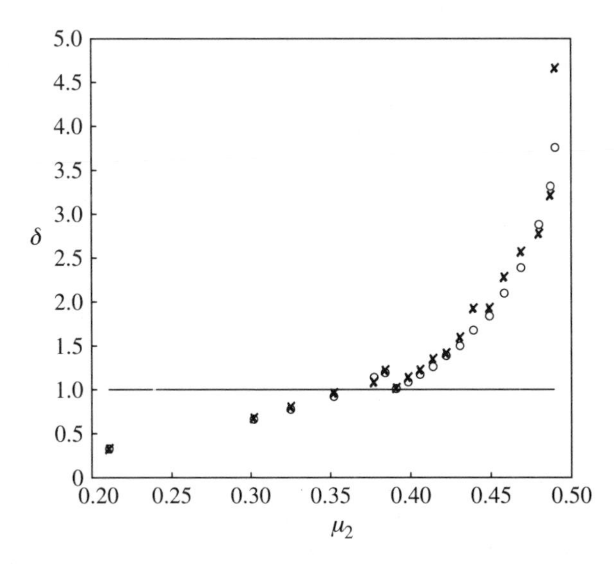

Fig. 3.15. Experimental determination of the parameter δ as a function of the control parameter μ_2.

parameter curve, as shown in Fig. 3.14(b), so that multiple states will arise and alternate ones will show forward and reverse period-doubling cascades to chaos.

Healey *et al.* [6] were able to obtain estimates of the parameter δ from the experimental data sets. This was done by constructing a linear model of the flow on the attractor near the unstable fixed point in the parameter space region, where the orbits are of saddle focus type. The results, which are presented in Fig. 3.15, are the estimates of δ for a range of fixed values of μ_2 when μ_1 is such that the orbit is closest to homoclinicity. It can be seen that δ does indeed change from less than to greater than 1, and so we should expect to see a monotonic approach to homoclinicity in some parameter

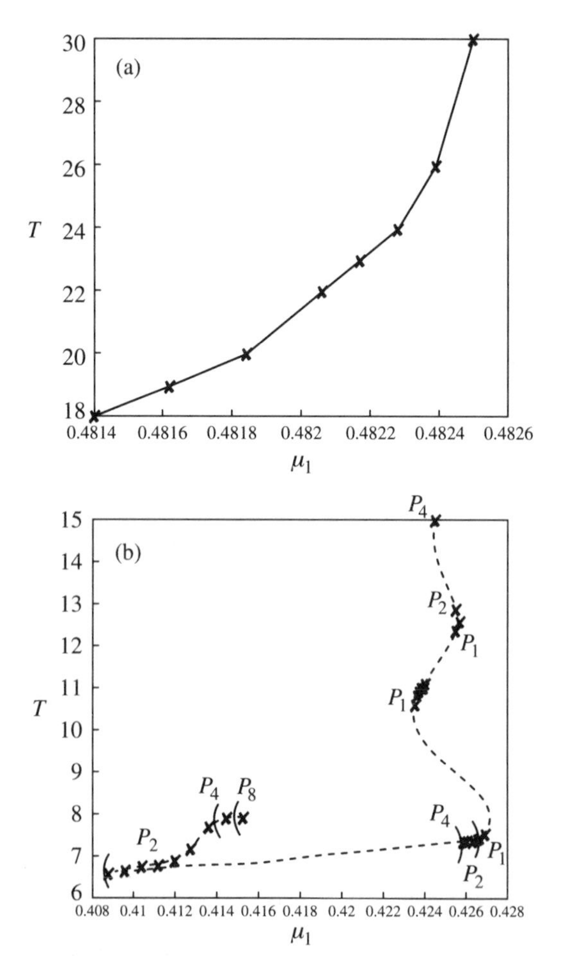

Fig. 3.16. Experimental observations of the two limiting cases of (a) monotonic and (b) folded approach to homoclinicity.

ranges, while, in others, multiple states and forward and inverse period-doubling cascades with intermediate bands of chaos will be found.

Next we turn to the experimental observations presented in Fig. 3.16. These represent the two extremes of the behaviour predicted in Fig. 3.14. Figure 16(a) shows the predicted monotonic approach to homoclinicity as μ_1 is varied; on the other hand, in Fig. 3.16(b) we see the experimental evidence for the predicted period-doubling cascades and chaos. This is the first direct experimental observation of this mechanism which had only previously been considered in abstract and in mathematical models. It is a very powerful result because it shows that the layers of period-doubling sequences and chaos, which pervade a wide range of parameter space of the oscillator, fan out from a single point in the control plane. Thus, if one wished to remove or use the chaos in any particular application of a circuit of this type, then knowledge of this fact would be extremely useful.

3.6 Summary

A combination of experimental and theoretical work has uncovered the origins of chaos in an electronic oscillator circuit. The equations of motion for the system are ordinary differential equations and this fact not only makes the theoretical analysis straightforward but also permits predictions to be made from the nonlinear analysis of experimental time series. This dual approach led directly to the uncovering of an organizing centre for the observed chaos and a Silnikov structure which gives rise to the period-doubling sequences. Thus this familiar transition scenario doubling forms only part of a hierarchy of bifurcation sequences which are all organized by a particular codimension-2 event that has previously been studied in abstract. It now remains to be seen whether this general approach can add anything to the understanding of the much more difficult problem of the transition to disordered motion in fluid flows, the subject of the next chapter.

References

1. Zeeman, E. C. (1975). Euler buckling. In: *Lecture notes in mathematics 525* (eds. A. Dold and B. Eckmann). Springer, Berlin.
2. Healey, J. J., Broomhead, D. S. Cliffe, K. A., Jones, R., and Mullin, T. (1991). The origins of chaos in a modified Van der Pol oscillator. *Physica* D, **48**, 322–39.
3. Golubitsky, M. and Schaeffer D. G. (1985). *Singularities and group in bifurcation theory*, Vol. 1., Applied Mathematical Sciences 51. Springer, Berlin.
4. Wiggins, S. (1990). *Introduction to applied nonlinear dynamical systems and chaos.* Texts in Applied Mathematics 2. Springer, Berlin.

5. Silnikov, L. P. (1969). On a new type of multidimensional dynamical system. *Sov. Math. Dokl.*, **10**, 1368–71.
6. Glendinning, P. and Sparrow, C. (1984). Local and global behaviour near homoclinic orbits. *J. Statist Phys.*, **43**, 479–88.

4
Chaos in fluid dynamics

Tom Mullin

The application of the ideas which have been developed in the study of low-dimensional chaos to fluid dynamics has become a rapidly increasing research area in recent years. Some of the reasons for this are that fluid turbulence still remains one of the greatest challenges of classical physics and it is of immense practical importance. In fact, it is often written in popular accounts of the subject that chaos lies at the heart of an explanation for turbulence. We will discuss this possibility using an example which has been thoroughly studied both theoretically and experimentally, to see how much confidence we can have in this assertion.

Following on from some introductory remarks on the equations of motion for fluid flows, we will introduce the particular problem of flow between concentric cylinders called Taylor–Couette flow. This example has been selected because it changes from the smooth laminar to disordered state through a sequence of bifurcations as the stress on the system is increased. Thus we will adopt the approach of a detailed study of the sequences of bifurcations which was so successful in the cases of the pendulum and electronic oscillator discussed previously. We show how a good theoretical and practical knowledge of the bifurcation structure can be used to build up a deep understanding of the organization of the complicated motions that can arise in the nonlinear regime. This powerful combination of theory and experiment has given rise to results which are highly suggestive of a definite connection between chaos and irregular fluid flow. After a discussion of the application of ideas of low-dimensional dynamical systems to this specific problem, we end with a commentary on the wider implications of these results for the understanding of turbulence.

4.1 The equations of motion

The equations of motion which govern fluid flows have been known for over a century and are called the Navier–Stokes equations. They can be written in the following vector form:

$$\frac{\partial U}{\partial t} + (\mathbf{U}\cdot\nabla)\mathbf{U} = -\nabla p + \frac{1}{Re}\nabla^2\mathbf{U}, \qquad (4.1)$$

where $\mathbf{U} = (u, v, w)$ is the velocity vector, p is the pressure, and Re is the Reynolds number which is defined below. Two modern introductory texts on fluid mechanics which discuss fundamentals as well as applications are by Acheson [1] for mathematicians and Tritton [2] for physicists. We will only give the briefest of discussions of the salient points concerned with one particular field of study in fluid dynamics, i.e. hydrodynamic instabilities. The interested reader is referred to [1] or [2] for a discussion of the broader issues involved in hydrodynamics research.

The Navier–Stokes equations are essentially Newton's equations of motion for a continuum, i.e. they model the macroscopic properties of the fluid and are not derived from the microscopic equations for individual molecules. They are partial differential equations which depend on space and time. Thus they are unlike any of the examples we have discussed in the previous chapters, where the physical systems have all been adequately modelled by ordinary differential equations. The nonlinearity $(\mathbf{U}\cdot\nabla)\mathbf{U}$ arises in the Navier–Stokes equations because an 'element' of fluid can accelerate not only by rate of change with respect to time but also by translation to a region of different velocity.

The only dynamical parameter in the above equations is the Reynolds number Re, which is named after the British engineer Osborne Reynolds. It is essentially a ratio of inertial to viscous effects and is defined thus:

$$Re = \rho U_0 d/\mu,$$

where ρ is the density, U_0 is a typical velocity, d is a relevant length scale, and μ is the molecular viscosity. It may seem that the choice of a 'typical' velocity or 'relevant' length scale makes this definition somewhat arbitrary, but in practice they usually arise naturally. The magnitude of Re gives an immediate gross classification of the flow. As it tends to zero, the equations of motion are dominated by the viscous term, and so essentially the nonlinear term becomes negligible and the motion becomes governed by a linear equation. On the other hand, as as $Re \to \infty$, the classical Euler equations for inviscid flow are recovered. Turbulence arises at modest values of the Reynolds number and it is therefore present in most practical fluid flows.

Thus far the equations are not complete as there are three equations, viz. one for each velocity component u, v, and w, but four unknowns in u, v, w, and p, the pressure. Thus we need a fourth equation to enable us to obtain solutions, and this is provided by the equation for the conservation of mass. For an incompressible fluid, which is the case for a wide variety of practical situations, it is given by

$$\nabla\cdot\mathbf{U} = 0. \tag{4.2}$$

We also need boundary conditions for these equations; for example, it is usually assumed that there is no slip between the wall of a solid body and the immediately adjacent fluid or that the velocity field is prescribed far from a body. These are only two of the boundary conditions which could be employed and many others are discussed in [1] and [2].

Exact solutions are available to the Navier–Stokes equations for a few simple situations some of which are of significant practical importance. However, for the majority of situations, explicit solutions are not readily available. Moreover, it is not in general possible to solve the full Navier–Stokes equations if the flow is turbulent or at least highly disordered, even using the largest available computers. These types of flow are the most common in nature and are to be found in rivers, the atmosphere, central heating systems, and out of kitchen taps, to give but a few instances.

In many practical situations, progress can be made by using models derived from the full equations. Some of these models are highly successful in specific applications, yet the principal difficulty remains that we cannot solve the full equations other than for moderate values of Re. This is a very old problem, which the majority of physicists abandoned around the turn of the century, even though it is often described as probably the most important problem in classical physics. It has been left to engineers and mathematicians to make many of the advances in this fundamental problem. It is still true today that very few physics courses contain any discussion of hydrodynamics although the situation is beginning to change with the advent of chaos.

4.2 The transition to turbulence

One common flow which encapsulates the issues we wish to address is shown in Fig. 4.1. There, we see the flow from a laboratory tap which is in its smooth laminar flow regime when the tap is barely opened, as in Fig. 4.1(a). It is perhaps surprising to realize that it would take the power of a modern supercomputer to solve the Navier–Stokes equations for this problem. It could be argued that there is no need to solve the full equations for this situation since it is adequately modelled by simpler equations which can be justifiably reduced from the full equations of motion. However, the point at issue is that the flow must also be a solution of the Navier–Stokes equations and obtaining it using modern numerical methods is not a trivial exercise even for this apparently simple flow. Further, this flow is not of great practical interest because the more common situation is as shown in Fig. 4.1(b), i.e. turbulent flow. We can no longer solve the equations of motion directly for this situation. The idea behind hydrodynamic instability theory is to see whether an insight into the more complicated motion can be gained from a study of the first instability of the laminar state. This alternative approach to the direct attack on the full equations of motion by intensive computing thus tries to give an understanding of how the turbulent motion arises.

Loosely speaking, at small Reynolds numbers any disturbance in the

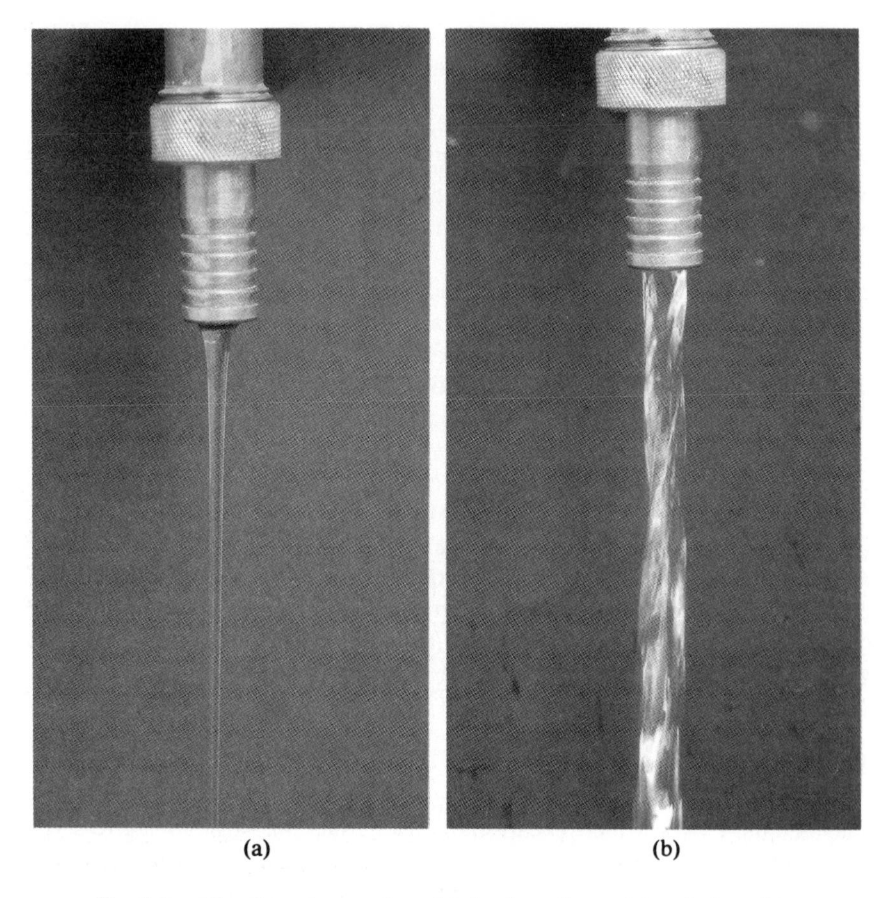

(a) (b)

Fig. 4.1. The flow from a tap: (a) laminar flow; (b) turbulent flow.

flow will be damped out by the action of viscosity. As the tap is opened and *Re* is increased, disordered motion sets in and the flow becomes turbulent. This appearance of disorder in the flow is not understood and, in this chapter we will see if the ideas of low-dimensional dynamical systems help shed any light on the process. We should remark here that it could be that the continuum approximation, which is the basis of the Navier–Stokes equations, breaks down, so that the equations may not be a good model for turbulent flow. This is generally not believed to be the case despite the fact that complete proof of existence, in the mathematical sense, of solutions to the equations in such regimes has not been achieved to date. Thus, if we accept that the Navier–Stokes equations model turbulent flows, then a study of the properties of these equations in a particular flow problem which progresses to turbulence ought to shed light on the processes involved.

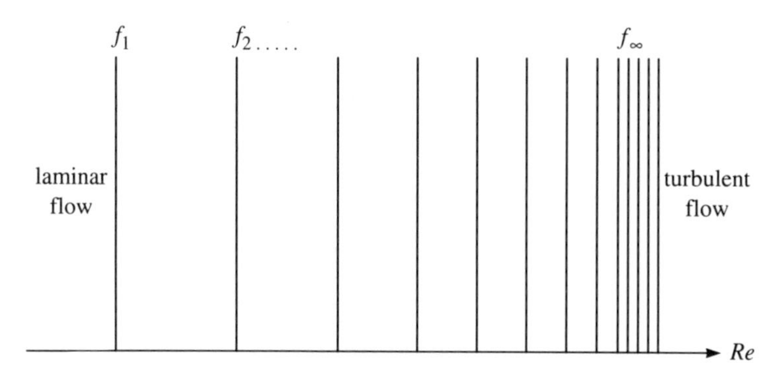

Fig. 4.2. Schematic representation of the transition from laminar to turbulent flow via the Landau route.

An important idea postulated by Landau in the 1940s is that if a fluid flow is in some laminar state and the stress parameter *Re* is increased then a simple periodic oscillation will be excited in the flow above a critical value Re_c. Further increase in *Re* excites another oscillation whose frequency is incommensurate with the first one and the process continues so that more and more modes are excited sequentially as *Re* is increased. Each of the new modes is introduced at smaller and smaller increments of *Re* so that an accumulation point is rapidly reached where a continuum of discrete modes is excited. (Note that this should not be confused with the creation of an accumulation point in a period-doubling cascade where each new frequency is related to the others by factors of 2.) The sequence of events for the Landau scenario is shown schematically in Fig. 4.2. Thus turbulence may be thought of in these terms as an infinite sum of discrete incommensurate frequencies with random phase. These ideas are consistent with a statistical description of turbulence, which has been successfully used in the development of turbulence models, for example.

More recently, two theoretical physicists called Ruelle and Takens [3] suggested that perhaps this wasn't a generic way to get from a quiescent to a disordered state. If we recast the initial stages of the Landau scenario into modern dynamical systems parlance, then the laminar state can be thought of as an attracting fixed point in phase space. Thus, if the system is disturbed from this state, then any resulting trajectory will head back towards the fixed point, as shown in Fig. 4.3 (a). If *Re* is now increased, a time-periodic mode will be excited in the flow which can be represented by a limit cycle in phase space (Fig. 4.3(b)). Yet further increase in *Re* will lead to the appearance of a second frequency which is incommensurate with the first one. This will give a torus in phase space, as shown in Fig. 4.3(c). As the Reynolds number is increased yet further, a third frequency begins to appear which would form a three-frequency torus.

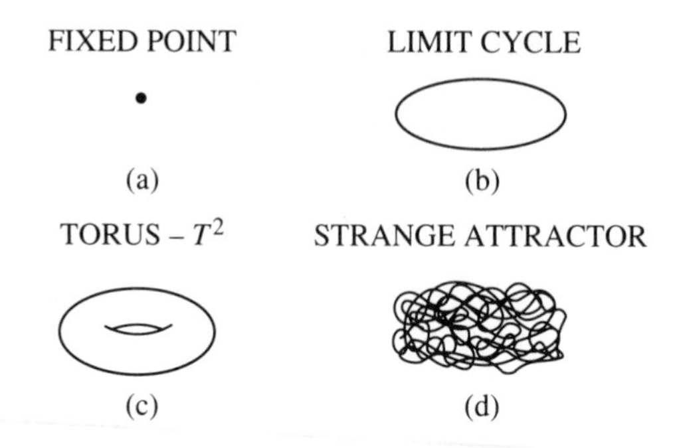

Fig. 4.3. Schematic representation of the transition to turbulence via the Ruelle–Takens route: (a) fixed point; (b) simple periodic flow on a limit cycle; (c) quasi-periodic flow on a 2-torus; (d) irregular flow as a strange attractor.

However, Ruelle and Takens suggest that this is a non-robust situation and it is more typical for a so-called strange attractor be in the solution structure at this stage. This situation is shown schematically in Fig. 4.3(d).

We have seen several examples of strange attractors in the preceding chapters where the physical systems have been adequately modelled by finite-dimensional ordinary differential equations. We now wish to extend the idea to a continuous, i.e. infinite-dimensional, system and although this step cannot yet be justified rigorously, in the mathematical sense, we will see that significant progress can still be made in practice. As a reminder, strange attractor behaviour means that trajectories from just outside the attractor are drawn into it, but once inside they display sensitivity to initial conditions with the concomitant stretching and folding. This is said to be typical behaviour for classes of equation for which the Navier–Stokes equations are believed to belong. Thus the temporal disorder associated with turbulence may be described in terms of strange attractor behaviour. This latter statement has not yet been proved but evidence is mounting that, at least in certain confined flows, it is a better descriptor than the Landau picture.

4.3 Taylor–Couette flow

In the case of flow out of a tap, there is a sudden transition from laminar to turbulent flow as the volume flow rate is increased. The experimental study of such a system is very difficult in practice as it is a situation which is not under good control. Taylor–Couette flow, which is concerned with fluid motion in the gap between concentric rotating cylinders, is an example of a fluid system which proceeds from laminar to irregular flow through a sequence of critical events and which may be operated under very close experimental control. In addition, it provides a good test for numerical studies of the full Navier–Stokes equations on realistic boundary conditions. Thus there is scope for progress to be made using a combination of theory and experiment in the study of some fundamental properties of the equations of motion.

A schematic diagram of a typical experimental arrangement is shown in Fig. 4.4. There are three control parameters which govern the Taylor–Couette system. These are the radius ratio $\eta = r_1/r_2$, the aspect ratio $\Gamma = l/d$, where l is the length of the annular domain and d is the gap width and the Reynolds number which is defined here as

$$Re = \Omega r_1 d/\nu,$$

where Ω is the angular speed of the inner cylinder whose radius is r_1, d is the width of the cylindrical gap, and $\nu = \mu/\rho$ is the kinematic viscosity of the fluid. The radius ratio is fixed in any experiment and the remaining two parameters define a control plane in the same way that the values of the two nonlinearities in the nonlinear electronic oscillator did. In the present problem these parameters are precisely maintained in the experiment by accurate construction of the apparatus, fine control of the motor speed, and fastidious regulation of the temperature. This is important

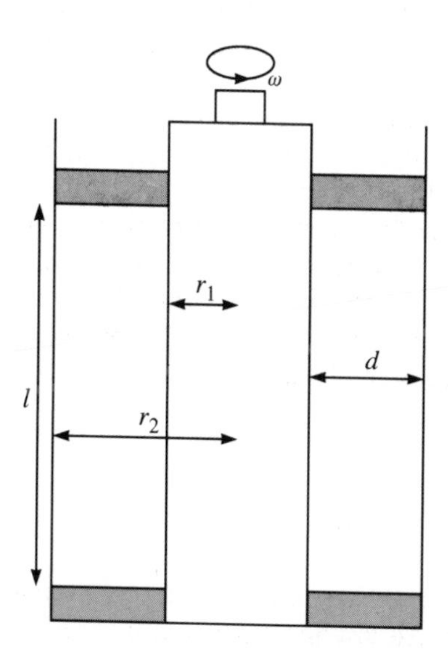

Fig. 4.4. Schematic diagram of Taylor–Couette apparatus. Inner rotating cylinder radius r_1 concentrically placed inside stationary outer cylinder radius r_2. Here, h is the working cylindrical gap between stationary end collars and ω is the angular frequency of rotation of the inner cylinder.

because critical bifurcation sequences can be covered by very narrow spans of control space. Observations can either be made using flow visualization, which is described below, or by a point measurement of a velocity component using a laser Doppler velocimeter. This latter technique uses the Doppler shift of laser light scattered by suitable particles from a small well-defined region in the flow field to obtain a 'point' measurement of a selected velocity component. An introductory book on the technique is by Drain [4].

The flow regimes we will consider are found when the inner cylinder rotates and the outer is stationary. Many complex instabilities are found when the cylinders are rotated differentially, and the reader is referred to the original article by Andereck, *et al.* [5] for a discussion of these phenomena. In the present problem, at small *Re* the fluid is dragged around by the motion of the inner cylinder so that there is a shear across the fluid layer. If particles are added to the flow which reflect the incident

Fig. 4.5. Front view of visualization of Taylor–Couette flow just below the first appearance of cells.

Fig. 4.6. Front view of visualization of Taylor cells.

light, then this flow regime will give an almost uniform reflectance as shown in Fig. 4.5. There are always some three-dimensional effects at the ends of the cylinders but these are too weak to be discernible in the photograph. When *Re* is increased above a critical range, a banded structure appears along the length of the cylinder, as shown in Fig. 4.6, and the flow has the appearance of a series of doughnuts stacked upon each other.

There is now a secondary circular motion superimposed upon the main azimuthal component so that the particles follow spiral paths around the inner cylinder. If we now take a sectional view through the fluid, as shown in Fig. 4.7, then the circular motion may be clearly seen. The photograph shows a view across the cylindrical gap illuminated by a sheet of light at right angles to the camera. We also show alongside for comparison the calculated streamline pattern which was obtained by K.A. Cliffe using the Cray supercomputer at AERE Harwell. The reason we see the banded structure in Fig. 4.6 is that the particles are in the form of small platelets. They are aligned by the shear in the flow and in the cellular regime follow

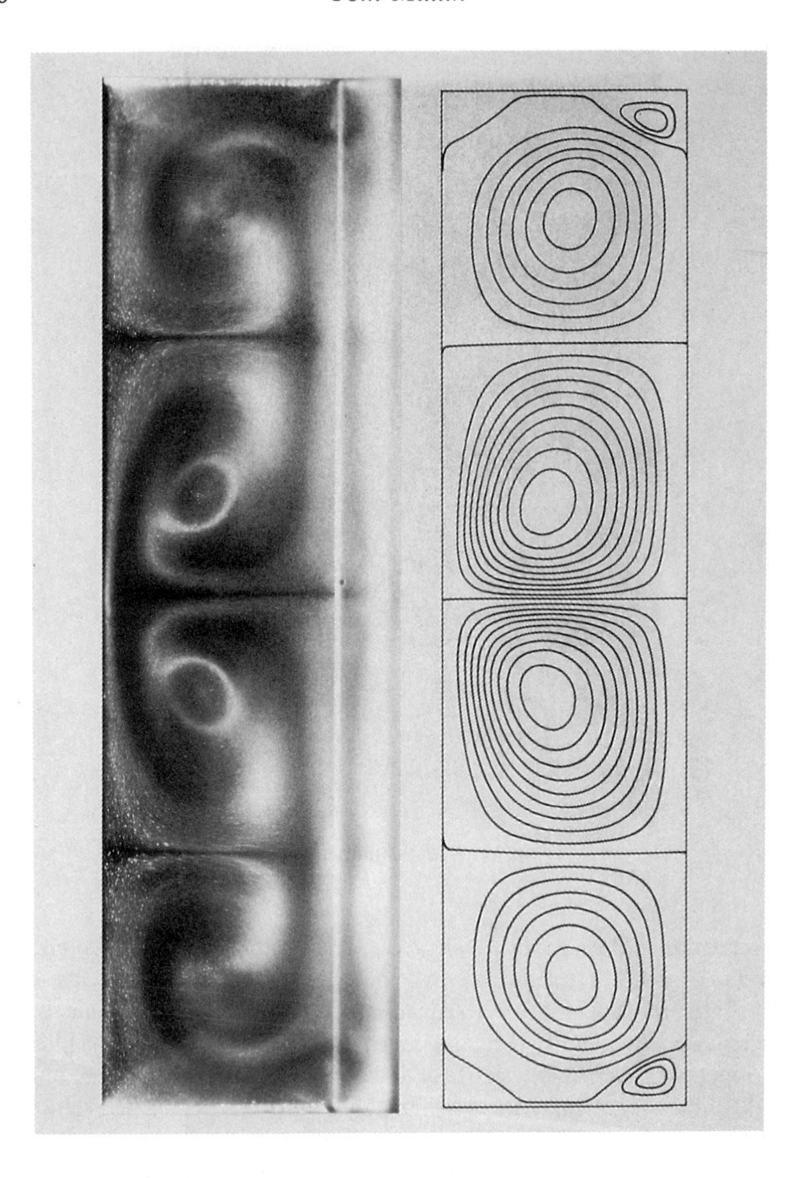

Fig. 4.7. Cross-sectional view of Taylor cells shown in comparison with calculated streamlines.

spiral paths which are a combination of the main azimuthal and secondary circular flow components. The incident light is thus reflected when the particles are face-on, but not when they are edge-on, to the field of view. Thus, when the flow field is viewed from the front, the regions where

Fig. 4.8. Front view of wavy Taylor vortices.

the secondary flow is mainly in towards or out from the inner cylinder will appear as dark bands.

If we now increase the Reynolds number by a small amount, then travelling waves appear above a critical value of *Re* and an example of this singly periodic flow is shown in Fig. 4.8. An integer number of waves travel around the annulus at some fraction of the speed of the inner cylinder. The number of waves and the speed at which they travel, usually between approximately 0.1 and 0.5 times that of the inner cylinder, are both dependent on the 'history' of the flow, i.e. there may be many different solutions to the problem at the same value of *Re*. This is yet another example of nonuniqueness of the solution in a nonlinear system and we will return to this feature below.

Thus far, we have reached the situation of a single excited mode in the Landau picture or a limit cycle in the Ruelle and Takens scenario. Now, if we increase the speed yet further, the simple wave becomes modulated as shown in Fig. 4.9. Here we have a doubly periodic state or the formation

Fig. 4.9. Front view of doubly periodic Taylor vortex flow. The second wave is just discernible in the darker regions.

of a torus in phase space. It may not be obvious that this is the case from Fig. 4.9, but this result is readily confirmed using laser Doppler velocimetry and was first done so by Gollub and Swinney [6].

Finally, another increase in Re leads to the appearance of the disordered state shown in Fig. 4.10. The flow is temporally irregular, so that the power spectrum of a single point measurement would contain a broadband component as well as discrete lines, but inspection of Fig. 4.10 shows that there is clearly spatial order retained on the lengthscale of the cells. The Reynolds number of the flow shown is approximately 10^3, and it is interesting to note that even at $Re \approx 10^9$ the spatial structure is maintained so that this is a rather special kind of disordered fluid motion. Nevertheless, the main point at issue is that the flow has become disordered through a sequence of bifurcations and we wish to know if it has followed the Landau or the Ruelle and Takens route. This question has been adressed

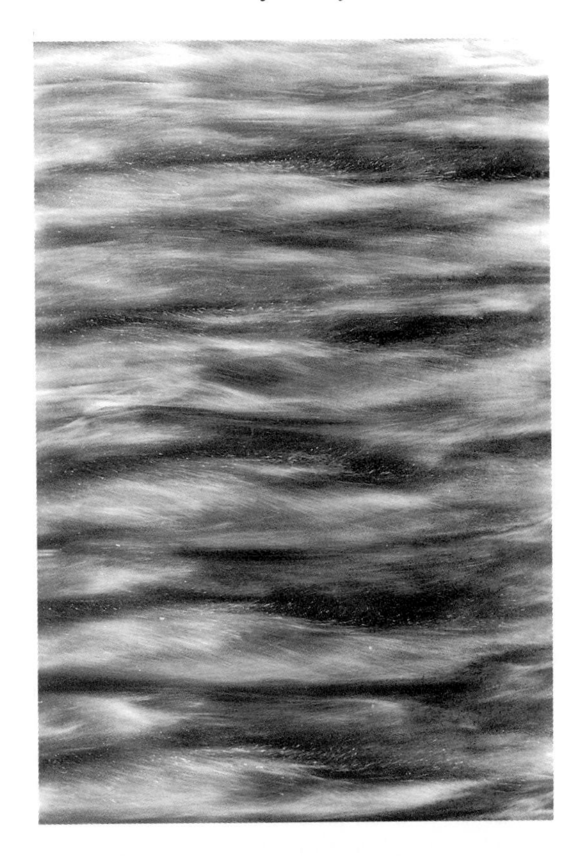

Fig. 4.10. Irregular Taylor vortices. Note that the spatial structure is still clearly visible.

in several studies (see [7] for a review of this approach), but we will now present an alternative view in which we attempt to retain a closer link with the Navier–Stokes equations than was attempted in the earlier investigations.

4.4 Steady bifurcations

Before we can address the issue of the appearance of disordered motion, there are certain other essential features of this problem we must first consider. The first of these is concerned with the steady cellular structure. In a brilliant combination of classical analysis on the Navier–Stokes equations and fluid experiment, G.I. Taylor [8] gave an explanation for the first appearance of the cells as a hydrodynamic instability. In his model he considered the cylinders to be infinitely long so that solutions can be

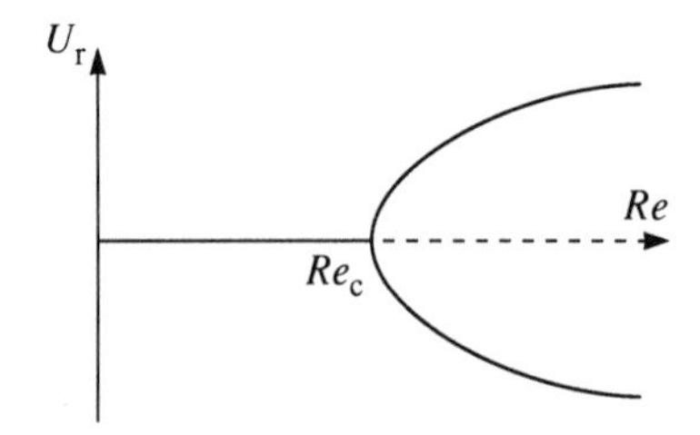

Fig. 4.11. Pitchfork bifurcation of the theoretical model for Taylor–Couette flow. The unique solution below Re_c is Couette flow which exchanges stability with cellular flows above Re_c.

sought which are periodic along the length of the cylindrical gap. Thus the cells can be represented mathematically as simple sine or cosine functions along the length of the cylinders. Using this approximation together with the assumption of a narrow gap between the cylinders makes the problem amenable to mathematical analysis.

In modern bifurcation parlance, the situation can be thought of as a pitchfork bifurcation from the trivial rotary shear state to an axisymmetric cellular flow, as shown in Fig. 4.11. Each branch of the pitchfork would correspond to a shift of one cell with respect to the other along the length of the cylinder as indicated. This is of course easily achieved on the periodic boundary conditions of the model where the periodic solution is unchanged by translation, but it is a situation which cannot be realized in the laboratory. Thus, the central portion of the flow may appear to have an approximately periodic structure, but there will always be departures from this at the ends of the physical system, no matter how large the apparatus.

We have already seen how the presence of physical imperfections can be accommodated in bifurcation theory in the example of the Euler strut in the previous chapter. Indeed, as mentioned above, there are three-dimensional flow effects at the ends of the flow domain which are precursors to the appearance of cells, i.e. the cellular structure spreads from the ends of the apparatus over a small but finite range of *Re*. Thus it might seem that we could represent the effects of the ends as a simple softening of the original pitchfork. We can investigate this possibility by measuring the effects of the imperfection. This is done by obtaining estimates of the lower limits of stability of the disconnected portion of the pitchfork. Benjamin [9] was the first to carry out such an investigation and it was later extended both numerically and experimentally in a number of studies (see [10] for a review of this work).

The main result of all of these studies is that the bifurcation is disconnected by a factor of approximately 2.5, as shown schematically in Fig. 4.12. Thus the physical realization of the bifurcation in this case

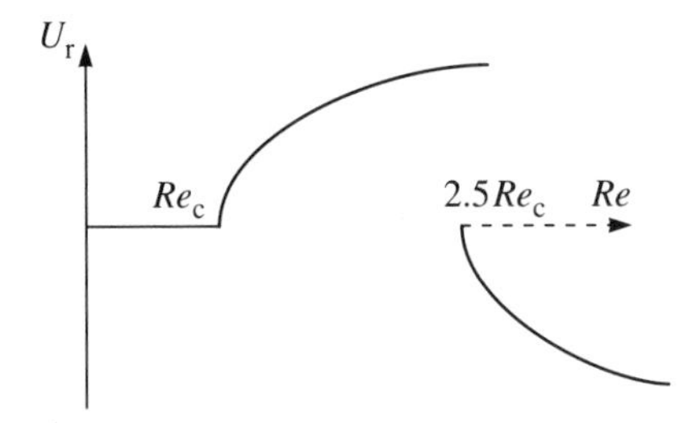

Fig. 4.12. Schematic diagram of disconnection of pitchfork in mathematical model by the application of physical boundary conditions.

is not an example of the simple softening of the model pitchfork by physical imperfections, such as we saw earlier for the Euler strut. The reason is that the essential symmetry of the model, which has been lost in the experiment, is that of translational invariance. In other words, we cannot simply translate the solutions by half a wavelength in the physical experiment as the end boundary conditions must always be matched. A heuristic argument for the large disconnection could proceed as follows. A simple physical model which is often used to explain the direction of rotation of the cells adjacent to the stationary end walls concerns the formation of Eckman boundary layers. It is argued that the centrifugal force gradient which drives the secondary circulation decays towards the stationary walls at the end of the cylindrical gap and so the circulation will always be inwards there. This is in agreement with the most common type of flow observed experimentally and thus it seems to be the one which is energetically most favoured. However, outward flow along the end wall is required for the state corresponding to the disconnected branch of the pitchfork. These flows are only observed at high values of the Reynolds number in practice, so that there is a huge difference between cellular states with opposite directions of rotation. In other words, the physical system is a major perturbation of the pitchfork in the model.

We stress the importance of this fundamental issue here because it has very important consequences for the dynamical behaviour. In Chapter 3 we saw how chaos can be organized by the interaction of different types of bifurcation for the example of the nonlinear electronic oscillator. Since the main bifurcation in the present case has to all intents and purposes been destroyed, we might expect that any associated finite-dimensional dynamics would also disappear. However, as we shall see below another type of bifurcation, which is not present in the model but exists in the physical system provides an organizing centre for the dynamics.

One other pertinent feature of nonlinear systems, which we have already discussed for the case of the pendulum, is that the solution set can be non-unique. By this we mean that there can be more than one outcome from a transient start of system to any given point in parameter space where there is a reasonable degree of nonlinearity. At small values of the Reynolds number, where the equations are essentially linear, the flow is unique. However, in the present problem, it was shown by Coles [11] that in the range of $Re \approx 10^2$ to 10^3 the solution set exhibits a high degree of nonuniqueness in the time-dependent flows. Further, it was demonstrated by Benjamin and Mullin [12] that with an aspect ratio or non-dimensionalized cylinder length of approximately 12 there were at least 39 steady solutions to the problem at one point in (Re, Γ) parameter space. Each solution was identified as having a different number of cells along the length of the cylinder. Twenty of these flows were realized in the laboratory and are therefore stable. However, as mentioned above, we know that the flow is unique at small values of Re and therefore there must be 19 unstable modes which coalesce with their stable counterparts when Re is decreased. These unstable flows cannot be observed directly but they are essential components of the whole solution structure and, as we have already seen in the case of the electronic oscillator, they can contribute to the dynamical behaviour. Thus in a very large system, where an attempt is made to mimic the theoretical model of an infinite cylinder, the solution set will be unwieldy. Therefore it will be very difficult to pin down the organizing centres for the dynamical behaviour and the dynamics themselves may be constantly flipping from one type to another. Further, if an attempt is made to establish a connection between theory and experiment, it will be very difficult in such a situation.

One way of thinking about this issue is to consider the solution surface as a tabletop on which there are lots of dimples with smooth sides. A ball is then cast onto the surface and it will eventually settle into one of the dimples. This we may think of as corresponding to one of the vortex states in the fluid dynamical problem. If the number of dimples is very large and they are not deep, then the basins of attraction for the various outcomes will be intricately interwoven so that ball may never settle in a specific location particularly when noise is present in the system. This continual hopping between states may not be a fundamental property of the system but may just be an indication of the practical limits of experimental control. Thus any potential for finite-dimensional dynamical behaviour in the fluid dynamical problem will be swamped by extraneous events caused by inevitable fluctuations in the experiment if the physical system is large.

4.5 Finite-dimensional chaos

One way to obviate the situation of high multiplicity outlined above is to make the aspect ratio of the system very small, which will in turn limit the number of steady solutions. Thus the number of fixed points in solution space will be restricted, but of course the fluid is still a continuum so that the number of temporal modes can in principle still be high. This was the approach that Pfister [13] adopted when he studied a Taylor–Couette system with an aspect ratio of approximately 1. Here the possible steady flow states are a two-cell flow and a pair of single cells. By restricting the flow in this way, he observed a period-doubling cascade to chaos in a very narrow range of *Re* and managed to follow it as far as a period-16 cycle. It is interesting to note that perhaps the most famous observation of period doubling in a fluid convection experiment by Libchaber and Maurer [14] also required this extremely tight control using a conducting fluid in a very small container with an external magnetic field to help pin down the steady cell structure. Perhaps one inference we can draw from these observations is that a generic or typical route to chaos in the restricted version of a continuous system, where the number of spatial modes is limited, is period doubling. We saw in the case of the nonlinear electronic oscillator that period doubling can form part of the so-called Silnikov mechanism for chaos, but this has not yet been shown for these continuous systems.

Now we will consider some of the events which are observed at larger aspect ratios in a modified version of the Taylor–Couette system. Here the plates which form the ends of the fluid column are rotated with the inner cylinder in an attempt to impose a strong symmetric forcing on the flow. The reason for the additional forcing is that Mullin *et al.* [15] had previously shown that the standard system with stationary ends gave rise to symmetry-breaking bifurcations. The flows which arise at these points break the up–down mirror symmetry of the flow domain, as shown in Fig. 4.13, i.e. one pair of cells is larger than the other. The symmetry-broken solutions must necessarily exist in pairs and are akin to the buckled beam states in the Euler strut. The idea behind the rotating-ends study was to use symmetric forcing to suppress the symmetry-breaking bifurcations but perversely they were found to occur over a wider range of parameter space.

It would therefore appear that spatial symmetry-breaking bifurcations are an essential feature of recirculating flow fields on symmetric boundary conditions. As we have pointed out above, the pitchfork bifurcation of the theoretical model on periodic boundary conditions is destroyed in the physical system. These new types of symmetry-breaking bifurcations are only slightly disconnected by the presence of imperfections and thus there is the potential for them to form organizing centres for low-dimensional

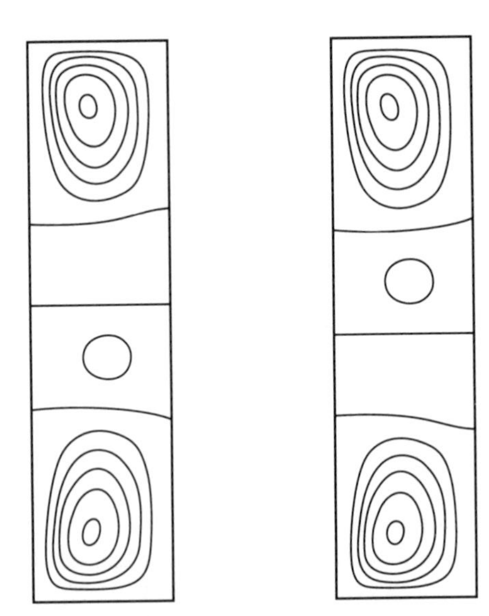

Fig. 4.13. Streamline plots of the pair of asymmetric states which arise at pitchfork bifurcation of the flow with physically realistic boundary conditions.

dynamical behaviour. This could proceed in an analogous manner to that found in the electronic oscillator since we know that there are also Hopf bifurcations present which give rise to the travelling wave solutions. However, the full picture is much more complicated here than in the case of the oscillator and the full details are beyond the scope of the present chapter; the interested reader is referred to [16] for full details. Therefore, we will concentrate on those features which are essential for the understanding of one particular route to chaos.

We show in Fig. 4.14 a comparison between experimental and theoretical results for two different types of bifurcation. The experimental results are shown as points and give a good absolute comparison with the theoretical curves. The theoretical results are determined numerically from the full Navier–Stokes equations using a bifurcation package called ENTWIFE, which was developed at AERE, Harwell. Not only does it allow the detection of bifurcation points but their paths can also be followed as the control parameters are varied. Thus we are able to map out the full steady and first time-dependent bifurcations for this set of partial differential equations, just as we did using AUTO for the ordinary differential set of the earlier examples.

The solid line is the locus of lower limit points for an asymmetric four-cell state which arises through a subcritical bifurcation. We have already discussed a subcritical bifurcation with reference to the parametrically

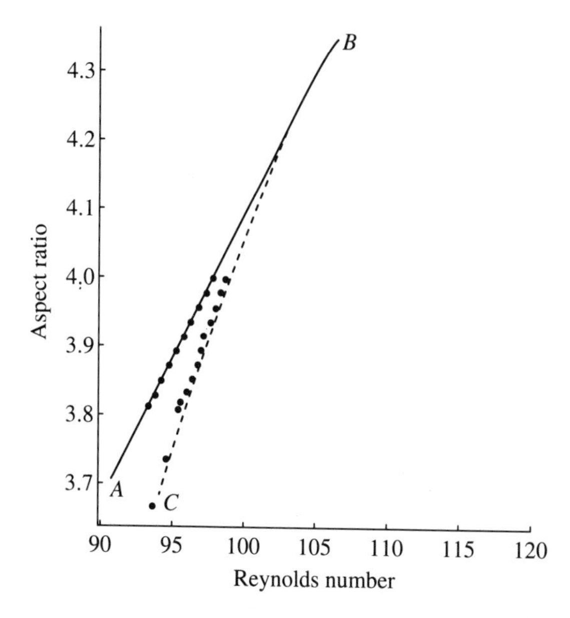

Fig. 4.14. Numerical and experimental results for the bifurcation set near the codimension-2 point. *AB* is the locus of limit points of the fold bifurcation and *CB* is the locus of Hopf bifurcations to travelling waves.

excited pendulum. There we saw that there was hysteresis in the appearance of the pendulum oscillation on the left-hand side of the instability tongue. In the present case there is hysteresis in the exchange of stability between the symmetric state with the asymmetric pair of solutions. In other words, one of the pair of asymmetric flows is reached suddenly at a critical value of *Re* and this parameter must then be reduced to a lower value before the symmetric flow is regained. The bifurcation point is far to the right on the present scale and so is omitted from the diagram. The locus therefore shows how the limit point of a particular asymmetric four-cell solution varies with external control parameters.

The chain-linked line shown in Fig. 4.14 is the numerically determined locus of Hopf bifurcations to a travelling wave which has wave number 1 in the azimuthal direction around the cylindrical gap. In the experiment this wave is observed to have the form of a tilt on the intercellular boundaries so that there is a π phase shift between opposite sides of the cylinder. The line of Hopf bifurcations crosses the fold line at A which is, therefore, a multiple bifurcation or codimension-2 point. Thus, analogous to the study of the oscillator in Chapter 3, there is potential for an organizing centre for chaos at this point.

The numerical results represent the limits of the calculations of the

Navier–Stokes equations using the Cray supercomputer. However, we can infer from these theoretical results the topological or geometrical form of the bifurcation structure and represent it using simple algebraic formulas. Using this information and the experimental results, we can construct a mathematical model which will hopefully describe the more complicated dynamical behaviour. The mathematical model can then be solved for various parameter values on a small personal computer.

One experimental fact we need to include in the model, and which is not available from the numerics, is that the singly periodic flow undergoes a secondary Hopf bifurcation to motion on a torus in phase space. The second wave is axisymmetric, so that it is in phase all around the annular gap. One abstract model which contains all of these ingredients has been extensively studied by Langford [17]. It is

$$\dot{x} = xz - \omega y,$$
$$\dot{y} = \omega x + yx,$$
$$\dot{z} = p + z - \tfrac{1}{3}z^3 - (x^2 + y^2)(1 + qx + \varepsilon z),$$

The coefficients p, q, ε, and ω may be chosen with guidance from the experimental and numerical results, but the extremely difficult task of estimating them directly from a reduction of the full Navier–Stokes equations remains a formidable challenge. Nevertheless, it is interesting to see how relevant the model is to the present study. The connection is not straightforward, as we find five different routes to chaos in the experiment, only one of which can be seen in the model. One reason for this is that there are other bifurcations nearby in parameter space which we have not included in the present discussion for the sake of simplicity.

We will now concentrate on one route to chaos which is qualitatively the same as the model. Figure 4.15(a) contains the velocity time series measured at a single point in the flow using the laser system and at a parameter setting near the double bifurcation point. The measurement point was chosen so that data recorded there contains contributions from all of the time-dependent components in the flow. The most striking feature is the regular appearance of a wave followed by a quiescent period. The intervals between the waves correspond to eight times the fast wave period, so that the two waves appear to be locked in phase. If the second wave is perfectly axisymmetric, then there could of course not be locking of the two motions as there would be no phase reference point. However, there are inevitably some small eccentricies in the cylinders which give rise to a small parameter range over which locking occurs.

The reconstructed phase portrait obtained from this time series is presented in Fig. 4.15(b) and this shows that the motion takes place on a locked torus. The two frequencies which make up the torus are related to each other by a factor of eight and so the motion is really taking place

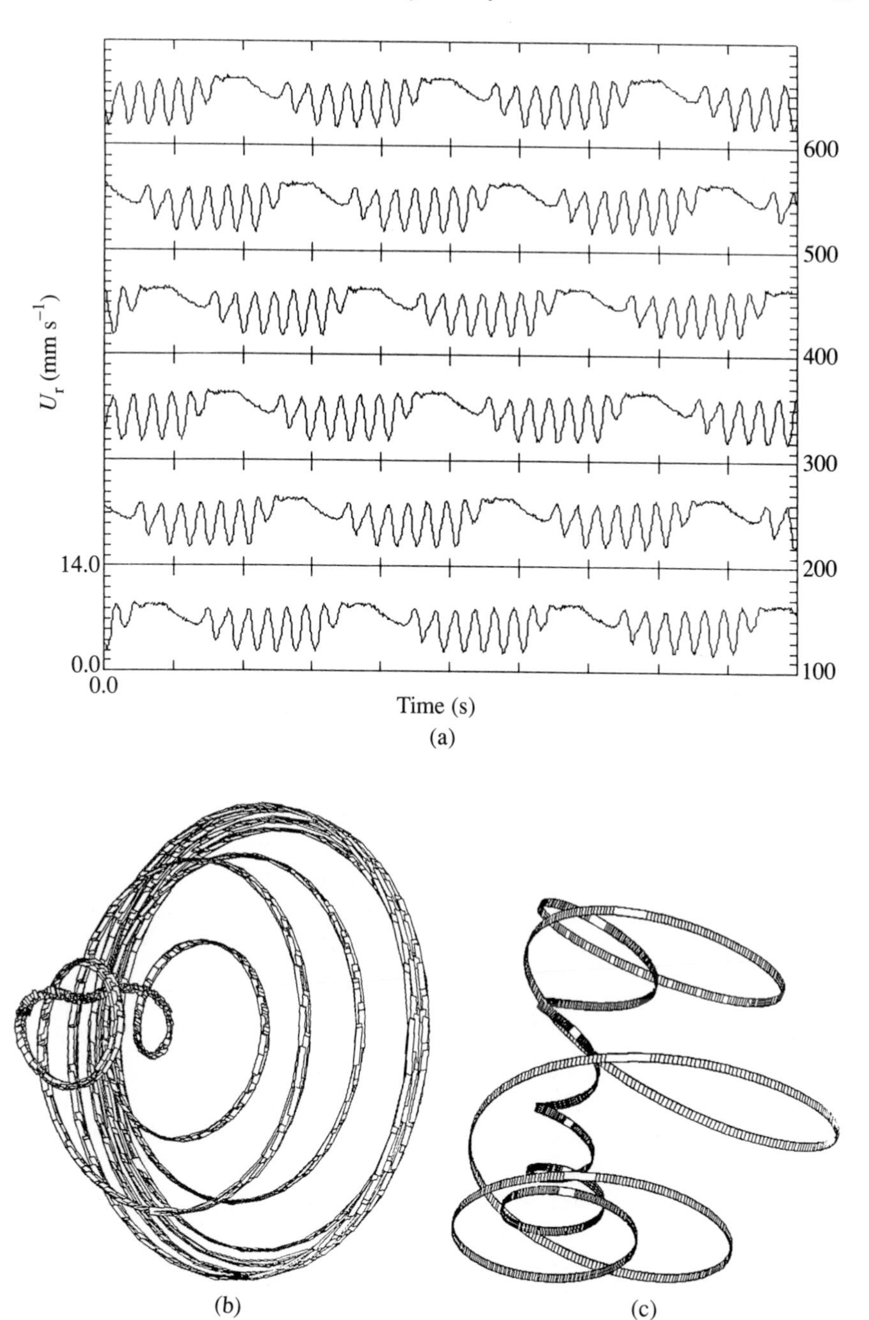

Fig. 4.15. Motion on a locked torus on one of the asymmetric four-cell states at an aspect ratio of 4.25: (a) velocity time series at $Re = 132.7$; (b) reconstructed experimental phase portrait; (c) model phase portrait.

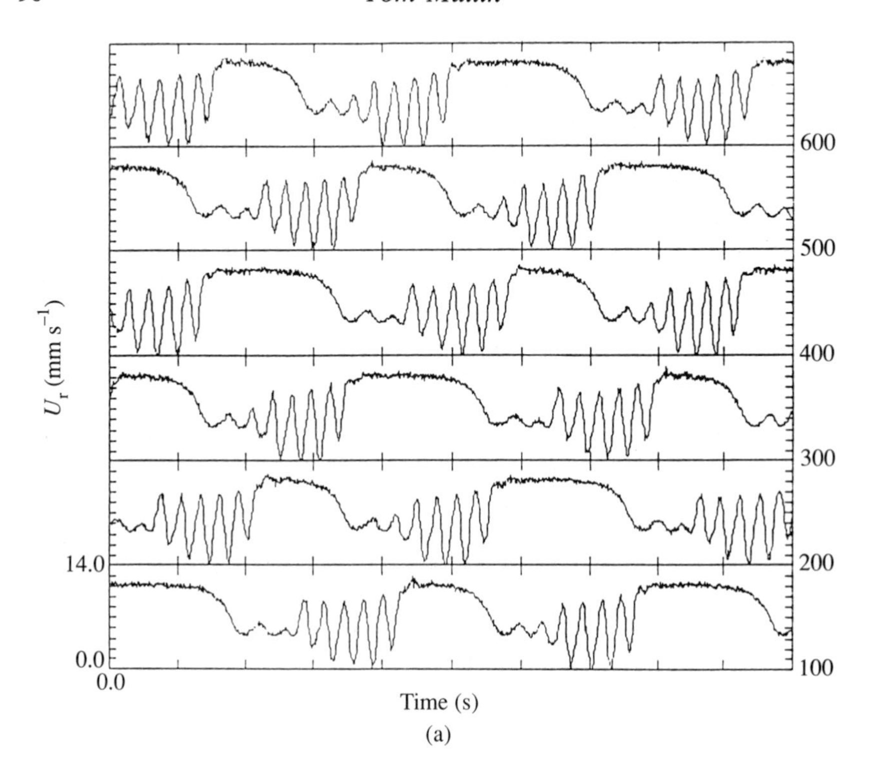

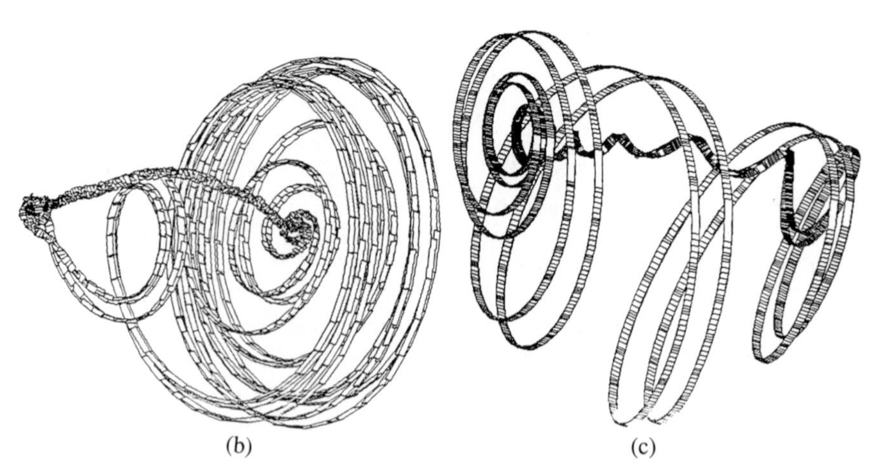

Fig. 4.16. Motion on a doubled locked torus: (a) velocity time series at $Re = 129.6$; (b) reconstructed experimental phase portrait; (c) model phase portrait.

on a complicated limit cycle. The torus comprises a central core which corresponds to the quiescent phase and the wavy phase is represented by the large-scale orbits round the outside which enter and leave at the ends of the core. We show in Fig. 4.15(c) the phase portrait obtained from the model and it can be seen that it is qualitatively the same as the experimental one.

If the Reynolds number is reduced by approximately 1% in the experiment, then the time series shown in Fig. 4.16(a) is obtained. Now the waves appear at alternately long and short intervals indicating a form of doubling. The reconstructed phase portrait shown in Fig. 4.16(b) displays a doubled limit cycle which is in close qualitative agreement with the model result shown in Fig. 4.16(c).

Finally, we show in Fig. 4.17(a) the time series for the chaotic motion found on further reduction of *Re*. The wave packets now arrive at irregular intervals and there is no obvious pattern to the time-series plot. The reconstructed experimental phase portrait shown in Fig. 4.17(b) begins to fill in and the trajectories no longer hit the end of the core but land at irregular points along it. The model result presented in Fig. 4.17.(c) also shows chaos and so is in apparent agreement with the experiment. However, close inspection of the experimental results shows self-intersection of trajectories, indicating that the full motion takes place in more than three dimensions. Since the model is only three dimensional, there cannot be a one-to-one correspondence between the model and the experiment. Nevertheless, this striking result shows strong evidence for finite-dimensional behaviour in the Navier–Stokes equations of more than a trivial kind.

Here, a note of caution must be added. The sequence of events described above appears to show good qualitative comparison between theory and experiment. However, the coefficients in the model have not been evaluated, so that a direct comparison is not yet possible. Moreover, the experimental results were taken at parameter values far away from the codimension-2 point. Therefore, the possibility of coincidental agreement cannot be ruled out at this stage. However, the quite remarkable similarities between the model and experimental results gives a clear indication that finite-dimensional dynamics leads to chaos in this flow.

As mentioned above, other routes to finite-dimensional chaos have been found in neighbouring parameter ranges for this flow problem. These include homoclinicity [18] and intermittency [19]. This latter route involves the transition from an oscillatory state with constant amplitude to one that exhibits large-amplitude infrequent bursts. A detailed and lucid account of three generic routes to chaos through intermittency is given in the book by Berge *et al.* [20]

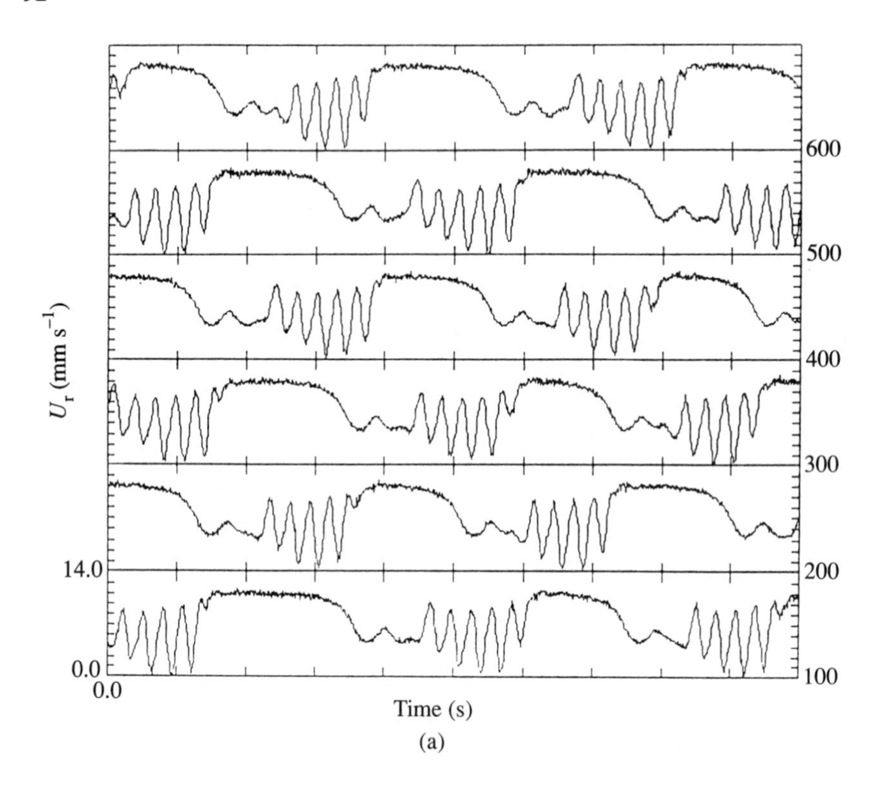

(a)

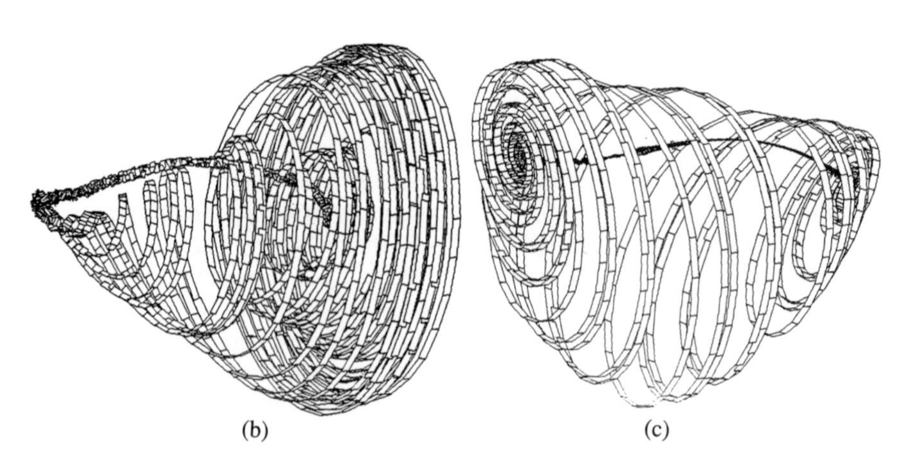

(b) (c)

Fig. 4.17. Chaotic motion: (a) velocity time series at Re = 127.9; (b) reconstructed experimental phase portrait; (c) model phase portrait.

4.6 Concluding remarks

We have shown that the notions of finite-dimensional chaos are relevant to the understanding of the transition to weakly irregular states in a carefully controlled closed fluid flow system. Further, the complicated motion which is found over a wide range of parameter space is organized by localized interactions of bifurcations of differing types. This observation makes the possibility of establishing a connection between the governing equations of motion and the considerably simplified dynamical models realistic.

The above observations appear to be more in accord with the Ruelle–Takens picture than that of Landau. However, despite the considerable successes of the present studies of the application of modern ideas on chaos to well-controlled fluid flows, they appear to have little relevance when applied to the more general problem of fluid turbulence. One reason for this is that turbulence is almost certainly an infinite (or at a least very large) dimensional phenomenon and so is not describable directly in terms of the drastic restrictions imposed in the small-scale experiments of the type discussed above. This is primarily because only a few spatial modes are allowed to participate in the interactions which produce the chaos. In addition, low-dimensional chaos is primarily concerned with the appearance of temporal disorder in finite-dimensional systems, whereas turbulence is a feature of continuous systems and is irregular in both space and time. A lively discussion of these issues from a variety of viewpoints is given in the book edited by Lumley [21].

Nevertheless, it should not be concluded that the ideas of low-dimensional chaos add nothing to our understanding of complicated fluid motions. There are many practical examples of fluid flows where there are large-scale coherent regions of fluid which move around in highly complicated ways, e.g. atmospheric flows. It is in these situations that the 'simple' models of low-dimensional chaos may have a significant role to play. Finally, and perhaps most importantly, the lessons to be learned about symmetries and multiplicities in carefully controlled experimental studies of the type outlined above yield new insights into highly complicated practical flow problems.

References

1. Acheson, D. J. (1990). *Elementary fluid dynamics.* Clarendon Press, Oxford.
2. Tritton, D. J. (1988). *Physical fluid dynamics.* Clarendon Press, Oxford.
3. Ruelle, D. and Takens, F. (1971). On the nature of turbulence. *Commun. Math. Phys.*, **20**, 167–92.

4. Drain, L. E. (1972). The laser Doppler technique. Wiley, New York.
5. Andereck, C. D., Liu, S. S., and Swinney, H. L. (1986). Flow regimes in a circular Couette system with independently rotating cylinders. *J. Fluid Mech.*, **164**, 155–83.
6. Gollub, J. P. and Swinney, H. L. Onset of turbulence in a rotating fluid. *Phys. Rev. Lett.*, **35**, 927–35.
7. DiPrima, R. C. and Swinney, H. L. (1981). Instabilities and transition in flow between concentric rotationg cylinders. In: *Hydrodynamic instabilities and the transition to turbulence* (ed. H. L. Swinney and J. P. Gollub), Topics in Applied Physics 45. Springer, Berlin.
8. Taylor, G. I. (1923). Stability of a viscous liquid contained between two rotating cylinders. *Phil. Trans. R. Soc.* A, **233**, 289–343.
9. Benjamin, T. B. (1978). Bifurcation phenomena in steady flows of a viscous fluid. *Proc. R. Soc. Lond.* A, **359**, 1–26, 27–43.
10. Mullin, (1991). Finite-dimensional dynamics in Taylor–Couette flow. *IMA J. Appl. Math.*, **46**, 109–19.
11. Coles, D. (1965). Transition in circular Couette flow. *J. Fluid Mech.*, **21**, 385–425.
12. Benjamin, T. B. and Mullin, T. (1981). Anomalous modes in the Taylor experiment. *Proc. R. Soc. Lond.* A, **377**, 221–49.
13. Pfister, G. (1985). Period doubling in Couette flow. In: *Lecture notes in physics 235*. Springer, Berlin.
14. Libchaber, A. Maurer J. (1982). A Rayleigh–Bénard experiment: Helium in a small box. In: Universality in chaos (ed. P. Cvitanovic). Adam Hilger, Bristol.
15. Mullin, T., Cliffe, K. A., and Pfister, G. (1987). Unsual time-dependent phenomena in Taylor–Couette flow at moderately small Reynolds numbers. *Phys. Rev. Lett.*, **58**, 2212–15.
16. Travener, S. J., Mullin, T., and Cliffe, K. A. (1991). Novel bifucation phenomena in a rotating annulus. *J. Fluid Mech.*, **229**, 438–97.
17. Langford, W. F. (1984). Numerical studies of torus bifurcations. *Int. Ser. Numer. Math.*, **70**, 285–93.
18. Mullin, T. and Price, T. J. (1989). An experiental observation of chaos arising from the interaction of steady and time-dependent flows. *Nature*, **340**, 294–6.
19. Price, T. J. and Mullin, T. (1991). An experimental observation of a new type of intermittency. *Physica* D, **48**, 29–52.
20. Berge, F., Pomean, Y., and Vidal, C. (1984). *Order within chaos*. Wiley, New York.
21. Lumley, J. L. (1990). Whither turbulence? Turbulence at the crossroads. In: *Lecture notes in physics 357*. Springer, Berlin.

5
Chaos and one-dimensional maps

David Holton and Robert M. May*

In our everyday experience we frequently encounter processes which exhibit chaotic or apparently random behaviour: the constantly changing weather pattern, the flow of a turbulent stream, or the fluctuations in the population of a species. These are dramatic examples of systems which display complicated and seemingly indeterminate behaviour. However, these are precisely the kinds of systems which hold most fascination and which we wish most to be able to rationalize, to model, and, using estimates based on the model, to predict future behaviour.

The traditional view of the complexity of nature is that there is little hope of unravelling the underlying reasons for randomness. The elements of cause and effect are believed to be so intricate and the interactions so complicated, appearing on so many spatial and temporal scales, that the net result is that the many influences act incoherently and statistically, to yield an unfathomable mess.

In the field of population dynamics and epidemiology it is certainly true to say that there is still a tendency of most entomologists, for example, to interpret apparently random population data as purely stochastic noise or Wienerian random experimental error. This view is also held by many researchers in other areas of the life sciences. This perception of the complexity underlying nature may well be partially accurate, but what a revelation it would be if there was an alternative scenario to the complexity that we perceive in nature, with mathematically simple roots.

The emerging view, challenging the traditional perspective, is the conjecture that low-dimensional chaotic dynamical systems hold the key to the description of a huge spectrum of evolutionary processes, and, further, that nonlinearity plays a central role in the description of such complicated dynamics. There is an abundance of theoretical and experimental evidence that the qualitative characteristics of the temporal evolution of many physical systems are the same as the complex solutions resulting from simple low-dimensional evolution equations of the form

$$\frac{d\boldsymbol{x}}{dt} = \boldsymbol{F}(\boldsymbol{x}(t); \lambda). \tag{5.1}$$

*Chapters 5–8 were written by David Holton, on the basis of my lectures along with notes and other material supplied to him. I am grateful to him. RMM.

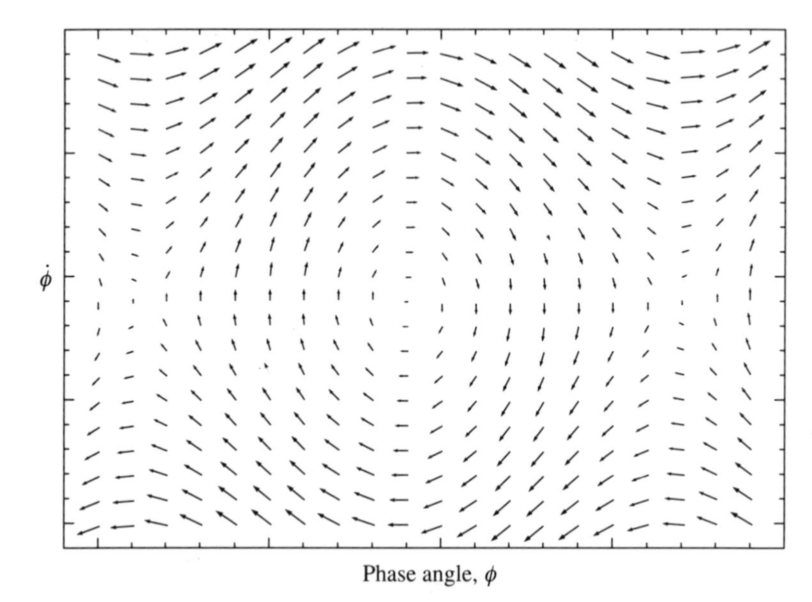

Fig. 5.1. A vector field generated by the equation for the simple pendulum, $\ddot{\phi} = \sin \phi$, where ϕ is the phase angle. The phase variables are ϕ and $\dot{\phi}$. At each point in phase space there is a vector pointing in the direction of the vector field.

Here each $\boldsymbol{x}(t) = (x_1, \ldots, x_n)$ represents a state of the system and may be interpreted as a point in an n-dimensional phase space P, and $\lambda = (\lambda_1, \ldots, \lambda_m)$ is a collection of stress parameters. The dimension n of P indicates the intrinsic number of degrees of freedom of the system. The function $\boldsymbol{F}(\boldsymbol{x}(t); \lambda)$ is in general a nonlinear operator that acts on points in P and specifies a vector field; Fig. 5.1 shows the vector field for the simple pendulum ($n = 2$). The parameters λ_i specify pertinent physical quantities. Equations such as this, and complex solutions arising from it, are aspects of *dynamical systems theory*.

The notion of a unique solution (and its existence), for a given a set of initial conditions $\boldsymbol{x}(0)$, is perhaps a familiar one when the function $\boldsymbol{F}$ is a linear function of the set of variables $\boldsymbol{x}$ [1]. When $\boldsymbol{F}$ is nonlinear, many distinct solutions can exist for the same set of parameters $\{\lambda_i\}$. Arguably, the most important characteristic of nonlinear systems is that they are able to generate temporal solutions that look for all the world like a sample function of a random process. This tumultuous behaviour is called simply *Chaos*, which nomenclature is due to Li and Yorke [2].

5.1 A brief history of chaos

The mathematics of dynamical systems theory, which encompasses the concepts of multiplicity of a solution set and chaoticness of a dynamical trajectory, has a relatively short but illustrious history. The origins of dynamical systems theory began over a century ago. The basic early components of the theory of multiplicity of solutions of nonlinear evolution equations were grouped into three distinct areas: critical solutions, stability, and structural stability. These formed a branch of mathematics known as bifurcation theory. The development of bifurcation theory has enhanced the understanding of how multiple solutions can exist in nonlinear systems and how the numbers and stability of solutions change as an experimental control, or stress parameter, is changed. The crucial question of stability of a solution, that is, whether or not a solution persists under an infinitesimal perturbation and how it changes as the number of solutions change, has occupied the time of many great mathematicians of the eighteenth and nineteenth centuries (e.g. Lagrange, Laplace, and Dirichlet). Their ultimate aim was to shed light on the age-old n-body problem and in doing so to determine the stability of the solar system.

Pierre Simon de Laplace wrote, in 1776:

> The present state of nature is evidently a consequence of what it was in the preceding moment, and if we conceive of an intelligence which at a given instant comprehends of all the relations of the entities of this universe, it could state the respected positions, motions, and general effects of all these entities at any time in the past or future.

Indeed, Laplace claimed that he had indeed proved, using a series expansion technique, that the solar system was stable. Perhaps this was a justifiably cautious conclusion, given the circumstances. The importance of Laplace's claim reinforced the opinion held by many scientists: that once an initial condition (or configuration) is specified the future is completely determined. Laplace's claim was eventually proved by Henri Poincaré to be fallacious. Poincaré stated, in 1903:

> A very small cause which escapes our notice determines a considerable effect that we cannot fail to see . . . even if the case that the natural laws had no longer secret for us, we could only know the initial situation approximately. . . . It may happen that small differences in initial conditions produce very great ones in the final phenomena.

Deterministic chaos, with its inherent sensitivity to initial conditions, provides a fresh insight into problems encountered in dynamics.

Towards the end of the nineteenth century the theory of bifurcations, due to Poincaré, was formulated. In Poincaré's seminal paper of 1880, Mémoire sur les courbes définies par les équations différentielles [3], he gave a mathematically precise criterion for a system to undergo fundamental qualitative change in nature as parameters of the system are changed. One of those conditions requires a change in the number of critical solutions.

However, the conceptual basis of bifurcation theory was begun by Leonhard Euler, predating Poincaré's effort with work published in 1774 on the buckling of loaded struts. Euler was able to demonstrate how, as the strut is loaded, it will deflect to the left or right. In other words, the initially stable, upright, solution becomes unstable to a solution with the strut either bent to the left or right, an example of a symmetry-breaking bifurcation.

Around about the turn of the century, King Oscar of Sweden, a great science advocate, offered a prize for what would have closed in a grand way the view of the world as a deterministic clockwork system. His prize, which was eventually offered to Poincaré, was for a rigorous mathematical proof to demonstrate that the solar system was stable. Poincaré [4] pointed out that the series expansions by Laplace diverged and hence stability could not be assured. His contribution provided fresh insight into the problem of stability, but in so doing it also ended the quantitative era. One of Poincaré's most important inceptions was the introduction of qualitative dynamics, introduced in a flurry of important papers in 1880 [4], 1890 [5] and 1899 [6]. Also, his contribution provided a basis for the development of the modern version of differential geometry. The Russian academician A. M. Lyapunov continued the spirit of Poincaré's work and made significant advances in the formulation of the notion of stability [6]; it is his definition of stability we often use today.

Poincaré was one of the earliest contributors to bifurcation theory and hence initiated the development of dynamical systems theory. Important contributions to the mathematics of dynamics occurred during the twentieth century. Birkoff's work in 1920 on the mappings of the circle to itself has greatly influenced our understanding of resonances [7]. Julia in 1917 contributed greatly to the inception and understanding of complex analytical maps; this work was substantially enhanced in recent times by Mandelbrot [8].

A strong Soviet school existed in the late 1940s (though Lyapunov had an influence dating from much earlier than this), led originally by Lefschetz. More recently Kolmogorov and Arnold opened up of a new chapter in dynamical systems theory with the proving of the so-called KAM theory, which is important in the understanding of so-called elliptic points in dynamical systems.

Much more recently, a significant development in chaos (some say its birth) occurred in 1963 with Lorenz's [9] seminal paper on deterministic nonperiodic flow, an article concerned with turbulence. Lorenz's name has now become synonymous with the discovery of chaos. He discovered aperiodic solutions of numerical integrations of a set of three ordinary nonlinear differential equations. Lorenz's equations were a crude caricature of the Navier–Stokes equations. However, it would seem remarkable that those three equations could give rise to complicated aperiodic behaviour. Could this complicated behaviour be a manifestation of a simpler version of turbulence? Lorenz's contribution marked a perceptual change which was to alter the face of dynamics.

The discrete analogue of Lorenz's initiative, pursued altogether independently, focused on difference equations. Difference equations or maps were advocated originally as models for ecological systems. A difference equation or map is an example of a dynamical system in which time is discrete. The evolution equation analogous to equation (5.1) is

$$\boldsymbol{x}_{t+1} = \boldsymbol{f}(\boldsymbol{x}_t; \lambda), \tag{5.2}$$

where t is the discrete time and $\boldsymbol{f}$ is an n-dimensional function of $\boldsymbol{x}$ and λ. The flourish of activity in the 1970s, by population biologists, gave much impetus to the understanding of chaotic dynamics in maps and in particular *period doubling*, in which a fixed point (when $\boldsymbol{x}_{t+1} = \boldsymbol{x}_t$) of the map given in equation (5.2) becomes unstable to a new solution whose period is double that previously, that is, $\boldsymbol{x}_{t+2} = \boldsymbol{x}_t$ is stable, but period-1 points $\boldsymbol{x}_{t+1} = \boldsymbol{x}_t$ are not. It is much easier to understand what is happening in one-dimensional discrete maps than in three-dimensional sets of differential equations (which are the lowest-dimensional such systems capable of exhibiting chaotic dynamics).

In the USA, the modern theory of dynamical systems was conceived by S. Smale [10] who invented a bizarre mathematical entity known as a 'Smale horseshoe'. Ruelle in France and Takens in Holland developed a new theory of turbulence [11] based on the theory of strange attractors, which challenged the Landau [12] view of turbulence as a confluence of quasi-periodic mayhem. Landau viewed a physical system becoming more and more turbulent as a result of the number of independent periodic motions progressively increasing. However, when there are nonlinear interactions between the excited oscillators, the net result is that there can be a complex finite-dimensional motion that need not be quasi-periodic. The important result suggested by Ruelle and Takens is due to the nonlinear coupling between oscillators corresponding to different frequencies; this tends to destroy quasi-periodicity and replace it by motion which is chaotic.

Lorenz's contribution and the work on mappings in the early 1970s by Yorke, May, Oster, and others provided two strands in a narrative which has exploded into a growing realization that the simplest deterministic rules can generate dynamical trajectories which not only look like the sample function of some random process but furthermore have the important property of sensitivity to initial conditions. This means that even if the rule defining evolution of a chaotic dynamical system were known exactly, long-term prediction would still be impossible, not simply because the notion was 'random' but because two nearly equivalent starting conditions could give distinctly different trajectories, which diverge at an exponential rate.

The development of dynamical systems theory in general, and Lorenz's contribution and May's [13] review in *Nature* in particular, triggered a change in perception: that a large number of complicated equations were not necessary (though could, of course, support complex structures) for solutions to be chaotic or turbulent-like. Dynamical systems theory has been driven to the forefront of many fields of science with an impressive number and variety of applications.

5.2 Model worlds

The study of a model problem is often precarious. The concept of a model engenders a degree of simplicity, a stripping away of all that is thought to be unnecessary. We often circumvent mathematical difficulties and can prove deeper results for the model equations than one could hope to achieve for the true equations, if they were known. Clearly, in reformulating the underlying equations, we have created a model world whose laws of interaction no longer precisely match those of nature. Our desire is that general behaviour of the solutions of the model equations have qualitative features similar to the kind of behaviour that occurs in nature. The mathematical reasoning for dealing only with qualitative results goes back to the time of Poincaré. We shall demonstrate that often solutions of model equations have behaviour that is common to a large number of problems.

Modelling involves the construction of an unnatural world that encapsulates or reflects the essence of the problem to be modelled. The model problem, however, has the advantage that it is unfettered by mathematical complexity. This was the essence of Poincaré's approach for studying the three-body problem, which involved difficulties inherent in the study of the dynamics of the solar system. In the life sciences we are not able to give the basic equations that reflect some deeper physical principle like Newton's equations of motion. We cannot unravel the true laws that govern evolution of a species from first principles. Instead we seek a pedagogical approach for studying model equations as metaphors for species interacting. This pedagogical approach will, it is hoped, be able to enhance our ability to understand global features of ecological and epidemiological interactions.

The classical equations of motion are well known for a variety of problems. In fluid mechanics, the governing equations of motion, the Navier–Stokes equations, are an excellent description of the motion of weakly compressible fluids. The problems encountered in seeking solutions of the Navier–Stokes equation reflect the profound difficulties in dealing with such complicated equations of motion. For example, to determine accurate time-dependent solutions of the Navier–Stokes equations is a formidable numerical task.

Early analysis of biological population data was based upon the assumption that if one could strip away environmental noise then the population would reside at an equilibrium population. We aim to demonstrate why this approach is not only fraught with difficulties but has no theoretical basis except in exceptional circumstances.

5.3 One-dimensional maps and population dynamics

In population dynamics, it is desirable to predict trends in populations due to external influences. One of the simplest population systems is a seasonally breeding organism whose generations do not overlap. Many natural populations, particularly among the temperate zone insects, including many important crop and orchard pests, are of this kind. We seek to understand how the size x_{t+1} of a population in generation $t + 1$ is related to the size x_t of the population in the preceding generation t. Often an adjustable parameter appears, accounting for, say, the net reproductive rate of the population. We may express such a scalar relationship in general form

$$x_{t+1} = f(x_t; \lambda). \tag{5.3}$$

We shall henceforth adopt the convention of referring to the variable x as 'the population'. The function $f(x_t; \lambda)$ is alternatively known as a 'density-dependent' relationship and is in practice often nonlinear. The function f is also called a mapping, or map for short. If we are interested in deriving information about the maximum, or the average, or the total annual population, we may consider describing the evolution of the population as a difference equation. We claim equation (5.3) is a general relationship for the dynamics of a fictitious (albeit metaphorical) species, but in doing so we have made some strong and as yet unjustified assumptions.

We have assumed that the species in equation (5.3) is self-interacting and that this is set against a constant and unchanging environment, typified by the constancy of the control parameter λ, which itself represents the propensity of a population to grow. Clearly, if the population interacts strongly with other populations which have a regulatory effect on it (such as a competing species or a finite food source), a single equation would not suffice. Competing populations could come with their own chain of interactions. The net result is a huge hierarchy of complex and largely unknown equations.

Despite this possibility, there is a rationale for constructing overly simplified models: to capture the essence of observed patterns and processes without being enmeshed in the details. Moreover, mathematical results do exist, in special cases, that can partially justify a general reduction of large systems of equations to a small number of important equations, by projecting the large-dimensional phase space onto a low-dimensional subspace [14, 15]. It must be stressed that these results are by no means general enough to justify a rigorous reduction for the problems we shall discuss.

For a continuously varying population, we could think of the function

f as a stroboscopic snapshot of the continuous dynamics at fixed time intervals. In general, however, the stroboscopic study of continuous dynamical systems is not useful, unless a natural frequency is apparent (e.g. when the system is forced at a fixed frequency ω). A more general, and more useful, mapping derived from a continuous dynamical system is commonly called a Poincaré map. It is formed by making a transverse cut through the n dimensions of phase space to give a set of points in $n - 1$ dimensions. The importance of the idea of reducing a continuous time system to a discrete time system is, not suprisingly, due to Poincaré. It offers the advantages of being able to reduce the dimension of the system by 1, thus enabling a lower-dimensional graphical interpretation of the dynamics. Also, the results for linear instabilities for continuous dynamical systems are most concisely stated in terms of the corresponding instabilities of their Poincaré maps.

5.3.1 *Examples of maps (through applications)*

A variety of model population problems have been the subject of studies by entomologists; some of these are catalogued in Table 5.1. Equation (ii), for example, has been used in the analysis of data pertaining to insect and fish populations by Moran [19] and Ricker [20], respectively, and equation (iii) is the basis for 'k-factor analysis' in pest control. The forms of equations (iii)–(vi) have been used in analyses of synoptic collections of data on insect populations. A number of equations of this type have been studied in various contexts and they share a number of special features.

Table 5.1. One dimensional 'maps' proposed in ecology studies of single populations.

Equation number	$f(x;\lambda)$	References
(i)	$x(1 + r(1 - x/k))$	[16], [17], [18], [2]
(ii)	$xe^{[r(1 - x/k)]}$	[19], [20], [21], [22]
(iii)	$\lambda x^{(1 - b)}$	[23], [24]
(iv)	$\lambda x[1 + \alpha x]^{-b}$	[25], [26]
(v)	$\lambda x[1 + (\alpha x)^b]^{-1}$	[27], [28]
(vi)	$xe^{[r(1 - (x/k)^\theta)]}$	[28], [29]
(vii)	$\lambda x[1 + \alpha x]^{-1}$	[30], [31], [32]
(viii)	$\lambda_+ x$; if $x < k$ $\lambda_- x$; if $x > k$	[33]
(ix)	$\lambda x[1 + e^{(Bx - A)}]^{-1}$	[34], [35], [36]

The maps given in Table 5.1 are all unimodal, that is to say they possess one critical point,

$$\frac{\mathrm{d}f(x_0;\lambda)}{\mathrm{d}x_t}=0, \tag{5.4}$$

for some isolated x_0 and the parameter λ fixed. Another common feature of the function $f(x_t;\lambda)$ is that it increases monotonically in the variable x_n in the range $0<x<x_0$, attaining a maximum at x_0, and decreases monotonically $x>x_0$. The general equation of the form given in equation (iii) is germane to the social sciences, for instance, and in elementary theories of learning, where x is the number of bits of information that can be remembered after an interval t. All of the examples given in Table 5.1 may be thought of as gross caricatures of the dynamical processes they hope to mimic.

5.3.2 *Bifurcation and stability analysis*

A typical starting point in any analysis of the dynamics of a map is to determine its fixed points. These are solutions of the map that do not depend on the discrete time t. A fixed point of the map, or the steady population of our model species, is sought when $x_t=x_{t+1}$. To evaluate the steady population we consider the *time-1 map*, that is, the map that relates the population x at generation t to that at generation $t+1$. The fixed points are found by requiring that $x_t=x_{t+1}=x^*$. Substitution of this into a one-dimensional map given in equation (5.3) requires

$$x^*=f(x^*;\lambda). \tag{5.5}$$

The solution to equation (5.5) can involve one of two approaches. A geometrical approach involves drawing the functional form of $f(x)$ as a function of x and fixed points occur as points of intersection with the line $x_t=x_{t+1}$; this is depicted in Fig. 5.2 for a unimodal map. We see for this generic example there are two points of intersection: the trivial solution $x^*=0$, and a nontrivial solution for finite x^*. For example, for the quadratic map, equation (i) of Table 5.1, the nontrivial solution or fixed point is given by $x^*=1+1/\lambda$.

It is important to determine whether the fixed points are stable, that is, whether an infinitesimal perturbation away from the fixed point grows or decays. According to Lyapunov's stability criterion [6], if the perturbation dies, the fixed point is stable; if it grows indefinitely, then it is unstable. Stability involves evaluating a linear map (the linearization) of f in a neighbourhood of the fixed point.

The linearization of f at the fixed point x^* is obtained by considering a perturbation δ_n such that

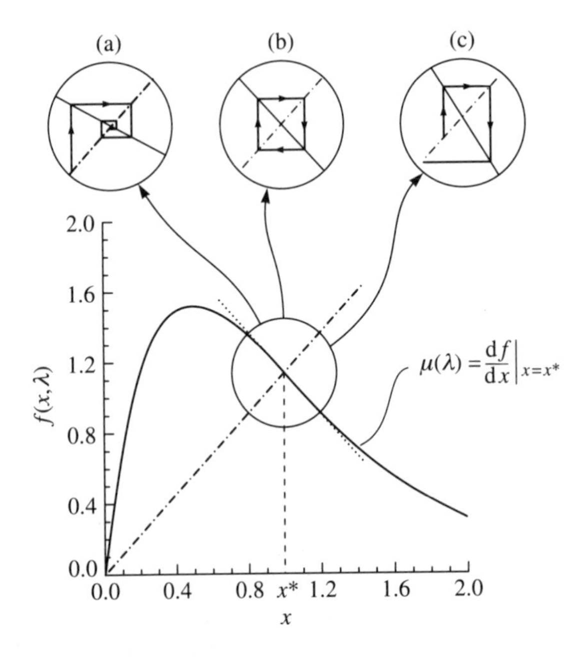

Fig. 5.2. The eigenvalue (slope) at x^* determines the local stability of the fixed point $f(x^*;\lambda) = x^*$: (a) $|\mu| < 1$, the fixed point is an attractor; (b) $|\mu| = 1$, the fixed point is neutrally stable; (c) $|\mu| > 1$, the fixed point is a repeller.

$$x_{n+1} = x^* + \delta_{n+1} \tag{5.6}$$

$$= f(x^* + \delta_n;\lambda) \tag{5.7}$$

$$= f(x^*;\lambda) + \delta_n \frac{df(x^*;\lambda)}{dx_n} + \delta_n^2 \frac{d^2 f(x^*;\lambda)}{2dx_n^2} + \cdots, \tag{5.8}$$

where we have expanded $f(x^* + \delta_n;\lambda)$ as a Taylor series. The linear map for δ_n is known as the equation for the *first variation* and is obtained by retaining the largest terms, that is, the first order terms in δ_n:

$$\delta_{n+1} = \frac{df(x^*;\lambda)}{dx}\delta_n = \mu^{(1)}(x^*;\lambda)\delta_n. \tag{5.9}$$

For the quadratic map, the slope $\mu^{(1)}(x^*;\lambda) = 2 - \lambda$. The slope of the map $\mu^{(1)}(x^*)$, at the fixed point x^*, gives the condition on λ for the stability of the fixed point x^*. It is locally stable for $1 \leqslant \lambda \leqslant 3$, since $|\mu^{(1)}(x^*;\lambda)| \leqslant 1$ for these values. For values of λ greater than 3 the equilibrium point becomes unstable, and a bifurcation occurs as the magnitude of the slope becomes greater than 1. What can we generically expect

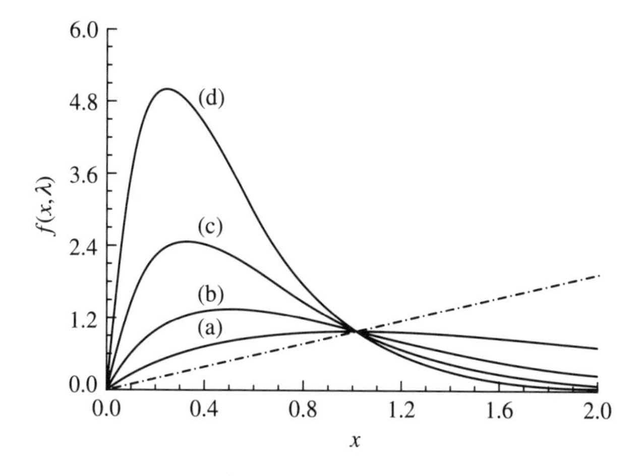

Fig. 5.3. The equation $x_{t+1} = x_t \exp\,[r(1 - x_t)]$ for various values of r: (a) $r = 1$; (b) $r = 2$; (c) $r = 3$; (d) $r = 4$.

to happen as the fixed point becomes unstable? Typically, either new fixed points occur, or the iterates of the map become unbounded, or iterates remain bounded but do not settle down to a stable periodic orbit.

To illustrate what happens when the parameters in f change, we use the one-parameter model, equation (ii) of Table 5.1:

$$x_{t+1} = x_t \exp[r(1 - x_t)] \equiv f(x_t; r). \tag{5.10}$$

Here we have for notational convenience defined $x_t \equiv x/k$. Figure 5.3 shows $f(x; r)$ for several values of the parameter r.

Although we use this particular example in order to give a detailed analysis, we emphasize that the general character of the dynamical behaviour we shall describe is not specific to this model, but rather is generic to any mapping with a single hump (provided it is sufficiently smooth) and whose steepness can be parametrically tuned. That is, our analysis will depend, in its essentials, only on the general shape of the function $f(x_t; r)$.

The nontrivial fixed point of equation (5.10) is at $x^* = 1$, and the slope of $f(x; r)$ at x^* is $\mu^{(1)}(x^*; r) = 1 - r$. Thus, the equilibrium is stable if $0 \leqslant r \leqslant 2$. However, as r increases past 2, the fixed point changes from an attractor to a repeller; this is shown in Fig. 5.4.

To see what is happening, we examine the relation between x_t and x_{t+2}:

$$x_{t+2} = f^{(2)}(x_t; r), \tag{5.11}$$

where

$$f^{(2)}(x; r) \equiv f(f(x; r); r). \tag{5.12}$$

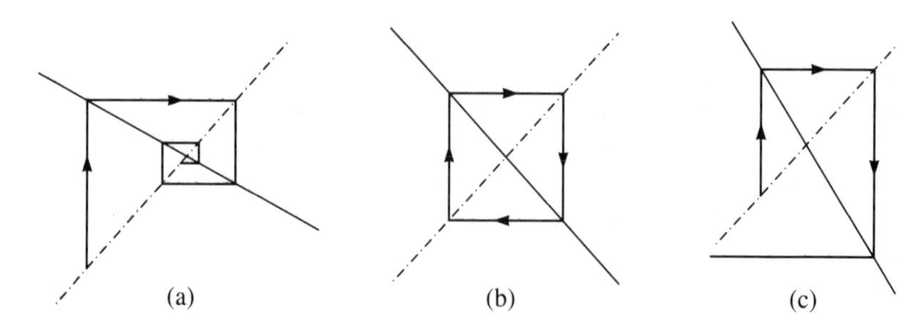

Fig. 5.4. See Fig. 5.2.

We now examine the fixed points $x^{*(2)}$ of period 2, that is, those points which are invariant under two iterations of the map f. These points can be located either by solving the appropriate algebraic equation,

$$x^{*(2)} = f^{(2)}(x^{*(2)}; r), \tag{5.13}$$

or by graphically iterating the map f.

We see that if the map f has one critical point then $f^{(2)}$ will have two critical points. However, if $r < 2$ (i.e. the fixed point of f is stable), then $f^{(2)}$ will only intersect the 45° line once, as shown in Fig. 5.5. At $r = 2$, $f^{(2)}$ is tangent to the 45° line, and the fixed point is neutrally stable: small oscillations about it will persist undamped. When $r > 2$, $f^{(2)}$ begins to develop a 'loop', intersecting the 45° line in three points. As r increases past 2, the original fixed point becomes unstable and splits into two new fixed

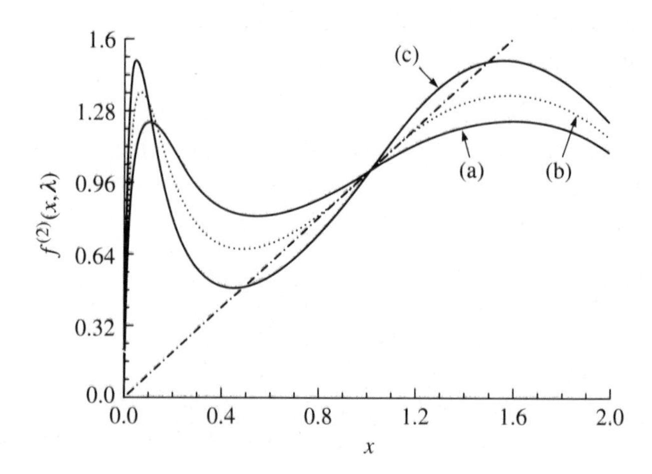

Fig. 5.5. The period-2 map: (a) $r < 2$, stable period-1 fixed point; (b) $r = 2$, the period-1 fixed point bifurcates; (c) $r > 2$, two period-2 attracting fixed points emerge from the original fixed point which is now a repeller.

points x_1^{*2} and $x_2^{*(2)}$, which move apart as the loop in $f^{(2)}$ becomes larger with increasing r. These new fixed points are initially stable, since the slope (eigenvalue) of $f^{(2)}$ at these points has magnitude less than unity. These stable oscillations of period 2 alternate between $x_1^{*(2)}$ and $x_2^{*(2)}$

As r continues to increase, the humps of $f^{(2)}$ become steeper, and the slopes at $x_1^{*(2)}$ and $x_2^{*(2)}$ increase until the period-2 points in turn become unstable, when $\mu^{(2)}(x^{*(2)};\lambda) \equiv (\mathrm{d}/\mathrm{d}x)f(x^{*(2)};\lambda) < -1$. Then they too will bifurcate, splitting into pairs of attracting fixed points of period 4. This can be easily seen by graphically composing $f^{(2)}$ with itself.

If we continue to increase r, the same process will repeat itself. The graph of $f^{(k)}$ when composed with itself will give rise to a graph of $f^{(2k)}$ which develops kinks. As the points of period k become unstable (which will happen when the slope $\mu^{(k)}$ of the intercept of $f^{(k)}$ graph on the 45° line steepens beyond $-45°$), the kinks in $f^{(2k)}$ grow so as to loop through the 45° line, corresponding to the period-k points bifurcating into pairs of fixed points with period $2k$ (whose stability depends on the eigenvalues $\mu^{(2k)}$ of $f^{(2k)}$).

This process can be best visualized by drawing a bifurcation diagram

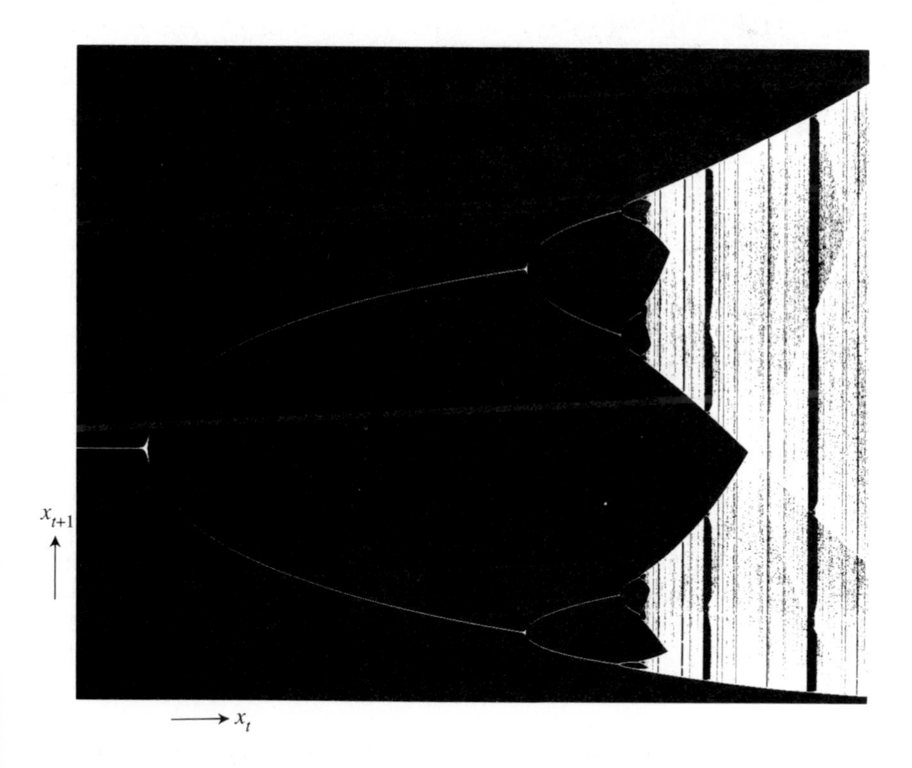

Fig. 5.6. Bifurcation diagram for the exponential map given by $x_{t+1} = x_t \exp[r(1 - x_t)]$ for $r > 1.9$.

showing the variable x versus r, as in Fig. 5.6. Figure 5.6 shows the way increasing r gives rise to a hierarchy of bifurcating stable points of period 1, 2, 4, 8, 16, ... Note that, although this bifurcation process produces an infinite sequence of cycles with period $2^n (n \to \infty)$, the window of r values wherein any one cycle is stable progressively diminishes, so that the entire process is a convergent one, being bounded above by some parameter value $r = r_c$ (which in this example is $r_c = 2.6924\ldots$).

So far, we have discussed the way the fixed points of $f(x; \lambda)$ give way to a hierarchy of stable cycles of periods 2^n. After first pausing to emphasize that the phenomena is a generic one, we go on to discuss what happens beyond r_c.

5.4 The generic character of period doubling

Later we shall see the universality associated with the period-doubling process. When a fixed point of period-k becomes unstable, it bifurcates to give two (initially stable) fixed points of period-$2k$.

The essential ingredient of the process consists of the observation that, at any fixed point $x_i^{*(k)}$ of period-k, the slope $\mu^{(2k)}(x_i^{*(k)}; r)$ of the intercept of $f^{(2k)}$ with the 45° line is simply the square of the slope of the intercept of $f^{(k)}$ with the 45° line:

$$\mu^{(2k)}(x_i^{*(k)}; r) = \{\mu^{(k)}(x_i^{*(k)}; r)\}^2. \tag{5.14}$$

Thus, if the period-k cycle is stable $(-1 < \mu^{(k)} < 1)$, we have $|\mu^{(2k)}| < 1$, and the kink in $f^{(2k)}$ does not yet intercept the 45° line, that is, we have the generic behaviour, much in the same way the first period-doubling took place. Conversely, once the points of period-k become unstable $(\mu^{(k)} < -1)$, we have $|\mu^{(2k)}| > 1$, and the kink in $f^{(k)}$ does intercept the 45° line, so that the period-k cycle bifurcates into one of period-$2k$.

5.4.1 *Period 3*

It must be noted that not every new solution must arise from bifurcation from a pre-existing fixed point. This can happen when a 'valley' of some iterate of f deepens and eventually touches the 45° line, or when a 'hill' rises to touch it; a pair of fixed points will then appear on the bifurcation diagram (one stable, one unstable). This is illustrated in Fig. 5.7(a) and is the crucial difference between bifurcation from $\mu(r) = 1$ (the so-called tangent bifurcation), and those from $\mu(r) = -1$ (the period-doubling bifurcation).

Higher-order cycles can be created by a tangent-type bifurcation for equation (5.10). Having appeared, such period-k cycles will in turn become

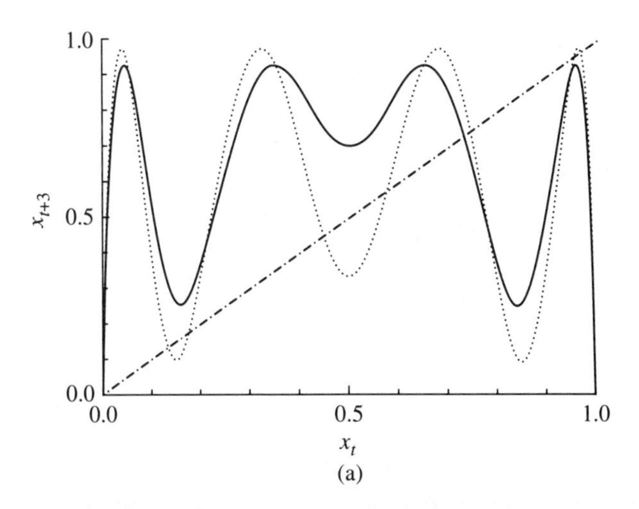

(a)

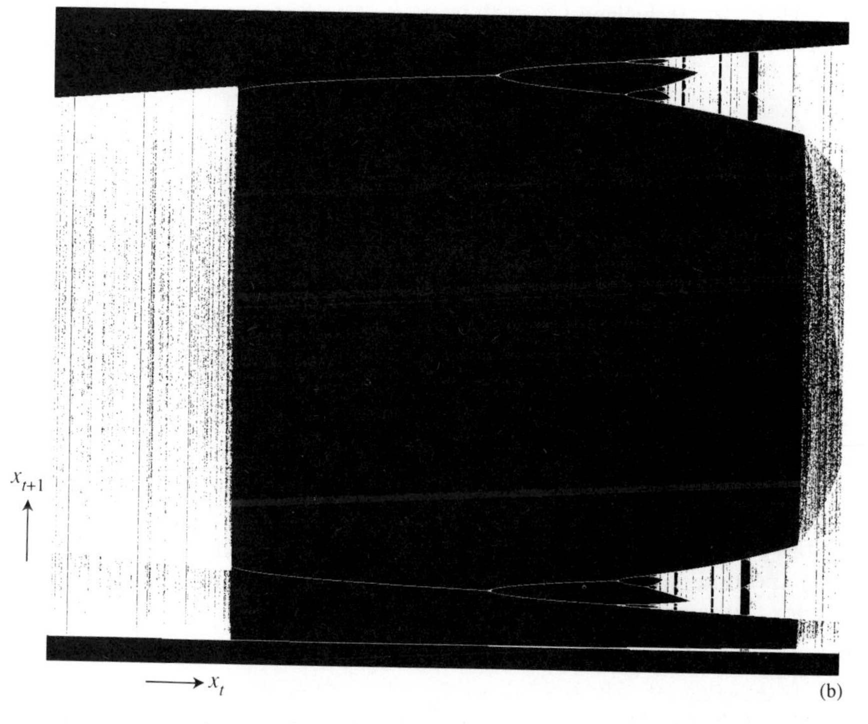

(b)

Fig. 5.7. The formation of the period-3 points, the (dashed) loops shown in (a) cut the diagonal to form a tangent or saddle-node bifurcation (solid lines), as λ increases. The full bifurcation diagram for the stable orbits is shown in (b).

Table 5.2. Numbers of periodic orbits up to period 10, for a general unimodal map f.

k	1	2	3	4	5	6	7	8	9	10
Possible total number of points with period k	2	4	8	16	32	64	128	256	512	1024
Possible total number of points with nondegenerate period k	2	2	6	12	30	54	126	240	504	990
Total number of cycles of period k	2	3	4	6	8	14	20	36	60	108
Total number of nondegenerate cycles	2	1	2	3	6	9	18	30	56	99
Total number of nondegenerate, stable cycles, including harmonics	1	1	1	2	3	5	9	16	28	51
Total number of nondegenerate, stable cycles, excluding harmonics	1	0	1	1	3	4	9	14	28	48

unstable, giving rise by period-doubling bifurcations to cycles of period $2k$, $4k$, etc. Some specific examples (e.g. a sequence of stable cycles of periods 3, 6, 12, 24, ... inhabiting the region $3.196 > r > 3.102$) for equation (5.10) are noted in Table 5.2. A bifurcation diagram showing the period-3 window is shown in Fig. 5.7(b).

In dimensions higher than 1, the existence of bifurcations to periodic orbits does not even require the presence of a critical point.

5.4.2 *Beyond r_c: chaos*

Beyond the limiting value r_c in Fig. 5.6, there are an infinite number of periodic points. Furthermore, there are an uncountable number of points (initial conditions) whose trajectories are totally aperiodic; no matter how long the time series generated by $f(x; r)$ is run out, the pattern never repeats, although it is bounded both above and below.

5.4.3 *Folding*

We may establish, for a given map, the conditions for the existence of an infinite number of periodic orbits, and for the existence of asymptotic aperiodic behaviour. For a critical point b, trace out the post-images c and d and the pre-image a. If $a \leqslant d$, then in addition to an infinite number of periodic points, the mapping can generate aperiodic motions. Li and Yorke [2] proved in 1975 that if there exists any cycle with an odd period, and if there is a sequence of points (a, . . ., b, c) ascending towards the critical point, followed by a crash back to a point (d) below the starting point (a) then chaos can ensue.

An alternative interpretation of the behaviour can be give in terms of what the map f does to subintervals of the unit interval. If the circumstances prevalent to Fig. 5.8 are the same, then there is a subinterval (a, b) which is *folded* under the action of the map; that is, the interval (a, b) is mapped to (b, c), and (b, c) is mapped to (c, d), as shown in Fig. 5.8. If the map is iterated an infinite number of times, an infinite number of folds are created in the open interval $(0, c)$.

As a consequence, there must exist an infinite number of fixed points in the region of overlap. If the condition $d \leqslant a$ holds, then there is an interval which completely covers itself under f. It can be shown that the union of domains of attraction of the stable fixed point will not fill up the interval $(0, c)$; in addition to repelling fixed points, the complementary set must contain aperiodic points.

This complex behaviour was demonstrated and christened as being chaotic by Li and Yorke [2], who showed that if a periodic orbit of period 3 is present all other periods are also present. Earlier Myrberg [37], Šarkovskii [38], and others had proved a more general result that if an odd periodic orbit of period-k is present then all periods greater than period-k are present; this earlier work, however, was purely of a mathematical character, and shows no sign of having recognized the wider implications.

As we have seen, we may have an infinite number of fixed points for a given parameter value, but there can be only a finite number of stable fixed points. A rigorous proof of this is to be found in a paper by Guckenheimer

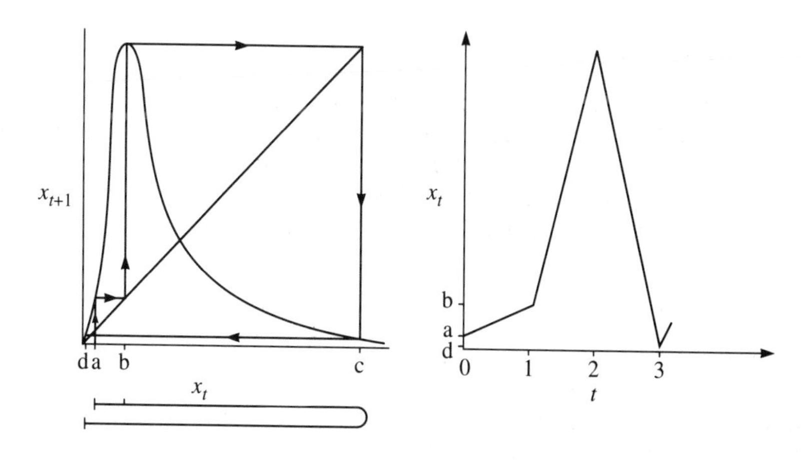

Fig. 5.8. If $d \leqslant a$ (a 3-point cycle exists), then there are initial conditions which generate aperiodic trajectories.

and Holmes [15] and the book of Devaney [39]. In essence, the proof follows from the observation that each successive iterate of f crowds twice as many humps into the same interval, so that the slopes of f^n at its intersections with the 45° line must steadily increase with each iteration. Thus it becomes harder for the eigenvalues at the periodic cycles to remain stable. Consequently, there can only be a finite number of stable cyclic points amongst the infinite number of possible periodic points; exceptions will occur at limiting values, and such exceptions will have measure zero. That is, after a finite number of iterates of f, all slopes at the periodic points will have magnitude in excess of unity, and so successive iterates produce only repelling periodic points.

We find the orbits of the map f can be partitioned into two distinct sets: a set containing the domain of attraction of a finite number of periodic points, and a domain containing an infinite number of unstable periodic points and an uncountable number of points that wander chaotically.

5.5 Numbers of periodic orbits

Suppose we wish to catalogue the numbers of periodic orbits arising from, say, the quadratic map. Perhaps the most elementary approach consists of first answering the question: how many period-k points can there be? In other words, how many solutions are there to

$$x_k^* = f^{(k)}(x_k^*)? \tag{5.15}$$

A simple-minded approach is as follows [13]. Once the parameter λ becomes sufficiently large, each mapping doubles the number of hills. The map $f^{(k)}(x)$ has 2^{k-1} hills; all of the hills and valleys will intersect the line $x_n = x_{n+1}$, consequently producing 2^k fixed points of period k. Included in our counting are those degenerate points of period k whose period is a submultiple of k. Table 5.2 gives a catalogue of periodic orbits.

For example, there are $2^9 = 512$ points of period 9. These include the stable and unstable pair of triplets of points of period 3, for a total of eight points whose basic period is a submultiple of 9; this leaves 504 points whose basic period is 9, corresponding to 28 pairs (stable and unstable) of distinct period-9 orbits.

The 2^k period-k points are arranged into various cycles of period k and submultiples of k. They appear in a succession of period-doubling bifurcations or as a tangent bifurcation, as parameters are varied. Table 5.2 catalogues, amongst other things, the total number of distinct cycles of period k which appear. Also, it catalogues the stable subharmonics which arise by period-doubling bifurcations, and pairs of stable–unstable pairs from the tangent or saddle-node bifurcation. The number of stable cycles are found by substracting out the cycles which are unstable from birth. The final row appearing in Table 5.2 lists the basic cycles whose period is k,

Table 5.3. Stable cycles for the quadratic map.

Period of basic cycle	λ value at which: Basic cycle first appears	Basic cycle becomes unstable	Subsequent cascade of subharmonics with period $2k^n$ all become unstable	Width of the range of λ values over which the basic cycle, or one of its harmonics, is attractive
1	1.0000	3.0000	3.5700	2.5700
3	3.8284	3.8415	3.8495	0.0211
4	3.9601	3.9608	3.9612	0.0011
5(a)	3.7382	3.7411	3.7430	0.0048
5(b)	3.9056	3.9061	3.9065	0.0009
5(c)	3.99026	3.99030	3.99032	0.00006
6(a)	3.6265	3.6304	3.6327	0.0062
6(b)	3.937516	3.937596	3.937649	0.000133
6(c)	3.977760	3.977784	3.977800	0.000040
6(d)	3.997583	3.997585	3.997586	0.000003

Table 5.4. Stable cycles for the exponential map given by equation (5.10)

Period of basic cycle	λ value at which: Basic cycle first appears	Basic cycle becomes unstable	Subsequent cascade of subharmonics with period $2k^n$ all become unstable	Width of the range of λ values over which the basic cycle, or one of its harmonics, is attractive
1	0.0000	2.0000	2.6924	2.6924
3	3.1024	3.1596	3.1957	0.0933
4	3.5855	3.6043	3.61153	0.0298
5(a)	2.9161	2.9222	2.9256	0.0095
5(b)	3.3632	3.3664	3.3682	0.0050
5(c)	3.9206	3.9295	3.9347	0.0141

and is calculated by subtracting out those stable cycles which appear as a result of a period-doubling bifurcation. Tables 5.3 and 5.4 show when the first few stable cycles appear for the quadratic map (equation 1 of Table 5.1) and the exponential map (equation (5.10)), respectively.

5.6 Universality

The fact that the patterns of bifurcations, and associated cycles, are generic to one-dimensional maps with one critical point (not being dependent upon the functional form of the map) was recognized by Myrberg [37], Šarkovskii

[38], Metropolis *et al.* [40], and others. They recognized the mathematical facts, but the broader implications—essentially, deterministic chaos—were not appreciated until these models were studied independently, yet again in ignorance of previous work by researchers motivated by practical problems [2, 41].

May and Oster [42] went further to recognize universal properties in how the quantitative details of the scaling of successive period-doublings depended on changes in parameters of the map. They, however, failed to recognize the real interest of this finding, using it merely as an aid to computing the kinds of numbers given in Tables 5.3 and 5.4. Feigenbaum [43, 44] independently discovered these properties by numerical studies, and recognized their wider significance (in possible implications for the onset spectrum of turbulence, and so on).

Specifically, Feigenbaum drew attention to the fact that certain quantitative features of the bifurcation process are independent of the particular map. For example, when the map cascades through period doubling to chaos, certain measures of this process are always invariant, such as the geometrical rate in which chaos sets in. This produces the now-famous Feigenbaum numbers. Feigenbaum's delta is one of these universal quantities:

$$\delta = \lim_{n \to \infty} \frac{\lambda_{n-1} - \lambda_n}{\lambda_n - \lambda_{n+1}} = 4.669201\ldots, \tag{5.16}$$

where λ_n is the critical parameter value for the period-$2n$ point.

Another universal quantity is the size α of successive branches:

$$\alpha = \lim_{n \to \infty} \frac{x_{n-1} - x_n}{x_n - x_{n+1}} = 2.5029\ldots, \tag{5.17}$$

where x_n is the corresponding x value at the critical period-$2n$ point.

Universality in chaos appears as a useful unifying principle for understanding the critical dynamics of the period-doubling phenomenon for one-dimensional maps. However, more general unifying principles become clear when period-doubling appears as a common route to chaos in many systems of equations, both discrete and continuous, from such disparate natural sources as hydrodynamics and chemical kinetics.

We conclude this discussion of universality by sketching a more detailed, although intuitively orientated, derivation of universality in general and the result of equation (5.16) in particular.

Let $f^{(k)}(x_n; \lambda)$ by the k-times composed mapping,

$$x_{n+k} = f^{(k)}(x_n; \lambda). \tag{5.18}$$

A period-k cycle will touch the points $x_i^{*(k)}$ (with $i = 1, \ldots,$ k) which are the fixed-points of this mapping, obtained by solving $x^{*(k)} = f^{(k)}(x^{*(k)}; \lambda)$. Let $\mu^{(k)}(\lambda)$ be the eigenvalue (slope) of the map $f^{(k)}$ at these fixed points, as a function of λ. Any particular stable cycle of period k is born with $\mu^{(k)} = 1$, and becomes unstable, giving birth to a period-doubled cycle with period $2k$, when $\mu^{(k)} = -1$. As $\mu^{(k)}$ sweeps through this range from 1 to -1, the change in the parameter λ may be denoted $\Delta\lambda(k)$. If we were interested in the asymptotic value of the ratio of the parameter separations from one period doubling to next, then we should consider

$$\delta = \lim_{k\to\infty} \frac{\Delta\lambda(k)}{\Delta\lambda(2k)}. \tag{5.19}$$

In principle, we could define $\lambda_0^{(k)}$ as the λ value at which the period-k cycle is born $(\mu^{(k)}(\lambda_0^{(k)}) = 1)$, and write subsequent λ values as $\lambda = \lambda_0^{(k)} + \varepsilon$. A Taylor expansion then gives

$$\mu^{(k)}(\lambda) = 1 + \varepsilon A_0^{(k)} + O(\varepsilon^2). \tag{5.20}$$

Here $A_0^{(k)}$ is defined as

$$A_0^{(k)} = \frac{d\mu^{(k)}}{d\lambda} = \frac{\partial^2 f^{(k)}}{\partial x \partial\lambda} \tag{5.21}$$

evaluated at the point where the first period-k orbit first appears. On this basis, the quantity δ can be calculated as the asymptotic ratio between $A_0^{(2k)}$ and $A_0^{(k)}$. The difficulty in this procedure is that $A_0^{(k)}$ cannot in general be evaluated analytically.

An alternative method, based on the Taylor series expansion of equation (5.18), is available if we can relate the stability-determining slope $\mu^{(2k)}$ of the $2k$-times composed map at the fixed points of period $2k$ to the corresponding slope $\mu^{(k)}$ of the period-k points. The stable orbit of period k is born when $\mu^{(k)} = 1$, and becomes unstable when $\mu^{(k)} = -1$, at which point the succeeding stable cycle of period-$2k$ is born, with $\mu^{(2k)} = 1$; this in turn becomes unstable when it attains some negative value, $\mu_c^{(k)}$ say, corresponding to $\mu^{(2k)} = -1$. Substituting in equation (5.20), and ignoring second-order terms, we obtain

$$\Delta\lambda(k) = -2/A_0^{(k)} \tag{5.22}$$

and

$$\Delta\lambda(2k) = (\mu_c^{(k)} + 1)/A_0^{(k)}. \tag{5.23}$$

Hence, we arrive at the result

$$\delta = \lim_{k\to\infty} \frac{-2}{\mu_c^{(k)} + 1}. \tag{5.24}$$

It remains to find the asymptotic relation between $\mu^{(2k)}$ and $\mu^{(k)}$, in order to determine $\mu_c^{(k)}$. We may approximate such a relation by using a cubic to approximate the way the map $f^{(2k)}$ exhibits bifurcation as the period-k cycle becomes unstable.

Suppose we expand $f^{(2k)}$ in a Taylor series to third order about the point $x^{*(k)}$:

$$f^{(2k)}(x^{*(k)} + \xi) \approx x^{*(k)} + A\xi + \tfrac{1}{2}B\xi^2 + \tfrac{1}{6}C\xi^3 + O(\xi^4). \qquad (5.25)$$

We need to carry out the expansion to at least order ξ^3 to account for the qualitative shape of $f^{(2k)}$ in the neighbourhood of $x^{*(k)}$, that is, to describe the bifurcation into a pair of points, one on each side of $\xi = 0$. Recall that

$$\begin{aligned} f^{(2k)}(x^{*(k)} + \xi) &= f^{(k)}((f^{(k)}(x^{*(k)} + \xi)) \\ &= f^{(k)}(x^{*(k)} + \mu^{(k)}\xi + \cdots). \end{aligned} \qquad (5.26)$$

We can express the coefficients of the expansion A, B, etc., in terms of derivatives of the lower-order map $f^{(k)}$. In this case we find that

$$A = (\mu^{(k)})^2 \quad \text{and} \quad B = \mu^{(k)}\left[(1 + \mu^{(k)})\,\frac{d^2f}{dx^2}\right]_{x = x^{*(k)}}. \qquad (5.27)$$

Notice that the fixed points of period-k bifurcate at $\mu(k) = -1$, so for λ just beyond this value $A \approx 1$ and $B \approx 0$.

For the fixed points of period $2k$, we write $x^{*(2k)} = x^{*(k)} + \xi^*$. The values of ξ^* are then determined by

$$\xi^* \simeq A\xi^* + \tfrac{1}{2}B(\xi^*)^2 + \tfrac{1}{6}C(\xi^*)^3. \qquad (5.28)$$

Apart from the degenerate period-k solution ($\xi^* = 0$), there is a pair of solutions given by

$$0 \simeq (A - 1) + \tfrac{1}{2}B\xi^* + \tfrac{1}{6}C(\xi^*)^2 \qquad (5.29)$$

Note that these solutions are real if and only if $A \geqslant 1$ (i.e. $|\mu^{(k)}| \geqslant 1$), in accord with the detailed discussion given earlier. Next, the slope at these period-$2k$ points is

$$\mu^{(2k)} = \left[\frac{df^{(2k)}}{d\xi}\right]_{\xi = \xi^*} \simeq A + B\xi^* + \tfrac{1}{2}C(\xi^*)^2 \qquad (5.30)$$

Using equation (5.29), this can be simplified to

$$\mu^{(2k)} \simeq (3 - 2A) - \tfrac{1}{2}B\xi^* \qquad (5.31)$$

The quantity ξ^*, which represents the distance between the unstable period-k point and the initially stable period-$2k$ point, will be small for even moderate values of k. In addition, the coefficient B is zero when the bifurcation first occurs. Thus, to a first approximation, $\tfrac{1}{2}B\xi^*$ may be neglected in equation (5.31).

Hence we arrive at an expression for the value of A at the point where $\mu^{(2k)} \to -1$ (where the period-$2k$ cycle becomes unstable): $2A \simeq 4$, or $A \simeq 2$. Returning to equation (5.27), we thence find that the corresponding value of $\mu^{(k)}$, which we have called $\mu_c^{(k)}$, is

$$\mu_c^{(k)} \simeq -\sqrt{2} + O(\delta^{-2}) \tag{5.32}$$

Substituting this result into equation (5.24), we arrive at the analytical approximation to Feigenbaum's ratio

$$\delta \simeq 2(1 + \sqrt{2}) = 4.828\ldots. \tag{5.33}$$

Comparing this result with the exact numerical result, we see there is an error of about 3%. An improved estimate could be obtained by going to the next order but the effort seems hardly worth it, considering we have demonstrated a universal quantity can result from general properties of the map f.

References

1. Hirsch, M. W. and Smale, S. (1974). *Differential equations, dynamical systems and linear algebra*. Academic Press, New York.
2. Li, T.-Y. and Yorke, J. A. (1975). Period three implies chaos. *Amr. Math. Monthly*, **82**, 481–5.
3. Poincaré, H. (1880). *Mémoire sur les courbes définies par les équations différentielles I–VI*. Oeuvre I. Gauthier-Villars, Paris.
4. Poincaré, H. (1899). *Les methodes nouvelles de la mechanique celeste*. Gauthier-Villars, Paris.
5. Poincaré, H. (1890) Sur les équations de la dynamique et le problème de trois corps. *Acta Math.*, **13**, 1–270.
6. Lyapunov, A. M. (1947) Problème général de la stabilité du mouvement. *Ann. Math. Study*, **17**.
7. Birkoff, G. D. (1960). *Dynamical systems*, Colloq. Publ. IX, 2nd edn. AMS, Providence, RI.
8. Mandelbrot, B. B. (1975). *Les objets fractals:forme, hasard et dimension*. Flammarion, Paris.
9. Lorenz, E. N. (1963). *J. Atmos. Sci.*, **20**, 130–41.
10. Smale, S. (1967) Differentiable dynamical systems. *Bull. Am. Math. Soc.*, **73**, 747–817.
11. Ruelle, D. and Takens, F. (1971). On the nature of turbulence. *Commun. Math. Phys.*, **20**, 167.
12. Landau, L. D. and Lifschitz, E. (1959). *Fluid mechanics*. Pergamon, New York.
13. May, R. M. (1976). Simple mathematical models with very complicated dynamics. *Nature*, **261**, 459–69.
14. Carr, J. (1981). *Applications of centre manifold theory*, Applied Mathematical Sciences 35. Springer, New York.
15. Guckenheimer, J., and Holmes, P. (1983). *Nonlinear oscillations, dynamical systems, and bifurcations of vector fields*. Springer, New York.
16. Maynard-Smith, J. (1968). *Mathematical ideas in biology*. Cambridge University Press.

17. May, R. M. (1972). On relationships among various types of population models. *Am. Nat.*, **107**, 46–57.
18. Krebs, C. J. (1972). *Ecology: The experimental analysis of distribution and abundance*. Harper and Row, New York.
19. Moran, P. A. P. (1950). Some remarks on animal population dynamics. *Biometrics*, **6**, 250–8.
20. Ricker, W. E. (1954). Stock and recruitment. *J. Fish. Res. Bd. Can.*, **11**, 559–623.
21. Cook, L. M. (1965). Oscillation in the simple logistic growth model. *Nature*, **207**, 316.
22. May, R. M. (1974). Ecosystem patterns in randomly fluctuating environments. In: *Progress in theoretical biology* (ed. R. Rosen and F. Snell), pp. 1–50. Academic Press, New York.
23. Varley, G. C. and Gradwell, G. R. (1960). Key factors in population studies. *J. Anim. Ecol.*, **29**, 339–401.
24. Stubbs, M. (1967). Density dependence population studies. *J. Anim. Ecol.*, **29**, 339–401.
25. Hassell, M. P. (1974). Density dependence in single species populations. *J. Anim. Ecol.*, **44**, 283–96.
26. Hassell, M. P., Lawton, J. H., and May, R. M. (1976). Patterns of dynamical behaviour in single-species populations. *J. Anim. Ecol.*, **45**, 471–86.
27. R. Maynard Smith, J. and Slatkin, M. (1973). The stability of predator–prey systems. *Ecology*, **54**, 384–91.
28. Bellows Jr, T. S. (1981). The descriptive properties of some models for density dependence. *J. Anim. Ecol.*, **50**, 139–56.
29. Thomas, W. R., Pomerantz, M. J., and Gilpin, M. E., (1980). Chaos, asymmetric growth and group selection for dynamical stability. **Ecology, 61**, 1312–20.
30. Utida, S. (1967). Damped oscillation of population density at equilibrium. *Res. Pop. Ecol.*, **9**, 1–9.
31. Skellam, J. G. (1951). Random dispersal in theoretical populations. *Biometrika*, **38**, 196–218.
32. Leslie, P. H. (1957) An analysis of the data for some experiments carried out by Gause with populations of the protozoa. *Paramecium aurelia* and *P. candatum. Biometrika*, **44**, 314–27.
33. Williamson, M. (1974). The analysis of discrete time cycles. In: *Ecological Stability* (ed. M. B. Usher and M. Williamson) pp. 17–23. Chapman and Hall, London.
34. Ullyett, G. C. (1950). Competition for food and allied phenomena in sheep blowfly populations. *Phil. Trans. Ry. Soc.* **B, 260**, 77–174.
35. Pennycuik, C. J., Compton, R. M., and Beckingham, L. (1968). A computer model for simulating the growth of a population. *J. Theor. Biol.*, **18**, 316–29.
36. Usher, M. B. (1972). Developments in the Leslie matrix model. In: *Mathematical models in ecology* (ed. J. N. R. Jeffers), pp. 29–60. Blackwell, Oxford.
37. Myrberg, P. J. (1958). *Ann. Akad. Sc. Fennicae*, A, I, No. 259.

38. Šarkovskii, A. N. (1964). Coexistence of cycles of a continuous map of a line into itself. *Ukr. Mat. Z.*, **16**, 61.
39. Devaney, R. (1986). *Introduction to chaotic dynamics*. Benjamin/Cummings, New York.
40. Metropolis, M., Stein, M. L., and Stein, P. R. (1973). On finite limit sets for transformations of the unit interval. *J. Combinatorial Theory* A, **15**, 25.
41. May, R. M. and Oster, G. F. (1976). Bifurcations and dynamical complexity in simple ecological models. *Am. Nat.*, **110**, 573–99.
42. May, R. M. and Oster, G. F. (1980). Period doubling and the onset of turbulence: An analytic estimate of the Feigenbaum ratio. *Phys. Lett.* A, **78**, 1–3.
43. Feigenbaum, M. J. (1978). Qualitative universality for a class of nonlinear transformations. *J. Statist. Phys.*, **19**, 25.
44. Feigenbaum, M. J. (1979). The universal metric properties of nonlinear systems. *Los Alamos Sci.*, **1**, 4.

6

Models of chaos from natural selection

David Holton and Robert M. May

Isaac Newton wrote, in 1667:

> In experimental philosophy we are able to look upon propositions obtained by general induction from phenomena as accurately or very nearly true . . . till such time as other phenomena occur, by which time they may be accurate or liable to exceptions.

The principles of general induction upon which Newton formulated the celestial mechanical equations of motion began the interpretation of dynamics in terms of mathematics. This early development spawned infinitesimal mathematics or calculus. A complete understanding of the solutions of these equations of motion, however, has remained out of reach.

It was Poincaré who first realized that, for the three-body problem, the occurrence of separatrices in phase space, the crossing of expanding and contracting trajectories, could have a profound influence on whether or not the dynamics are integrable (we can take this to mean that a closed solution can be found). In two-dimensional phase space, with solution curves which are one dimensional, there cannot be self-intersection of trajectories, as this would violate the uniqueness of solution (except where the vector field goes to zero). In three dimensions, a one-dimensional orbit can avoid another orbit by exploring the third dimension, thus avoiding self-intersection; consequently the phase space dynamics can, in this case, be remarkably complex.

6.1 Sensitivity to initial conditions

Suppose we set out to predict the orbit of a chaotic dynamical system, such as the quadratic map. We have seen that the orbit it traces out is complex. As we have an exact model description of the equation of motion, we may reasonably expect that we can predict its orbit indefinitely (a similar mistake was made by Laplace). However, such intuitive reasoning does not always lead to the correct conclusion; the problem is that orbits can be extremely sensitive to initial conditions.

Consider two initial conditions x_1 and x_2 which are close to one another. We allow the initial conditions to evolve under the iterates of the map. At first, as the initial conditions are close, the orbits trace out almost indistinguishable orbits. As the orbits evolve in discrete time it becomes noticeable that they are diverging. After a finite number of iterates, in this case about 10 (as shown in Fig. 6.1), the orbits have diverged sufficiently to become entirely different. Their divergence increases exponentially with time.

This property of sensitivity to initial conditions has far-reaching implications. If there is the slightest imperfection in specifying a starting condition, then this would prevent us from determining the long-term future of the chaotic system. In other words, because in practice we can never precisely specify a starting configuration, we can, in turn, never succeed at predicting the long-term behaviour of the system.

Suppose you had difficulty resolving the small thickness of the inkline in Fig. 6.1 at time t, which resulted in a small error, ε, in estimating where we started at time t. If it was sufficiently small, as in the case of fixed points, we could justify expanding the map f in a Taylor series:

$$f(x_t + \varepsilon) = f(x_t) + \varepsilon \frac{\mathrm{d}f}{\mathrm{d}x}(x_t) + O(\varepsilon^2) \tag{6.1}$$

The slope $(\mathrm{d}/\mathrm{d}x)f(x_t)$ will, in general, vary as t varies. Suppose that the logarithmic average of the slope is e^{λ}. If initially we make an error ε, that error will have grown, on average, to $\varepsilon \mathrm{e}^{\lambda}$ after one iteration. After t time steps, the difference between the two orbits will be $\varepsilon \mathrm{e}^{\lambda t}$. If λ is negative, the error becomes smaller. For $\lambda = 0$, the error is maintained. For $\lambda > 0$, the numerical value of the error becomes asymptotically large. Moreover, no matter how small ε may be, the error will grow to have the magnitude of unity after a time $\simeq [\ln(1/\varepsilon)]/\lambda$ has elapsed; hence an arbitrarily small error will grow to become significant after a time $t \sim 1/\lambda$ has passed.

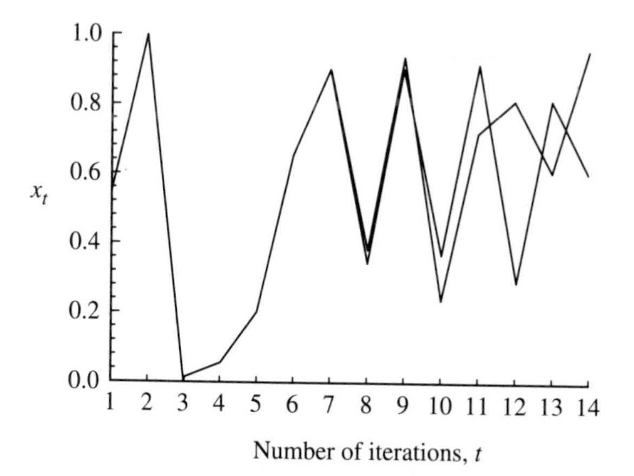

Fig. 6.1. Divergence of a chaotic trajectory: two almost indistinguishable initial conditions are exponentially magnified by a chaotic map.

6.2 Stability and the Lyapunov exponent

As the reader will now be aware, the exponential map, among an array of other dynamical systems, exhibits period doubling and chaotic behaviour. There appear to be windows of seemingly chaotic behaviour; within these windows there is a mosaic of unstable and stable cycles interspersed with truly chaotic behaviour. For all practical applications, the chaotic regime has an effectively random character which even the most accurate numerical investigation cannot disentangle, as periodicity within one of these windows can exist with arbitrary large period. We have seen that the quadratic map, like many other systems, exhibits sensitivity to initial conditions. Is it possible to quantify the sensitivity of an orbit to initial conditions? Can we quantify chaotic behaviour by computing an index of the motion from, say, the computed orbit? The index, it turns out, which is natural to use is due to Lyapunov, and called appropriately the Lyapunov exponent. The Lyapunov exponent is analogous to calculating eigenvalues in Floquet theory for the linear stability of periodic orbits. In Floquet theory we look at what happens after one period of a closed orbit. The Lyapunov exponent measures, as eigenvalues do, something of the expanding and contracting phase space. However, there are subtle differences and the analogy must stop there. The main difference is that eigenvalues are local quantities; Lyapunov exponents are global averaged quantities.

For a one-dimensional difference equation $x_{t+1} = f(x_t)$, the Lyapunov exponent λ is given by

$$\ln \lambda = \lim_{n \to \infty} \left\{ \frac{1}{n} \sum_{t=0}^{n} \ln \frac{\mathrm{d}f(x_t)}{\mathrm{d}x} \right\}. \tag{6.2}$$

In 1968, Oseledec [1] proved an important theorem given by Lyapunov: that the limit on the right-hand side exists and is finite.

We are able to calculate the Lyapunov exponent for the maps we have listed in Table 5.1. We wait until the map 'settles down', by throwing away, say, the first ten thousand iterates. Then we calculate the average of the slope of the map: the Lyapunov exponent. For the quadratic map,

$$x_{t+1} = ax_t(1 - x_t) \tag{6.3}$$

shown in Fig. 6.2 (due to Shaw [2]), below $a = 3$ there is a stable fixed point, and the Lyapunov exponent is consequently negative. As we enter the first region of period doubling (period 2, 4, 8, . . .), they all are discernibly attractive, and they all attract any (to measure zero) initial points into the periodic orbits; consequently the Lyapunov exponents are negative. As we move into the chaotic region, we observe that there appears to be a positive-valued region with some wiggles. The Lyapunov exponent appears to be positive in parameter intervals of positive measure.

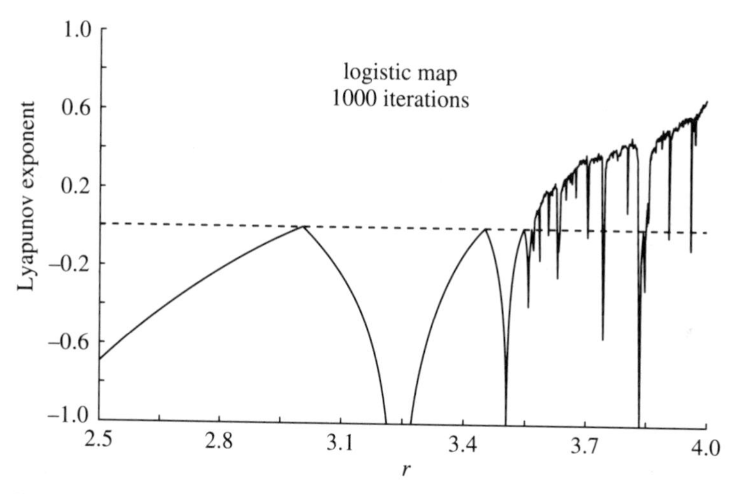

Fig. 6.2. Lyapunov exponent plotted against the parameter a for the logistic (quadratic) map.

In fact, it can be shown that the Lyapunov exponent tends to ln 2 in the limit as a $a \to 4$. As this is positive, one can estimate on average how many iterates the example given in Fig. 6.1 takes to diverge, so that the error is of order unity. The initial conditions are six parts in ten thousand apart. Therefore,

$$1 \approx \varepsilon e^{\lambda t} = \frac{6}{10000}\, e^{\lambda t} \tag{6.4}$$

$$\lambda t = \ln\left(\frac{5000}{3}\right) \approx 8. \tag{6.5}$$

For $a = 4$, we have $\lambda = \ln 2 = 0.69$. It thus follows that $t \simeq 11$.

This numerical computation of the Lyapunov exponent for the quadratic map is, in a strict mathematical sense, nonsense. In the so-called chaotic regime, there is—for almost all values of the parameter a of equation (6.3)—some unique attractor (a cycle of period 1137, for instance). Associated with this basic cycle is a cascade of period doublings (2, 274; 4, 548; etc.), just as for the basic 'period-1 point'. But these unique attractors change with infinitesimal changes in the basic parameter a in the 'chaotic' region, $4 > a > 3.57\ldots$. And also, although for most values of a there is some unique attractor, it may take 100 million or more iterations before the orbit discernibly settles to it. Thus, although Fig. 6.2 may be mathematically incorrect, for all practical purposes it gives a correct impression of the dynamical behaviour of the quadratic map. In Picasso's words (although here for dynamics not art), Fig. 6.2 is 'a lie that tells the truth'.

In more detail, we note that Jakobson [3] in 1981 was able to prove that there is a Cantor set of positive measures of chaotic values for a between 3.57 and 4. In other words, if we chose the a value between these limits, then we would be able to pick out, with finite probability, a chaotic orbit for which the Lyapunov exponent will be truly positive. This, of course, will be surrounded by a values where there is a unique attractor, so that a correctly computed Lyapunov exponent would be negative. The true version of Fig. 6.2 is thus ultimately not possible to draw. But for practical purposes Fig. 6.2 suffices.

We recall from Chapter 5 that the equation for the first variation δ_t is given by

$$\delta_{t+1} = \frac{\mathrm{d}f}{\mathrm{d}x_t}(x_t; \lambda)\delta_t, \tag{6.6}$$

for a one-dimensional map, where $\mathrm{d}f/\mathrm{d}x_t$ is the slope of the map at the fixed point x_t. There is a straightforward generalization to n dimensions (n first-order difference equations) in which the most appropriate formulation is using the Jacobi matrix J; the equation for the variation $\delta = (\delta_1, \ldots, \delta_t)$ is

$$\delta_{t+1} = J(x_t, \lambda)\delta_t, \tag{6.7}$$

where a multivariate Taylor series is used to derive this equation. The matrix J consists of a matrix of partial derivatives of the multivariate function $\mathbf{f} = (f_1, \ldots, f_n)$, with

$$J = \begin{bmatrix} \frac{\partial f_1}{\partial x_1} & \cdots & \frac{\partial f_1}{\partial x_n} \\ \vdots & & \vdots \\ \frac{\partial f_n}{\partial x_1} & \cdots & \frac{\partial f_n}{\partial x_n} \end{bmatrix}. \tag{6.8}$$

Equation (6.7) is the linearization of the dynamics about the fixed point x^*.

The stability of the n-dimensional fixed points is determined by the roots (the eigenvalues μ) of the characteristic equation:

$$\det(J - \mu I) = 0. \tag{6.9}$$

For the $n \times n$ matrix J, equation (6.9) is a polynomial of degree n. There can be n roots to equation (6.9), some of them complex. For any fixed parameter value, the n-eigenvalues can be plotted in the complex plane. As a parameter is smoothly changed, the n roots also smoothly change. If an eigenvalue (or complex conjugate pair) crosses the unit circle in the complex plane, as a parameter is smoothly varied, an instability or bifurcation can occur. Based on a genericity argument, we can show that a bifurcation can occur in one of three distinct ways for a one-parameter family. This is an extremely important result, as it tells us exactly how a system can become unstable. What happens after the fixed point becomes unstable is, however, not so easily determined. In a two-parameter family, combinations of the three basic types of bifurcation can give rise to yet more complex. bifurcations. A large amount of literature exists on this topic (see e.g. [4]).

The eigenvalues describe how, near to the fixed points, phase space is contracted and stretched under the action of the dynamics.

6.3 One-dimensional maps with two critical points

In many natural situations, organisms interact with their surroundings and other organisms in such a way that certain genotypes are favoured. Evidence has been marshalled to the effect that rarer genotypes have a selective 'frequency-dependent' advantage important in promoting and preserving genetic polymorphisms in natural selection [5–12].

The conventional mathematical caricature of this situation consists of one gene locus with two alleles A and a (with gene frequencies p_t and q_t, respectively, in the tth generation; $p + q = 1$). If the forces of natural selection are frequency dependent, it may be that allele A enjoys a selective advantage when rare, and a concomitant disadvantage when common. As a result, the corresponding gene frequency p will tend to increase from one generation to the next when it has a low value, and to decrease when it has a high value. This general situation is depicted in Fig. 6.3; the variable p is replaced for notational convenience by the variable $X = 2p - 1$.

In this paradigm, it is clear that if the advantage possessed by A at low frequencies and the disadvantage suffered at high frequencies are both relatively modest, then the system will settle down to a stable equilibrium at an intermediate gene frequency value. A stable polymorphism can thus be maintained by frequency-dependent natural selection. But what happens when the gradient of the map relating p_{t+1} to p_t steepens at a fixed point? What happens to the dynamics of the map described by Fig. 6.3 as the hill at low p values rises and the valley at high p values deepens? To answer these questions, we begin considering a general map of the form

$$X_{t+1} = F(X_t), \tag{6.10}$$

where the mapping $F(X)$ is defined on the interval $[-1, 1]$, and can have two critical points, a hill and a valley, as shown in Fig. 6.3.

Quite apart from possible biological applications, the dynamical behaviour generated by maps with two critical points is of intrinsic mathematical interest as the next step up from that generated by 'one hump' maps.

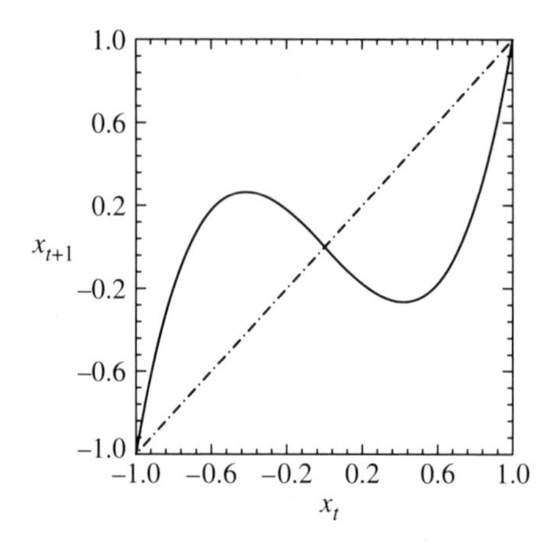

Fig. 6.3. First-order difference equation with two critical points. Here, x is defined on the interval $[-1, 1]$, and the map is specifically the cubic map with $a = 2$. The fixed points of period 1 (at $x = 1, 0, -1$) are at the intersection of the map and the 45° line.

6.3.1 *Properties*

For maps with one critical point, the maximum number of fixed points with period k is 2^k: each new iteration of the map doubles the number of possible intersections with the 45° ($X_k = X_{k+1}$) line, so that the k-times composed mapping can have a maximum 2^k fixed points. For a map with two critical points, as illustrated in Fig. 6.3, each new folding can triple the number of intersections with the 45° line, and there can consequently be a maximum of 3^k fixed points of period k. For the map with one critical point, various techniques have been used to elucidate the generic order in which the different cycles of period k appear, and in general to classify how the totality of 2^k fixed points of period are organized. The task of giving a generic classification of how the 3^k fixed points of period k are organized, for maps with two critical points, is more difficult. A beginning to this classification is given below.

For maps with one critical point, there can be at most one stable attractor; even in the chaotic regime, where there are infinitely many different periodic orbits and an uncountable number of aperiodic orbits, there is in general a unique cycle that attracts almost all initial points. One can show [4], for one-dimensional maps, that the maximum number of periodic attractors is equal to the the number of critical points. Thus, for maps with two critical points, there can be domains of parameter space in which the system possesses two distinct periodic attractors. We shall see below how this phenomena emerges from the bifurcation structure.

In what follows, there is first a generic discussion of the bifurcations that occur in the second and higher iterations of an (antisymmetric) map with two critical points, as the 'hill' and 'valley' steepen. Second, these processes are illustrated by a concrete example:

$$X_{t+1} = aX_t^3 + (1 - a)X_t. \tag{6.11}$$

Equation (6.11) is the canonical exemplar of a map with two critical points.

Let us consider a map, as defined by equation (6.10), with X defined on the interval $[-1, 1]$. In discussing the case where $F(X)$ has two critical points, it is convenient further to assume $F(X)$ is antisymmetric, so that

$$F(-X) = -F(X). \tag{6.12}$$

This is illustrated in Fig. 6.3.

We further assume the map is 'anchored' to three fixed points at $X = 1$, 0, and -1. We first focus on the stability of the fixed point at $X = 0$. From Chapter 5, we found the stability of the fixed point is determined by the slope $\mu^{(1)}(0)$ of the map at the fixed point:

$$\mu^{(1)}(0) = \left[\frac{\mathrm{d}F}{\mathrm{d}X}\right]_{X=0}. \tag{6.13}$$

So long as $|\mu^{(1)}| < 1$, the fixed point at $X = 0$ is attracting. However, as we tune the parameter a, the hill and valley in Fig. 6.3 will steepen, $\mu^{(1)}$ will steepen towards -1, and the fixed point at $X = 0$ will become unstable once $\mu^{(1)}$ steepens beyond -1.

To see what happens as this fixed point of period 1 ($X = 0$) becomes unstable, we turn to the map for the second iterate:

$$X_{t+1} = F^{(2)}(X_t). \tag{6.14}$$

As before, $F^{(k)}$ denotes the k-times composed mapping of F. The period-1 point at $X = 0$ is obviously a degenerate period-2 point, and the slope of the $F^{(2)}$ map at this point is

$$\mu^{(2)}(0) = \left[\frac{\mathrm{d}F^{(2)}}{\mathrm{d}X}\right]_{x=0}. \tag{6.15}$$

That is, explicitly writing $F^{(2)}(X) = F[F(X)]$,

$$\mu^{(2)}(0) = \left[\frac{\mathrm{d}F}{\mathrm{d}X}\right]^2_{x=0} \tag{6.16}$$

$$= [\mu^{(1)}(0)]^2. \tag{6.17}$$

The observation that $\mu^{(2)}(0)$ is the square of $\mu^{(1)}(0)$ is the key to the cascading bifurcations that arise in maps with one critical point: so long as the period-1 point at $X = 0$ is stable, which implies $|\mu^{(1)}| < 1$, then $|\mu^{(2)}| < 1$ and the $F^{(2)}$ map intersects the 45° line only once in the neighbourhood of $X = 0$; as the period-1 point at $X = 0$ becomes unstable, which implies $|\mu^{(1)}| > 1$, then $|\mu^{(1)}| > 1$ and the $F^{(2)}$ map steepens to make a loop which intersects the 45° line three times in the neighbourhood of $X = 0$. In this way, two new fixed points of period 2 are born by a symmetry-breaking bifurcation, at the same time as the period-1 fixed point becomes unstable. For the antisymmetric map under consideration here, we can label these two fixed points of period 2 as $X = \pm \Delta$:

$$\pm \Delta = F(\mp \Delta). \tag{6.18}$$

The stability of this new fixed point of period 2 depends, in turn, on the slope of $F^{(2)}$ at these fixed points $X = \pm \Delta$:

$$\mu^{(2)}(\Delta) = \left[\frac{\mathrm{d}F^{(2)}}{\mathrm{d}X}\right]_{X=\Delta}. \tag{6.19}$$

Again, writing $F^{(2)}$ explicitly,

$$\mu^{(2)}(\Delta) = \left(\frac{\mathrm{d}F}{\mathrm{d}X}\right)_{X=-\Delta}\left(\frac{\mathrm{d}F}{\mathrm{d}X}\right)_{X=\Delta}. \tag{6.20}$$

That is,

$$\mu^{(2)}(\pm\Delta) = \mu^{(1)}(\Delta)\mu^{(1)}(-\Delta). \tag{6.21}$$

Of course, the slope $\mu^{(2)}$ is the same at each of the two points $X = \Delta$ and $X = -\Delta$.

The antisymmetry of $F(X)$ necessarily implies that $\mu^{(1)}(\Delta) = \mu^{(1)}(-\Delta)$, whence

$$\mu^{(2)}(\pm\Delta) = [\mu^{(1)}(\Delta)]^2. \tag{6.22}$$

This is a second key relationship. The eigenvalue (or slope) of the time-2 map $F^{(2)}$ at these fixed points of period 2 is necessarily positive. This is in contrast to what happens for maps with one critical point, where the period-2 cycle is born with $\mu^{(2)} = +1$, and then as the hump steepens $\mu^{(2)}$ decreases to zero, beyond which $\mu^{(2)}$ becomes negative, so that $\mu^{(2)}$ steepens beyond -1, at which point the period-2 orbit becomes unstable to a stable period-4 orbit.

Here the 2-cycle always has $\mu^{(2)}$ positive: $\mu^{(2)}$ begins at $+1$, decreases to zero, then increases back to $+1$, at which point a *pair* of new 2-cycles appear. This situation is illustrated in Fig. 6.4. The system now has two alternative stable states, each a stable cycle of period 2. Initial conditions originating in the intervals $[-1, -\Delta]$ or $[0, \Delta]$ will be attracted to one cycle; points in the intervals $[-\Delta, 0]$ or $[\Delta, 1]$ to the other cycle: this is shown in Fig. 6.5. This now accounts for all $3^2 = 9$ fixed points of period 2.

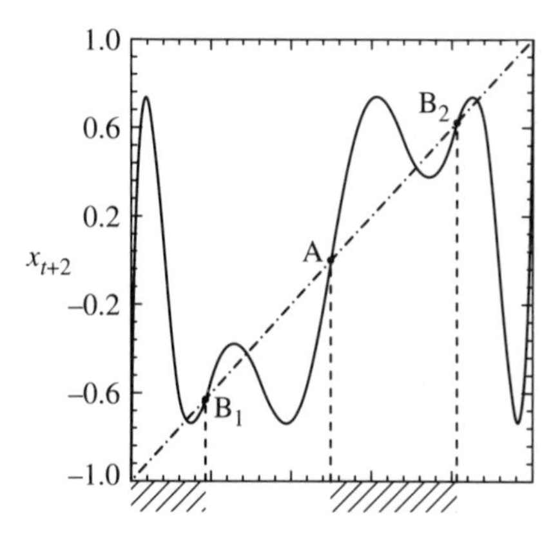

Fig. 6.4. This figure shows the two-times-iterated mapping $x_{t+2} = f^{(2)}(x_t, \lambda)$, for the cubic mapping of equation (6.11), with $a = 3.3$. The map intersects the 45° line at nine points, which are the fixed points of period 2. Here, the symmetric period-2 cycle between the points B_1 and B_2 is unstable, and there are two distinct periodic attractors: one attracts points originating in the hatched regions between -1 and B_1, and between A and B_2; the other attracts points originating in the two unhatched regions.

For each of these 'second generation' cycles of period 2, the stability-setting slope $\mu^{(2)}$ at the two fixed points will evolve in the manner made familiar by the map with one critical point: $\mu^{(2)}$ will decrease from $+1$, through zero, towards -1, beyond which point the period-2 cycle will become unstable, giving rise to an initially stable cycle with period 4. Thus each of the two domains of attraction will exibit the cascading bifurcation process, successively stable cycles of periods 2, 4, 8, 16, . . . , 2^n as shown in Fig. 6.5.

In summary, we observe that, for first-order difference equations with one critical point, the basic process is a bifurcation hierarchy of cycles of period $1 \to 2 \to 4 \to 8 \to \cdots \to 2^n$: other higher-order cycles echo this theme, with cycles of basic period k cascading through 'harmonics' of period $k2^n$. For one-dimensional maps with two antisymmetric critical points, the basic process is cycles with periods $1 \to 2 \to$ two distinct 2's, each with its own domain of attraction, and each of which then goes $\to 4 \to 8 \to \cdots \to 2^n$.

Higher-order cycles arise by 'tangent' bifurcation, and are similarly complicated. In general, there will be two stable periodic attractors for each parameter value. Consider, for example, period-3 points. For maps with one critical point, there are eight (2^3) such points: two period-1 points, and a stable and unstable pair of period-3 cycles. For maps with two critical points, there are potentially $3^3 = 27$ such points. Three of the 27 are period-1 points, the remaining 24 may be split equally into two; each set originates as two pairs of 3-cycles, one stable and one unstable. The task

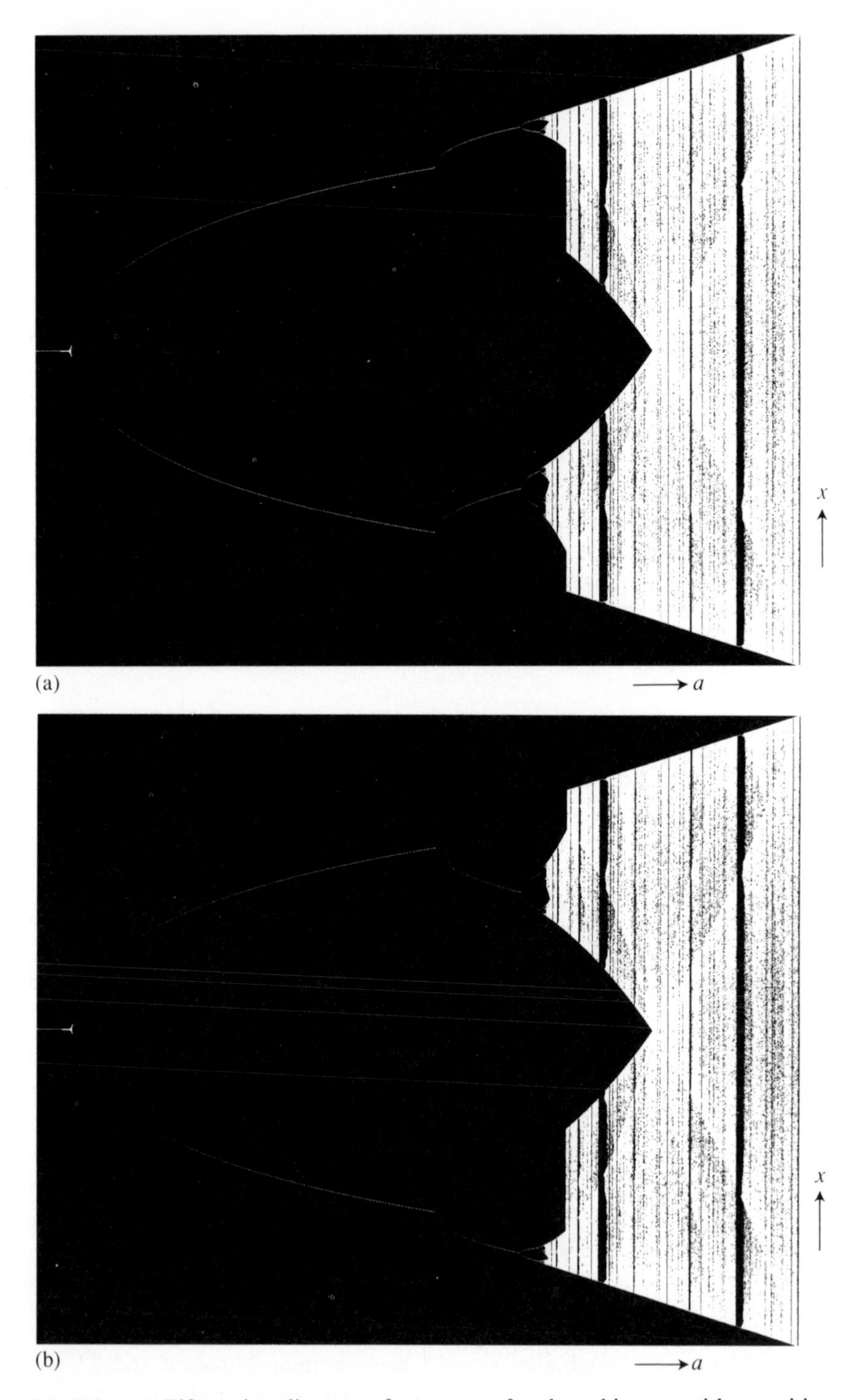

Fig. 6.5. (a) Bifurcation diagram of x versus a for the cubic map with a positive initial condition for x. (b) Bifurcation diagram of x versus a for the cubic map with a negative initial condition for x.

of cataloguing the way in which the 3^k fixed points of period k are organized, and the order in which the various cycles or period k originate, has been treated by several approaches.

6.4 The cubic map

Equation (6.11) has non-trivial behaviour for $0 < a < 4$; for $a < 0$, the end points at $X = \pm 1$ are attractors, and for $a > 4$ the hill top and valley bottom lie outside $[1, -1]$. The slope of the map (6.11), $\mu^{(1)}(0)$, at $X = 0$ is

$$\mu^{(1)}(0) = 1 - a. \qquad (6.23)$$

Thus the period-1 orbit is stable for $0 < a < 2$. For $0 < a < 1$ the damping is monotonic; for $1 < a < 2$ the damping is oscillatory.

The $F^{(2)}$ map bifurcates, to give symmetric fixed points of period 2, for $a > 2$, at $X = \pm \Delta$. We obtain these points from equation (6.18):

$$-\Delta = a\Delta^3 + (1 - a)\Delta. \qquad (6.24)$$

This gives

$$\Delta = \pm\, [(a - 2)/a]^{1/2} \qquad (6.25)$$

From equation (6.22) we can determine the eigenvalues (slope) of the $F^{(2)}$ map:

$$\mu^{(2)}(\pm\, \Delta) = (2a - 5)^2. \qquad (6.26)$$

At $a = 2$, we have $\mu^{(2)} = +1$; the slope $\mu^{(2)}$ decreases to zero at $a = 2.5$, and then increases to $\mu^{(2)} = 1$ at $a = 3$. For $a > 3$, this period-2 cycle becomes unstable.

The subsequent nonsymmetric fixed points of period 2 are found from solving

$$X = X[aX^2 + 1 - a]\,[aX^2(aX^2 + 1 - a)^2 + 1 - a]. \qquad (6.27)$$

Introducing $y = aX^2 + 1 - a$, we may, after some manipulation, write this as

$$[y^2 - 1]\,[y^2 + (a - 1)y + 1] = 0. \qquad (6.28)$$

The solution $X = 0$ has been discarded. Similarly, the pair of solutions $y = \pm 1$ corresponds to the fixed point solutions already found: $X = \pm 1$ and $X = \pm \Delta$. The remaining quadratic in y in equation (6.28) leads routinely to four other period-2 fixed points at $X = \pm\alpha$, $\pm\beta$, with α and β defined by

$$\alpha, \beta = \left(\frac{(a - 1) \pm [(a - 1)(a - 3)]^{1/2}}{2a}\right)^{1/2}. \qquad (6.29)$$

It is easy to verify that these four fixed points correspond to two distinct period-2 cycles, as illustrated in Fig. 6.5, with $\alpha \to -\beta \to \alpha$ and $\beta \to -\alpha \to \beta$, as illustrated in Fig. 6.4. The stability of these cycles depend on the slope of $F^{(2)}$ at these fixed points:

$$\mu^{(2)}(\pm\alpha, \pm\beta) = \mu^{(1)}(\pm\alpha)\mu^{(1)}(\mp\beta) \tag{6.30}$$

$$= 7 + 4a - 2a^2. \tag{6.31}$$

The cycles first appear at $a = 3$, with $\mu^{(2)} = +1$. They become unstable once $\mu^{(2)} < -1$, for $a > 1 + \sqrt{5}$. Beyond $1 + \sqrt{5}$, there are two distinct bifurcating hierarchies of cycles of periods 4, 8, and so forth.

There are two distinct attractors for all values of a in the range $3 < a < 4$, as can be seen most clearly in Fig. 6.5.

6.5 Frequency-dependent natural selection

Let us consider a subset of maps generated by simple models for frequency-dependent natural selection. For clarity, we replace the variable X of equation (6.10) by the variable p for the gene frequency, so that the maps are of the form

$$p_{t+1} = F(p_t), \tag{6.32}$$

with p defined on the interval [0, 1].

Specifically, we consider one locus with two alleles (A with frequency p; a with frequency $q = 1 - p$) in a diploid population where the genotypes AA, Aa, and aa have frequency-dependent fitnesses $w_{AA}(p)$, $w_{Aa}(p)$, and $w_{aa}(p)$, respectively.

Two assumptions are now made. First, we assume that the heterozygotes have fitness equal to the geometric mean of those of the homozygotes. We refer all fitnesses to that of the heterozygotes, taken to be unity ($w_{Aa} = 1$): this implies $w_{AA} \times w_{aa} = 1$. Second, the frequency-dependent selective advantage under discussion is assumed to be symmetric in the sense that $w_{aa}(1 - p) = w_{AA}(p)$. If we write $w_{AA} = f(p)$, the first of these assumptions implies that the relative fitnesses of the genotypes AA, Aa, and aa can be written as $f(p)$, 1, and $f(p)^{-1}$, respectively, and the second assumption implies the fitness function $f(p)$ obeys the relation

$$f(p) = 1/f(1 - p). \tag{6.33}$$

A reasonable frequency-dependent fitness for the three diploid genotypes is illustrated in Fig. 6.6. The above assumptions are made to keep the problem manageable.

The change in gene frequency between one generation (t) and the next ($t + 1$) is given by the standard expression

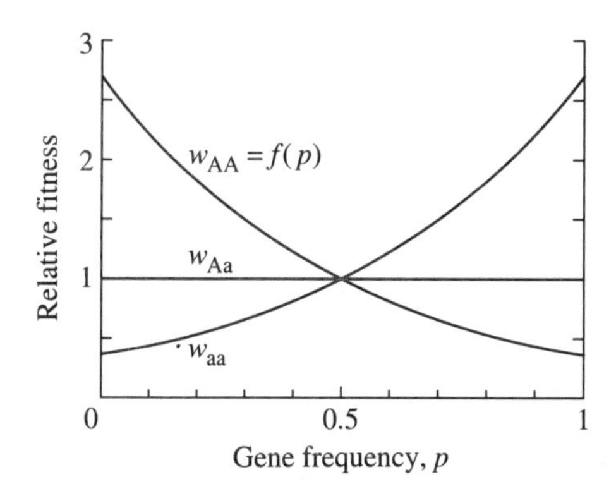

Fig. 6.6. This figure illustrates the assumptions made about the relative fitnesses w_{AA}, w_{Aa}, w_{aa} of the genotypes AA, Aa, aa, as functions of the gene frequency p. Note that the fitness function $w_{AA} = f(p)$ decreases monotonically.

$$p_{t+1} = \frac{w_{AA}p_t^2 + w_{Aa}p_t q_t}{w_{AA}p_t^2 + 2w_{Aa}p_t q_t + w_{aa}q_t^2}. \tag{6.34}$$

Under the above assumptions, this reduces to

$$p_{t+1} = \frac{p_t f(p_t)}{p_t f(p_t) + q_t}. \tag{6.35}$$

This has precisely the form of equation (6.32), with $F(p)$ given as

$$F(p) = \frac{pf(p)}{pf(p) + 1 - p} \tag{6.36}$$

As is intuitively obvious, equation (6.33) implies that $F(p)$ has the antisymmetric property of equation (6.12).

The slope of this map, $\mu^{(1)} = \mathrm{d}F/\mathrm{d}p$ at a point p is then

$$\mu^{(1)}(p) = \frac{f + p(1-p)(\mathrm{d}f/\mathrm{d}p)}{(pf + 1 - p)^2}. \tag{6.37}$$

In particular, the stability of the equilibrium point at $p = p_{\mathrm{I}}^* = \frac{1}{2}$ depends on the slope $\mu^{(1)}(\frac{1}{2})$, which is

$$\mu^{(1)}(\tfrac{1}{2}) = 1 + \tfrac{1}{4}[\mathrm{d}f/\mathrm{d}p]_{p=1/2}. \tag{6.38}$$

If the slope of the fitness function $f(p)$ is sufficiently steep, $\mu^{(1)}$ can steepen beyond -1, and the equilibrium point becomes unstable, bifurcating to give a stable 2-point cycle.

The gene frequency p can thus settle a stable oscillation between $p = p_{\mathrm{II}}^*$

and $1 - p^*_{\text{II}}$; the nontrivial solutions of the equation $(1 - p) = F(p)$, with $F(p)$ defined by equation (6.36). It follows, after some algebraic manipulation, that p^*_{II} and $1 - p^*_{\text{II}}$ are the two solutions of the equation

$$(1 - p)^2 = p^2 f(p). \tag{6.39}$$

The stability of this cycle depends on the slope of the period-2 map $F^{(2)}$ at these two points, and thence, via equation (6.22),on the slope $\mu^{(1)}(p^*_{\text{II}})$. Using the relations (6.39) and (6.38), we can write the slope of the $F(p)$ map at the period-2 points as

$$\mu^{(1)}(p^*_{\text{II}}) = 1 + \eta, \tag{6.40}$$

with η defined as

$$\eta = p^*_{\text{II}}(1 - p^*_{\text{II}})\,[d(\ln f)/dp]_{p = p^*_{\text{II}}}. \tag{6.41}$$

To proceed further, we make the assumption that $f(p)$ is a monotonically decreasing function of p. This assumption, which is as depicted in Fig. 6.6, seems to follow from any biologically plausible mechanism producing the frequency-dependent selective advantage to rare genotypes. The quantity η is now constrained to be negative: $df/dp < 0$ implies $\eta < 0$.

The 2-cycle is born near $p = \frac{1}{2}$, precisely when the slope at this point attains the value $\mu^{(1)} = -1$. As the map $F(p)$ steepens, the two period-2 points move away from the midpoint towards the critical points and for any monotonic curve $f(p)$, η will increase smoothly from $\eta = -2$ to $\eta = -1$ as the 2-cycle and critical points coincide. As the map $F(p)$ steepens further, the two period-2 points will move beyond the critical points, and η will continue to increase from $\eta = -1$ toward $\eta = 0$. However, for monotonic decreasing $f(p)$, we have $\eta < 0$ for all p, and consequently $\mu^{(1)}$ cannot surpass the value $+1$. Hence, the bifurcation at $\mu^{(2)} = 1$, whereby the two new distinct 2-cycles are born, is not attainable.

In short, the gene frequency in symmetrical diploid systems with frequency-dependent natural selection can exhibit either a stable point or a stable 2-point cycle. The richer range of dynamics available to the cubic map is not possible for the restricted class of maps implied by equation (6.36).

6.5.1 *An example*

Suppose we choose the fitness function $f(p)$ of the previous section in the form

$$f(p) = e^{\beta(1 - 2p)}. \tag{6.42}$$

This has the properties indicated in Fig. 6.6: it is monotonic decreasing for p on the interval [0, 1] and it has the symmetry property that $f(p)f(1 - p) = 1$.

Substituting equation (6.42) into equation (6.38), we see that the slope of the map $F(p)$ at the equilibrium point p is

$$\mu^{(1)}(\tfrac{1}{2}) = 1 - (\tfrac{1}{2}\beta). \tag{6.43}$$

There will be a stable equilibrium gene frequency provided $\beta < 4$. For frequency-dependent selective forces so strong that $\beta > 4$, this equilibrium point gives way to a stable 2-point cycle between two values of p determined from equation (6.39). Equation (6.39) now has the explicit form

$$\ln\left(\frac{1-p}{p}\right) = \beta(\tfrac{1}{2} - p). \tag{6.44}$$

The stability of the 2-point cycle depends on the slope $\mu^{(2)}(p^*_{\text{II}}) = [\mu^{(1)}(p^*_{\text{II}})]^2$, where $\mu^{(1)}(p^*_{\text{II}})$ is given by equations (6.40) and (6.41) as

$$\mu^{(1)}(p^*_{\text{II}}) = 1 - 2\beta p^*_{\text{II}}(1 - p^*_{\text{II}}), \tag{6.45}$$

with p^*_{II} from equation (6.44). It is easy to see that, as β increases from 4 toward ∞, $\mu^{(1)}(p^*_{\text{II}})$ increases from -1 to approach $+1$. For $\beta \gg 1$, we have

$$p^*_{\text{II}} \approx \mathrm{e}^{-\beta/2}, \tag{6.46}$$

and

$$\mu^{(1)}(p^*_{\text{II}}) \approx 1 - 2\beta\mathrm{e}^{-\beta/2}. \tag{6.47}$$

The value $\mu^{(1)} = 1$ is, however, never attained, and the 2-cycle never becomes unstable. Figure 6.7 bears out the above remarks, by showing the dynamical behaviour of the gene frequency p_t for various values of β.

In this example there is a stable equilibrium at $p = \frac{1}{2}$ unless $\beta > 4$. That is, there is a stable point unless the genotypes AA when very rare have a selective advantage of a factor $\mathrm{e}^4 \approx 55$ over the heterozygotes Aa, and of $\mathrm{e}^8 \approx 3000$ over the homozygotes aa. It is not easy to imagine ecological circumstances that will produce so large an advantage to rare alleles.

One would expect that the assumption in which an allele enjoys maximum selective advantage when it is rare, but not too rare, is realistic. However, p could perhaps equally be realized as a projection of some multilocus system with competing selection pressures; in such a circumstance the constraint $\eta < 0$ is lost. A much wider range of dynamical behaviour is possible with the loss of this constraint.

For example, the frequency-dependent fitness function

$$f(p) = \frac{1 + 2\alpha(1 - 2p)(1 - p)}{1 - 2\alpha(1 - 2p)p}, \tag{6.48}$$

(with $0 < \alpha < 4$) gives behaviour consistent with the cubic map of equation (6.11). But this $f(p)$ has the biologically strange feature that maximum

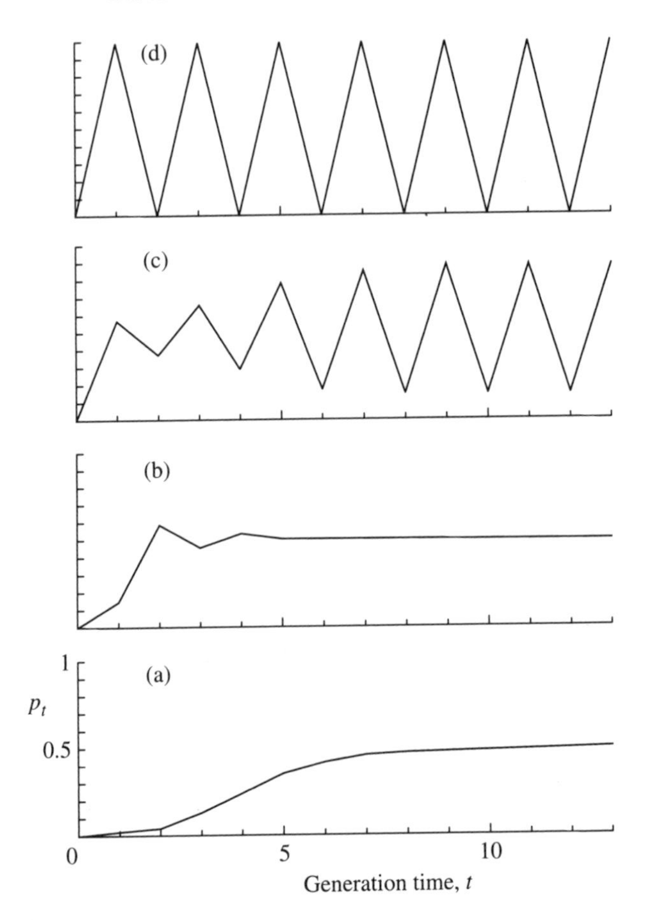

Fig. 6.7. The behaviour of the gene frequency p as a function of generation time t is shown for the specific frequency-dependent fitness assumptions. The parameter β, which measures the strength of the frequency-dependent effects, has the values: (a) $\beta = 1$; (b) $\beta = 3$; (c) $\beta = 5$; (d) $\beta = 10$.

selective advantage accrues to the genotype AA at intermediate rarity, rather than in the limit $p = 0$, once $\alpha > 1$.

On the whole, the assumption that $f(p)$ decreases monotonically seems very reasonable. Other forms are likely to be the exception rather than the rule. Less affirmative things have to be said about the other two assumptions in the previous section.

The assumption that the heterozygous genotypes have fitness geometrically intermediate between the homozygotes is of a kind that is common in population genetics. As a consequence, it reduces the diploid genetic system to a haploid one; for a haploid population with alleles A and a, and with the genotype A having fitness $f(p)$ relative to the genotype a, equation (6.35) is obtained immediately. If we give up this assumption about the fitness of the heterozygotes, the relative fitness of the genotypes may be written as $w_{AA} : w_{Aa} : w_{aa} = f(p) : 1 : g(p)$. Here the assumption that the frequency-dependent selection affects the rarer allele in a symmetrical way is retained by requiring that $g(p) = f(1 - p)$. What is lost is the further requirement that $f(p)f(1 - p) = 1$.

The final assumption is that the selective advantages possessed by the allele A when it is rare are precisely the same as those pertaining to the allele a when it is rare. This is the assumption that gives the map $F(p)$ its antisymmetric properties. While it is true that the gene frequency p can, without loss of generality, be rescaled to bring the equilibrium point to the mid-point at $p = \frac{1}{2}$, it is not in general true that the overall antisymmetry of $F(p)$ can be so preserved. Once the fitness functions for the genotypes AA and aa are unrelated, it is hard to make any general statement.

If frequency-dependent natural selection is carictured by a single-locus-two-allele model, the gene frequency will obey a first-order difference equation that can have two critical points. Such systems with two critical points can, in general, have very complicated dynamical behaviour. But if we assume frequency-dependent selection in a diploid population where the heterozygotes are of geometrically intermediate fitness, where the selective advantage accruing to the rarer allele is symmetric, and where these selective advantages decrease monotonically as the allele becomes commoner, then the consequent mappings can only exhibit either a stable point or a stable 2-point cycle.

On the one hand, not only are the models dynamically tame in possessing only stable points or stable 2-point cycles, but even these cycles require that the rarer allele possess implausibly large selective advantages. This may be taken as an argument that the frequency-dependent natural selection typically produces a balanced polymorphism.

On the other hand, studies of systems of difference equations in various contexts in population biology suggest that chaotic behaviour arises more generically as the dimension of the system increases. This suggests that chaotic behaviour, possibly with several alternative domains of attraction, may arise with frequency-dependent selection in multilocus systems.

6.6 Two-dimensional maps

A two-dimensional map can be written as

$$x_{n+1} = f_1(x_n, y_n, \lambda), \tag{6.49}$$

$$y_{n+1} = f_2(x_n, y_n, \mu), \tag{6.50}$$

where λ and μ are parameters. If the pairs (x_n, y_n), found by iterating the map remain bounded, then the sequence approaches an asymptotic set, called the attractor. There are three possibilities for the attracting set: a set of zero dimension (a periodic set of points), a one-dimensional curve, and a strange attractor of noninteger dimension. The dimension of the attractor is less than the dimension of the phase space that it sits in. The noninteger dimension is commonly referred to as the fractal dimension [13].

6.6.1 *Examples*

Populations with overlapping age classes

A population of plants and animals corpposed of several overlapping age classes can be described by a set of coupled first-order difference equations. One such example is a model used by the International Whaling Commission (IWC), which has the general form

$$N_{t+1} = (1 - \mu)N_t + R(N_{t-T}). \tag{6.51}$$

Here N_t is the population of mature adult whales in year t. The corresponding population N_{t+1}, one year later consists of the surviving fraction $(1 - \mu)N_t$ in addition those born T years ago, $R(N_{t-T})$, where T is the time taken to attain sexual maturity and R is a nonlinear function. The IWC model for the recruitment term R is

$$R(N) = \tfrac{1}{2}(1 - \mu)^T N\{P + Q[1 - (N/K)^z]\}. \tag{6.52}$$

Here K is the assumed unharvested equilibrium density of the population; P is the per capita fecundity of females at the equilibrium point $N = K$; Q is the maximum per capita increase of which the whales are capable as population densities fall to low levels; and z measures the severity with which this density-dependent response is manifested. If we choose $z = 1$, this corresponds to the usual logistic assumption, in which the density-dependent increase in fecundity is linear. The factor $(1 - \mu)^T$ measures the fraction of newly born whales that survive, with probability $(1 - \mu)^T$ for each of the T years, to adulthood. Finally, the factor $\frac{1}{2}$ arises because exactly half of the population is female, so that the per capita fecundity of females is to be multiplied by $\frac{1}{2}N$, the total number of females. The equilibrium population in the unexploited system is, by assumption, $N_{t+l} = N_t = N_{t-T} = K$; substituting this in equation (6.51) and (6.52) gives the identity

$$\mu = \tfrac{1}{2}(1 - \mu)^T P. \tag{6.53}$$

Hence, putting equations (6.51) and (6.53) together with equation (6.52) gives the baleen whaling equation

$$X_{t+1} = (1 - \mu)X_t + \mu X_{t-T}[1 + q(1 - X_{t-T}^z)]. \tag{6.54}$$

Here, $q = Q/P$; X is the rescaled population variable, $X = N/K$. Early studies of this equation were made by Clarke [14] and Levin and May [15].

The example we shall investigate more closely is perhaps the simplest example of a difference equation that contains time lags: the time-delayed logistic equation [16]

$$x_{t+1} = \lambda x_t(1 - x_{t-1}). \tag{6.55}$$

The time lag may, for example, account for the effect of the over-consumption of a primary food source. Equation (6.55) may be reformulated as a papir of coupled first-order difference equations:

$$x_{t+1} = \lambda(1 - y_t), \tag{6.56}$$

$$y_{t+1} = x_t. \tag{6.57}$$

Fixed points of the time-delayed logistic equation

The fixed points of the system (6.56, 6.57) occur when $x_{n+1} = x_n$ and $y_{n+1} = y_n$. In which case there are two fixed points at

$$(x_0, y_0) = (0, 0) \quad \text{and} \quad (x_1, y_1) = \left(1 - \frac{1}{\lambda}, 1 - \frac{1}{\lambda}\right). \tag{6.58}$$

To establish the local stability of the requires the analysis of the linearization of the nonlinear system (6.56, 6.57) at the fixed point. The pertinent system of linear equations, with new variables $\hat{x}_{n+1}$ and $\hat{y}_{n+1}$ is

$$\begin{bmatrix} \hat{x}_{n+1} \\ \hat{y}_{n+1} \end{bmatrix} = \begin{bmatrix} \dfrac{\partial f_1}{\partial x} & \dfrac{\partial f_1}{\partial y} \\ \dfrac{\partial f_2}{\partial x} & \dfrac{\partial f_2}{\partial y} \end{bmatrix} \begin{bmatrix} \hat{x}_n \\ \hat{y}_n \end{bmatrix}. \tag{6.59}$$

Here the matrix of partial derivatives (a constant matrix) is found by evaluating the partial derivatives at a fixed point $\mathbf{x}_0 = (x_0, y_0)$. For the trivial fixed point x_0, the eigenvalues μ of the matrix are found from the characteristic equation

$$\begin{vmatrix} \mu - \lambda & 0 \\ 1 & \mu \end{vmatrix} = 0, \tag{6.60}$$

where $|\cdot|$ is a determinant. This calculation gives two eigenvalues: $\mu_0 = 0$ and λ.

Therefore, we may conclude, the trivial fixed point $\mathbf{x}_0$ is stable for $0 < \lambda < 1$.

The behaviour of the fixed point $\mathbf{x}_1$ is perhaps more interesting. The characteristic equation for the fixed point $\mathbf{x}_1$ is

$$\mu(\mu - 1) + (\lambda - 1) = 0. \tag{6.61}$$

At $\lambda = 0$, the linearization indicates the local dynamics are saddle-like; that is, it has a stable ($\mu < 0$) and an unstable ($\mu > 0$) eigen-direction. At parameter value $\lambda = 1$, the unstable eigenvalues become critical ($\mu = 0$) and both directions become stable; contrast this behaviour with the linear map for the fixed point $\mathbf{x}_0$. At the parameter value $\lambda = \frac{5}{4}$, the eigenvalues coalesce to become a degenerate pair, $\mu = \frac{1}{2}$. Increasing the parameter λ further results in the eigenvalues becoming complex. A bifurcation occurs at $\lambda = 2$ when the moduli of the complex pair μ_c and $\bar{\mu}_c$ are unity, in which case we can write

$$\mu_c = \exp(i\theta_c), \tag{6.62}$$

with $\theta_c = \frac{1}{3}\pi$.

In general, bifurcation of this type (a Hopf-type bifurcation, in which the bifurcation occurs via a complex pair of eigenvalues becoming unstable) results in the ratio θ/π being either rational or irrational. If $\theta/\pi = l/m$, then resonance can occur, and, provided the resulting orbit is stable, a periodic orbit of period m will result. For an irrational value of the ratio θ/π, an invarient circle will result, the exact form of which is determined by the nonlinearities. For the time-delayed logistic equation, a period-3 orbit results.

Equations (6.56) and (6.57) have been studied by Pounder and Rogers [17], Aronson *et al.* [18] and Rogers and Clarke [19]. Pounder and Rogers showed that the trajectories of points (x_t, y_t) are attracted to an invariant curve, which upon close inspection has an extremely complicated shape. This occurs via a homoclinic bifurcation, in which the invariant circle from the Hopf resonance comes close to an unstable fixed point. Very briefly, this curve has infinitely many loops or folds issuing from the origin; the loops are successive images of the bottom arc of the curve [20]. Once λ exceeds a critical value, the bottom arc of the curve can cross the x-axis, enabling the system to 'escape' to negative values (corresponding, biologically, to extinction). Interior loops of the invariant curve can also lead to 'escaping' trajectories and extinction. The net result is that, for a range of λ values, there are initial values which lead to extinction (possibly after very long times), while others can lead to cycles of high period or chaotic fluctuations; the phase plane has a complicated filamentary structure, in which arbitrarily close initial points can undergo qualitatively different fates.

6.6.2 *Two interacting populations (host–parasitoid)*

Roughly 10% of all metazoan species are insect parasitoids: flies or hymenopterans (wasps, bees) which lay their eggs in the lava usually of lepidopterans (butterflies and moths). One of the simplest kinds of prey–

predator systems involves insect parasitoids. Studies of temperate zone insects and their predators or parasites, and of host parasite systems in general, often lead to two coupled first-order maps, one for the changes in the host population and the other for the predator or parasite population.

When one considers two or more interacting populations, two- and higher-dimensional difference equations can result. A two-dimensional example, corresponding to interacting prey and predator populations with discrete nonoverlapping generations, was propounded by the parasitologist Crofton [21] as a description of a host–parasite interaction:

$$x_{t+1} = \lambda x_t(1 + y_t)^{-k}, \qquad y_{t+1} = x_t y_t(1 + y_t)^{-(k+1)}. \tag{6.63}$$

Here, x_t and y_t are essentially the numbers of hosts and parasites respectively in generation t. The parameter λ represents the growth factor for the host population, and the parameter k measures the degree of parasite aggregation.

Dynamics and global stability

A fixed point occurs for the system (6.63) at

$$(\bar{x}, \bar{y}) = (\lambda^{1+1/k}, \lambda^{1/k-1}). \tag{6.64}$$

It is locally stable for all λ if $k < 2$, and for

$$\lambda < \left(\frac{k-2}{k+2}\right)^k \tag{6.65}$$

if $k > 2$. Figure 6.8 shows the asymptotic fate of the system for various values x_0 and y_0 of the initial populations. The initial condition x_0 ranges from 10^{-6} to 10^4 and y_0 from 10^{-4} to 10^2, on a logarithmic scale. The light region denotes points that are attracted to the equilibrium point. The dark area indicates points that lead to diverging oscillations which carry the variable y to zero. The initial condition $(x_0, y_0) = (10^{0.75}, 10^{-3.5})$ is special in that it is attracted to a stable 11-point cycle.

We observe the domain of attraction is not without structure. All points within an order of magnitude or so of the equilibrium point are attracted to it, and for large x_0 there is a tendency for points lying along two 'spiral arms' to converge to the equilibrium point.

An alternative approach to mapping out the domain of attraction of the equilibrium point is to run the equation backwards in time, starting from points in the neighbourhood of the equilibrium point. The pair of equations (6.63) are, moreover, easily inverted, to give

$$y_t = \frac{\lambda y_{t+1}}{x_{t+1} - \lambda y_{t+1}} \tag{6.66}$$

$$x_t = \frac{x_{t+1}}{\lambda}\left(\frac{x_{t+1}}{x_{t+1} - \lambda y_{t+1}}\right)^k. \tag{6.67}$$

Fig. 6.8. The global stability of the pair of first-order coupled difference equations (6.63), with $\lambda = 2$ and $k = 5$. The locally stable equilibrium values of x and y are marked by the cross, and those initial points (x_0, y_0) in the light region do indeed give trajectories that converge to this equilibrium point. All other points in the dark region lead to oscillations that diverge until y is less than 10^{-99}. The point at $x_0 = 10^{3/4}$, $y_0 = 10^{7/2}$ is exceptional, being attracted to a stable 11-point cycle.

When run backwards in time, the trajectories are not confined to the positive quadrant. The procedure does not indicate any more coherent a domain than is shown in Fig. 6.8.

Studies of the dynamics of host-parasitoid systems have aided empirical studies since the early work of Nicholson and Bailey in the 1930s. This is because each host produces either another host or a parasitoid in the next generation, which means simple models are also relatively realistic. The basic host-parasitoid system obeys the equations

$$x_{t+1} = \lambda x_t f(y_t), \qquad y_{t+1} = x_t - x_{t+1}/\lambda. \tag{6.68}$$

Here, x and y are the population densities of hosts and parasitoids respectively. The dynamics of the system (6.68) depends on the single parameter λ. We observe there is a stable fixed point at $y = y^*, x = x^* = y^*\lambda/(\lambda - 1)$, where $\lambda f(y^*) = 1$. A linear stability analysis of this fixed point leads to the characteristic equation for the eigenvalues μ:

$$\mu^2 - \mu(1 + \alpha) + \lambda\alpha = 0. \tag{6.69}$$

Here, for convenience, we have defined

$$\alpha \equiv -x^* \left(\frac{\mathrm{d}f}{\mathrm{d}y}\right)^*. \tag{6.70}$$

As the function f should be monotonically decreasing for increasing y, it follows that the parameter $\alpha > 0$. The fixed point (x^*, y^*) will be unstable as μ crosses the unit circle. At this Hopf bifurcation, we may write $\mu = \mathrm{e}^{\mathrm{i}\theta}$. The imaginary part of the characteristic equation (6.69) then gives

$$\sin 2\theta - (1 + \alpha)\sin\theta = 0 \tag{6.71}$$

The phase angle θ is given by

$$\cos\theta = \tfrac{1}{2}(1 + \alpha). \tag{6.72}$$

At the bifurcation, the parameters α and λ are related by $\lambda\alpha = 1$. We may use this to express α in (6.72) in terms of the more biologically familiar λ. We finally arrive at

$$\cos\theta = \tfrac{1}{2}\left(1 + \frac{1}{\lambda}\right). \tag{6.73}$$

As λ increases from unity to very large values, the phase angle increases from 0° to 60°. Consequently, we can obtain a 6-point cycle in the limit of $\lambda \to \infty$, but more generally a cycle of roughly 7 is the lowest likely to be found.

Hence, the functional form of the map given by equation (6.68) implies that stable periodic solutions have approximate periods of at least six generations. This accords with several numerical studies presented by Olsen and Degn [22]; see also Lauwerier [20].

6.7 Preturbulence

Yorke has shown that dynamical systems can exhibit preturbulence, in which an orbit wanders around for arbitarily varying periods of time, before suddenly converging to a point of equilibrium. Conceivably, the above two-dimensional systems can, for certain parameter values, exhibit this phenomenon. Long transient behaviour, in which the orbit wanders in phase space, can give a false impression of being a chaotic orbit. This, of course, exacerbates the difficulties involved in calculating measures such as the Lyapunov exponent.

6.8 Lyapunov exponents for two-dimensional maps

For a two-dimensional map, the Lyapunov exponent λ_1 and λ_2 are the average stretching factors for a infinitesimal circular area. To calculate the exponents we must compute the eigenvalues of

$$\lim_{n \to \infty} [J(x_n, y_n)J(x_{n-1}, y_{n-1}) \cdots J(x_1, y_1)]^{1/n}, \tag{6.74}$$

where J is the matrix of partial derivatives encounted previously. The exponents give the average stretching rate of the principle axes of the infiratesimal area. The map is area-preserving if $\lambda_1 \lambda_2 = 1$ and contracting if $\lambda_1 \lambda_2 < 1$. For a chaotic map, at least one of the exponents must exceed unity. When both or in higher-dimensional maps more than one of the exponents exceed unity, the mapping is considered to be hyperchaotic.

6.9 Delay–differential equations

For human populations, the annual death rates and birth rates are small; the maturation times are of the order of several years. Whale populations also fall naturally into this category. We then expect the whale population to be described in terms of many overlapping age classes and a well-defined breeding season. We may approximate the population growth by a delay–differential equation.

We can write the IWC equation (6.51) in terms of its continuous counterpart. Under such an approximation, equation (6.51) becomes

$$\frac{\mathrm{d}X}{\mathrm{d}t} = -\mu X + \mu X_{t-T}[1 - q(1 - X_{t-T}^Z)]. \tag{6.75}$$

Dynamical studies of related delay–differential equations have been performed by Mackey and Glass [23, 24] and Lasot and Wazewska [245]. Sparrow [26, 27] has investigated chaotic behaviour in simple feedback systems. In the context of human demography, Swick [28, 29] has studied integro-differential equations which reduce to equations of the IWC type if the kernels are reduced to a limiting δ-function form. Mackey and Glass have studied equations of the form

$$\frac{\mathrm{d}X}{\mathrm{d}t} = -\mu X + R(X(t - T)), \tag{6.76}$$

with several different 'one-humped' functions for $R(X(t-T))$, as models for specific physiological process.

6.9.1 *Linear analysis of equation (6.75)*

The linearized stability analysis of differential–delay equations such as equation (6.75) is straightforward, but the bewildering array of dynamical behaviour that can be displayed for certain parameter values is both fascinating and ill-understood.

A fixed point occurs for equation (6.75) at $X = 1$. It is natural to linearize

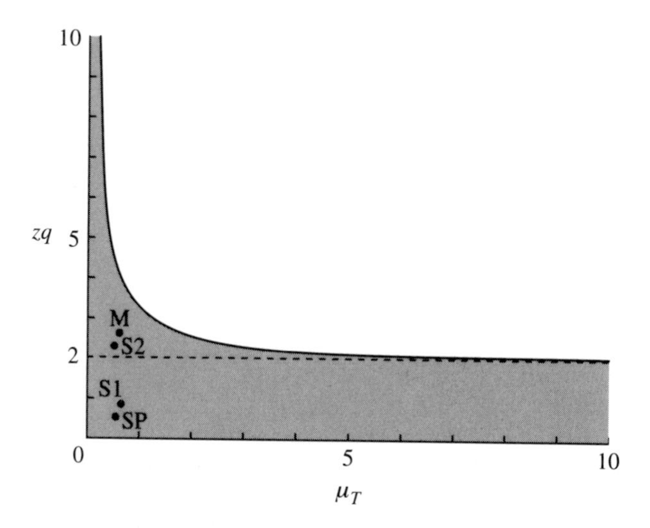

Fig. 6.9. The shaded area shows the domain of parameter space in which the IWC delay–differential equation has a locally stable equilibrium point. The IWC parameters μ, T, q, and z are as defined in the text. The four points correspond to the parameter values actually used by the IWC: the point labelled M is for minke whales; S1 is for sei whales with $z = 2.4$; S2 is for sei whales with $z = 1$; and SP is for sperm whales.

about the fixed point by writing $X(t) = 1 + x(t)$, where $x(t) = x_0 e^{\lambda t}$. A solvability condition on this approximation yields a transcendental equation for the eigenvalue λ:

$$\lambda = -\mu[1 + (zq - 1)e^{-\lambda T}] \tag{6.77}$$

The equilibrium point will be locally stable iff

$$\mu T < \frac{\pi - \cos^{-1}(1/b)}{(b^2 - 1)^{1/2}}. \tag{6.78}$$

Here $b = zq - 1$. We note that system has a locally stable equilibrium point if $zq < 2$. These results for local stability are illustrated in Fig. 6.9, which also gives estimates of the parameter values for real whole populations.

6.9.2 *Nonlinear behaviour of equation (6.75)*

When the fixed point is no longer locally stable there will be a Hopf bifurcation to a stable cycle. As b is increased further, the cycle undergoes a cascade of dynamical behaviour, eventually entering a regime of apparently chaotic dynamical behaviour.

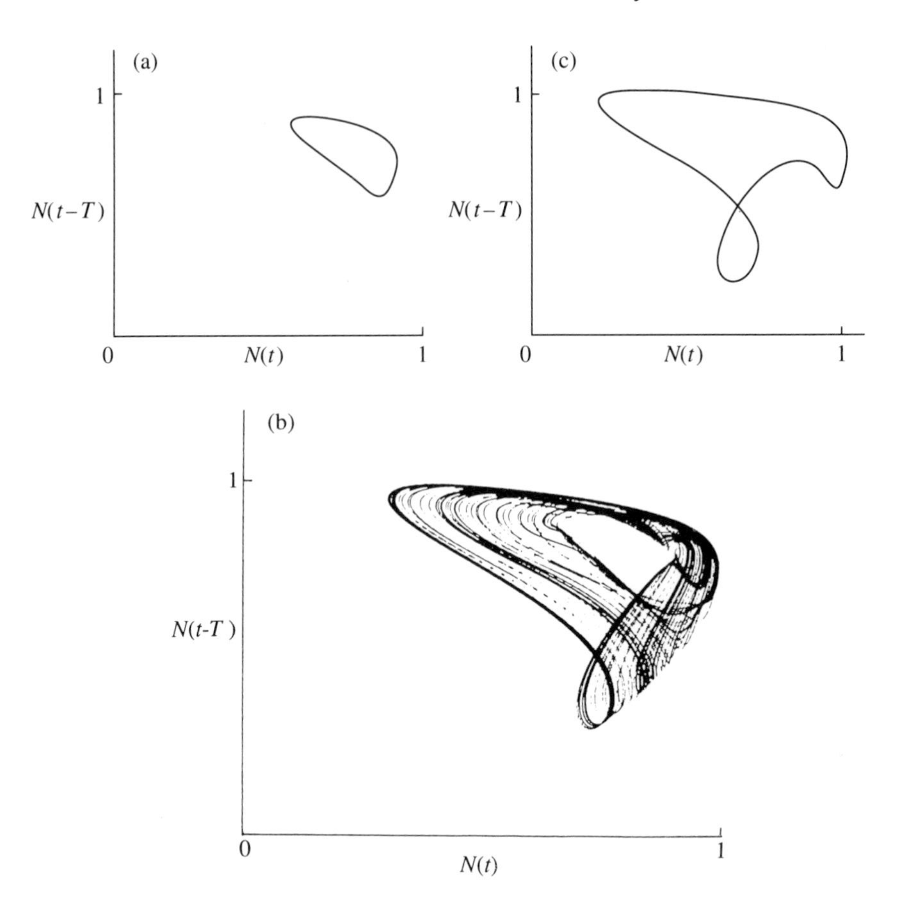

Fig. 6.10. The phase space plot of the asymptotic solution to the IWC delay-differential equation for an illustrative parameter choice, with $\mu = 1$, $T = 2$, $q = 1$, and various z: (a) $z = 3.0$ (stable cycle); (b) $z = 3.7$ (apparently chaotic trajectory); (c) $z = 4.0$ (a stable cycle).

The IWC has catalogued the values of P, Q, z, μ, and T that it has used for particular species of whales. Figure 6.9 shows the quantities z and μT for minke, sei, and female sperm whales. We note that in all cases the dynamic lies well within the domain of stable point behaviour.

Figure 6.10 shows a sequence of asymptotic phase space portraits, with $N(t - T)$ plotted against $N(t)$, after initial transients have died away. Figure 6.10(a) shows a simple cycle; this undergoes a sequence of period doublings to the complicated orbit shown in Fig. 6.10(b). Increasing the parameter further results in a reverse period doubing (inverse cascade) to a simple cycle, illustrated in Fig. 6.10(c).

Many other populations can plausibly be described by delay-differential equations. One example is Nicholson's blowfly data. Another laboratory study is by Halbach [30] on the dynamical behaviour of rotifer populations at different temperatures. As will be discussed in the next chapter, these observations can be successfully explained by using a delay-differential equation to model the population dynamics, and letting the intrinsic growth rate be an increasing function of temperature.

It must be noted that the above dynamical behaviour exhibited by equations (6.75) and (6.51) are of mathematical interest for their own sake, but they probably have little relevance to the dynamics of whale populations (see Fig. 6.10).

References

1. Oseledec, V. I. (1968). A multiplicative ergodic theorem: Liapunov characteristic numbers of dynamical systems. *Trans. Moscow Math. Soc.*, **19**, 197–231.
2. Shaw, R. (1981). Strange Attractors, chaotic behaviour, and information flow. *Z. Naturf.* **36a**, 80–112.
3. Jakobson, M. (1981). Absolutely continuous invariant measures for one-parameter families of one-dimensional maps. *Commun. Math. Phys.*, **81**, 39.
4. Guckenheimer, J., and Holmes, P. (1983) *Nonlinear oscillations, dynamical systems, and bifurcations of vector fields.* Springer, New York.
5. Clarke, B. C. (1964). Frequency-dependent selection for thr dominance of rare polymorphic genes. *Evolution*, **18**, 364–9
6. Clarke, B. C. (1969). The evidence for apostatic selection. *Heredity*, **24**, 347–52.
7. Clarke, B. C. (1975). Frequency dependent natural selection. In: *The role of natural selection in human evolution* (ed. F. M. Salzano), pp. 187–200. North-Holland, Amsterdam.
8. Wright, S. (1969). *Evolution and the genetic populations*, Vol. 2. Chicago University Press.
9. Kojima, K.-I. (1971). Is there a constant fitness value for a give genotype? No! *Evolution*, **25**, 281–5.
10. Endler, J. A. (1977). *Geographic, speciation and clines.* Princeton University Press.
11. Lewontin, R. C. (1958). A general method for investigating the equilibrium of gene frequency in a population. *Genetics*, **43**, 419–34.
12. Lewontin, R. C. (1974). *The genetic basis of evolutionary change.* Columbia University Press.
13. Grassberger, P. and Procaccia, I. (1983). Characterisation of strange attractors *Phys. Rev. Lett.*, **50**, 346.
14. Clarke, C. W. (1976). A delayed-recruitment model of population dynamics with application to baleen whale populations. *J. Math. Biol.*, **3**, 381–91.

15. Levin, S. A. and May, R. M. (1976) A note on difference-delay equations. *Theor. Biol.*, **9**, 178–87.
16. Maynard Smith, J. (1968) *Mathematical ideas in biology*. Cambridge University Press.
17. Pounder, J. R. and Rogers, T. D. (1980). The geometry of chaos: Dynamics of a nonlinear second order difference equation. *Bull. Math. Biol.*, **42**, 551–97.
18. Aronson, D. G., Chory, M. A., Hall, G. R., and McGehee, R. P. (1982). Bifurcations from an invariant circle for two-parameter families of maps of the plane: A computer assisted study. *Commun. Math. Phys.*, **83**, 303–54.
19. Rogers, T. G. and Clarke, B. L. (1981). A continuous planar map with many period points. *Appl. Math. Computat.*, **8**, 17–33.
20. Lauwerier, H. A. (1986). Two-dimensional iterative maps. In: *Chaos* (ed. A. V. Holden), pp. 58–95. Princeton University Press.
21. Crofton, H. D. (1971). A qualitative approach to parasitism. *Parasitology*, **63**, 179–93.
22. Olsen, L. F. and Degn, H. (1985). Chaos in biological systems. *Q. Rev. Biophys.*, **18**, 165–225.
23. Mackey, M. C. and Glass, L. (1977). Oscillation and chaos in physiological control systems. *Science*, **197**, 287–9.
24. Glass, L. and Mackey, M. C. (1979). Pathological conditions resulting from instabilities in physiological control systems. *Ann. N. Y. Acad. Sci.*, **316**, 214–35.
25. Lasota, A. and Wazewska, M. (1976) Mathematical models of the red blood cell system. *Math. Stosowana*, **6**, 25–40.
26. Sparrow, C. (1980). Bifuration and chaotic behaviour in simple feedback systems. *J. Theor. Biol.*, **83**, 93–105.
27. Sparrow, C. (1980). Chaos in single loop feedback systems. Ph. D. thesis, University of Cambridge. Chapter II.
28. Swick, K. E. (1980). Periodic solutions of a nonlinear age-dependent model of single species population dynamics. *SIAM J. Math. Anal.*, **11**, 901–10.
29. Swick, K. E. (1981). A nonlinear model for human population dynamics. *SIAM J. Appl. Math.*, **40**, 266–78.
30. Halbach, U. (1979). Introductory remarks: Strategies in population research exemplified by rotifier population dynamics. *Fortschr. Zologie*, **25**, 1–27.

7
Distinguishing chaos from noise

David Holton and Robert M. May

A central theme to dynamical systems is the recognition that simple deterministic mechanisms can generate random-looking data. However, can random-looking data, derived from an experiment, be attributable to a simple deterministic mechanism?

There are several sources of difficulty in attributing a random time series from a natural system to a deterministic source. The randomness may be a result of errors inherent in making the measurement, for example, in sampling errors in estimating sizes. Also, there may be fluctuations derived from unpredictable environmental changes; this is perhaps most insidious, since it suggests the system is strongly coupled to the environment and hence inherently of high dimension. Another difficulty could be that the system is intrinsically of high or infinite dimension with interactions on many temporal and spatial scales. This being the case, we can no longer hope for a low-dimensional dynamical description.

But, given these difficulties, an outstanding problem we wish to begin thinking about is: can one distinguish between deterministic chaos and noisy data generated from an experimental time series?

For all of the dynamical processes arising from natural population dynamics, an exact dynamical description is unknown. We have, however, seen in previous chapters an approach for obtaining a mathematical understanding of the natural world by setting up a model, involving a finite number of phase space variables. The analysis of the models may lead, it is hoped, to a deeper understanding of how these quantities vary. We wish to turn that approach around: given a sequence of data, can we derive a finite number of phase space variables which describe the dynamics of the time series?

7.1 Time series

In an experiment, one typically observes a finite number of variables (usually one) over a finite (often small) period of time. The sequence of data obtained, indexed by time, is called a *time series*. It is not clear how we can reconstruct the information given in the time series to give information about the whole system. This is a problem that has received a lot of attention from statisticians over the years; ideas derived from chaotic dynamical systems are beginning to bring new approaches to some old questions [1].

The simplest way of obtaining a multidimensional signal from a scalar time series is to use the method of delay [2], and was put on a firm mathematical basis by Takens [3]. Consider a time series $\{x_t\}$ of measurements (a sequence of data points), indexed by time t. We can construct a space in which the points of the multidimensional signal reside, called an embedding space E of dimension e, using a reconstructed trajectory vector $\boldsymbol{x}_t$:

$$\boldsymbol{x}_t = (x_t, x_{t+1}, \ldots, x_{t+(e-1)}). \tag{7.1}$$

The set of e-dimensional points we reconstruct using the available time series is a projection onto e dimensions of the true attracting set.

The approach we shall describe is designed for making short-term predictions about the trajectories of chaotic dynamical systems. The underlying idea is simple: if the random-looking time series is generated by a low-dimensional attractor, then short-term predictions should be possible (although long-term ones are not); if the time series is 'really random' (as, for example, sampling error), then both long- and short-term prediction are equally probabilistic.

To illustrate the method, we shall apply it to a variety of data, first from various mathematical models and then from naturally arising data on measles and chickenpox [4], and marine phyloplankton populations [5]. The overall aim is to show how apparent noise associated with deterministic chaos can be distinguished from sampling error and other sources of externally induced environmental noise.

The technique we shall use is to make short-term predictions which are based upon a library of past patterns in a time series (an idea also proposed for weather forecasting by Lorenz [6]). By comparing the predicted and actual trajectories, we can make a tentative distinction between dynamically generated chaos and uncorrelated noise. Although, for a chaotic time series, sensitivity to initial conditions makes long-time forecasting impossible, we can make reasonably reliable short-term predictions (within the Lyapunov time horizon). That is, for a chaotic time series, the accuracy of the nonlinear forecast falls off as the prediction time interval increases, at a rate which gives a rough estimate of the Lyapunov exponent. In contrast, forecasting with white (uncorrelated) noise is independent of the prediction interval.

7.2 Simplex projection

The first step in the simplex projection [7] method is to choose an embedding dimension e of an embedding space E. The data points sitting in this space represent a fixed-length section of the time series. For the reconstructed trajectory $\boldsymbol{x}_t$, we shall choose lagged coordinates such that

$$\boldsymbol{x}_t = (x_t, x_{t-\tau}, \ldots, x_{t-(e-1)\tau}). \tag{7.2}$$

Typically, we shall choose $\tau = 1$. More generally, we explore various values of τ and choose the one which leads to the best short-term predictability (as measured by some objective criterion). For our original time series $\{x_i\}$, each sequence for which we wish to make a prediction—each *predictee*—is now an e-dimensional point, consisting of the present value and the previous values, each separated by one lag time τ. We now locate all nearby e-dimensional points in the reconstructed phase space and choose a minimal neighbourhood, defined to be such that the predictee is contained within the smallest simplex formed from its $e + 1$ nearest neighbours. A simplex, as shown in Fig. 7.1, with $e + 1$ vertices is the smallest object which can contain an e-dimensional point as an interior point. The only complication occurs when there are points of the embedding on the boundary of the attracting set, in which case a lower-dimensional simplex of nearest neighbours is used.

The prediction is obtained by projecting the domain of the simplex into its range, that is, by keeping track of where the points in the simplex end up after p time steps. To obtain the predicted value, we compute where the original predictee has moved within the range of this simplex, giving exponential weight to its original distances from the relevant neighbours.

The merit of the method relies on the fact that it is nonparametrical, requiring no prior information about the model used to generate the time series, only the information in the output itself.

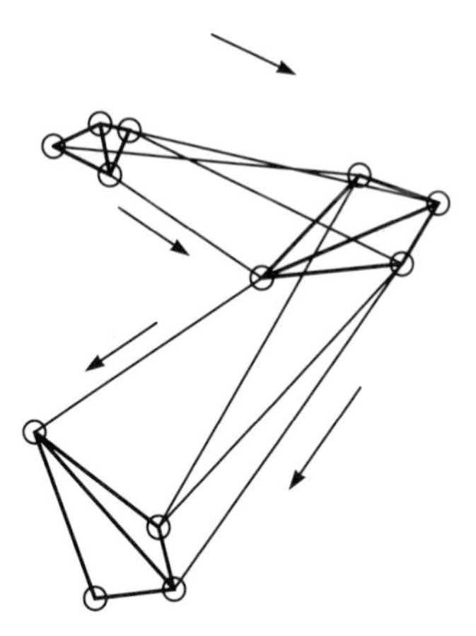

Fig. 7.1. A schematic representation of a simplex projection. A simplex (in a neighbourhood of a point in reconstructed phase space) is mapped, two time steps forward. The predicted point is estimated as a weighted average of the verticies of the mapped simplex.

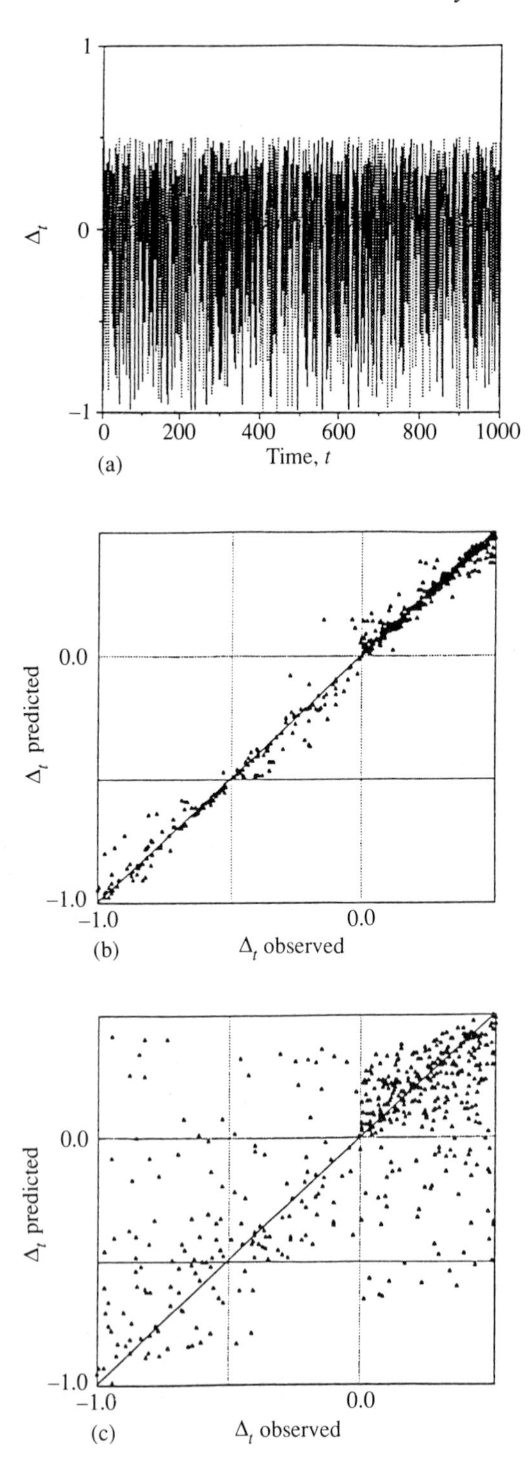
1
0
−1
Δ_t
0
200
400
600
800
1000
Time, t
(a)
0.0
−1.0
Δ_t predicted
−1.0
0.0
Δ_t observed
(b)
0.0
−1.0
Δ_t predicted
−1.0
0.0
Δ_t observed
(c)

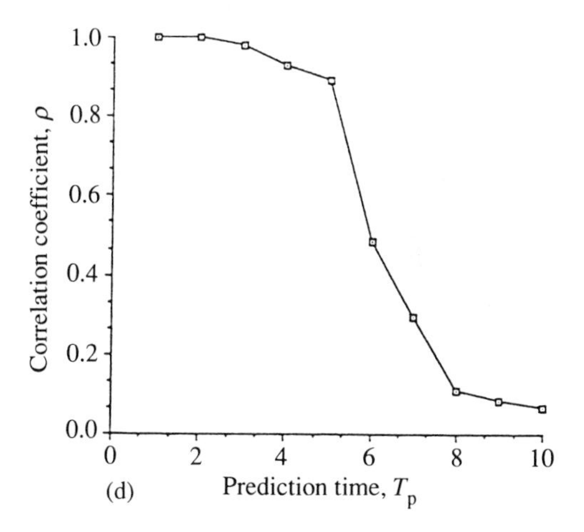

Fig. 7.2. (a) Time series of 1000 points generated by taking first-differences, $\Delta_t = x_{t+1} - x_t$, of the tent map: $x_{t+1} = 2x_t$ for $0.5 > x_t > 0$; $x_{t+1} = 2 - 2x_t$ for $1 > x_t > 0.5$. (b) Predicted values two steps into the future (T_p) versus observed values for the tent map Δ_t time series depicted in (a). Specifically, the first 500 points in the series were used to generate a library of patterns, which were then used as a basis for making predictions for each of the second 500 points. As described in the text, the predictions were made using a simplex projection method, and in this figure the embedding dimension and lag time are $e = 3$ and $\tau = 1$, respectively. Here the coefficient of correlation between predicted and actual values is $\rho = 0.997$ ($N = 500$). For comparison, we note that the corresponding correlation coefficient obtained using the first half of the series to predict the second half with an autoregressive linear model (where the predictions are based on the weighted average of three linear maps, one for each of the three different τ values that give the best results in such a linear scheme) is $\rho = 0.04$. (c) Exactly as (b), except here the predictions are five time steps into the future ($T_p = 5$). The correlation coefficient between predicted and actual values is now $\rho = 0.89$ ($N = 500$). (d) Summary of the trend between (b) and (c), by showing ρ between predicted and observed values in the second half (second 500 points) of the time series of (a) as a function of T_p. As in (b) and (c), the simplex projection method here uses $e = 3$ and $\tau = 1$. The prediction accuracy (as measured by the linear correlation coefficient between predicted and observed values) falls as the prediction extends further into the future: this is a characteristic of a chaotic attractor.

The method may be viewed as a simpler variant of techniques used by Farmer and Sidorowich [8] and by Casdagli [9]. It should be applicable to any time series showing stationary, periodic, quasi-periodic, or chaotic behaviour. The loss of predictive power can be measured by any one of a variety of statistics. We chose to use the conventional and familiar linear correlation coefficient ρ [10]. For pairs of quantities (x_i, y_i), with $i = 1, \ldots, N$,

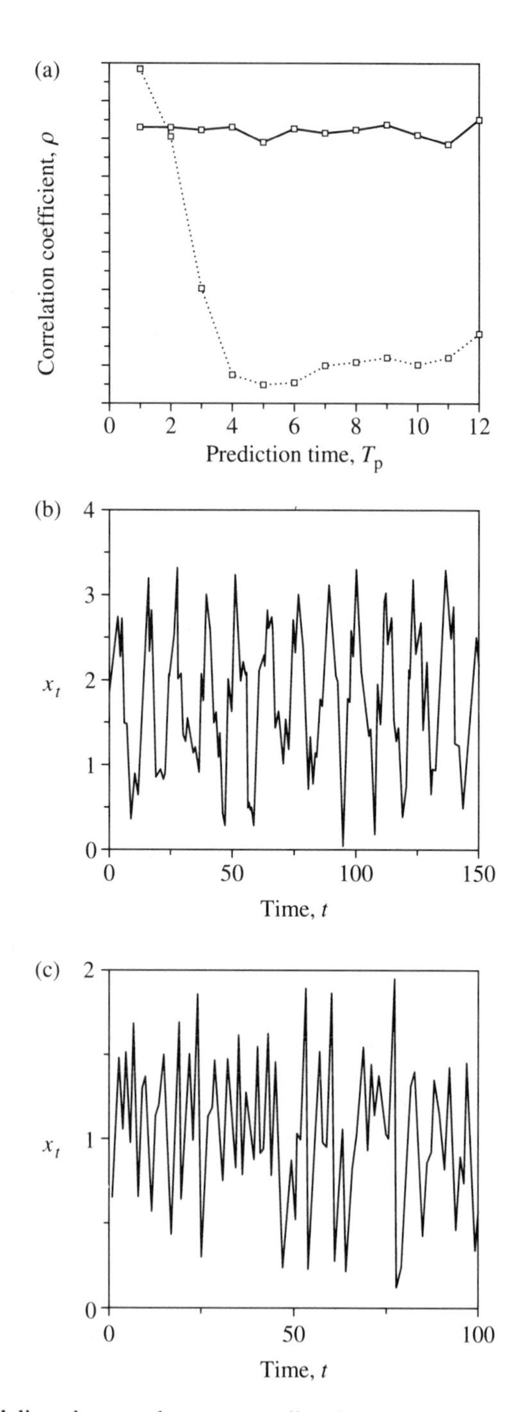

Fig. 7.3. (a) Solid line shows ρ between predicted and observed values for the second half of the time series defined in (b) (which is, in fact, a sine wave with additive noise) as a function of T_p. As discussed in the text, the accuracy of the prediction, as measured by ρ, shows no systematic dependence on T_p. By

$$\rho = \frac{\sum_i (x_i - \bar{x})(y_i - \bar{y}_i)}{\sqrt{\sum_i (x_i - \bar{x})^2}\sqrt{\sum_i (y_i - \bar{y})^2}}, \qquad (7.3)$$

where x_i is the observed variable, with mean $\bar{x}$, and y_i is the predicted variable, with mean $\bar{y}$.

A decrease in the linear correlation coefficient with increasing prediction time interval T_p may be interpreted a characteristic feature of chaos. Moreover, the rate of decay of ρ versus T_p gives a rough measure of the value of the positive Lyapunov exponent. Predictions with additive white noise have the same correlation with observation, regardless of how far, or close, into the future one attempts to predict. One should contrast this prediction with deterministic chaos which will deteriorate as one tries to forecast further into the future.

Farmer and Sidorowich [8] have derived asymptotic results, for long time series, which describe how this error propagates over time in a simple chaotic system. Other statistical measures of the relation between predicted and observed values could be adopted; the results obtained appear to be essentially identical [7].

7.2.1 *Forecasting*

The signature of ρ decreasing with the prediction time T_p does not arise when the erratic time series is noisy periodic motion. Here (shown in Fig. 7.3(a)) we have uncorrelated additive noise superimposed on a sine wave, like a noisy limit cycle. Such additive noise is reckoned to be characteristic of sampling variation. Here the error remains constant as the simplex is projected further into the future; past sequences of roulette-wheel numbers that are similar to the present ones can tell us as much or little about the next spin as the next hundredth spin. By contrast, the dashed line in Fig. 7.3(a) represents ρ as a function of T_p for a chaotic sequence generated as the sum of two independent runs of a tent map, that is, for the time series illustrated in Fig. 7.3(c). Although, at first sight, the two time series in Fig. 7.3(b,c) both look like the sample function of some random

---

constrast, the time series shown in (c) (which is the sum of two separate tent map series) does show the decrease in ρ with increasing T_p, as illustrated by the dashed line, that is characteristic of a chaotic sequence. Both curves are based on the simplex methods described in the text, with $e = 3$ and $\tau = 1$. (b) First 150 points in the time series generated by taking discrete points on a sine wave with unit amplitude ($x_t = \sin 0.5t$), and adding a random variable chosen (independently at each step) uniformly from the interval $(-0.5, 0.5]$. That is, the series is generated as a 'sine wave + 50% noise'. (c) Time series illustrated here is generated by adding together two independent tent map sequences.

process, the characteristic signatures in Fig. 7.3(a) distinguish the deterministic chaos in Fig. 7.3(c) from the additive noise in Fig. 7.3(b).

7.2.2 *Embedding dimension*

The predictions in Fig. 7.2 and 7.3 are based on an embedding dimension of $e = 3$. The results are, however, sensitive to the choice of e. Figure 7.4(a) compares the predicted and the actual results for the tent map two time steps ahead (T_p), as in Fig. 7.2(b), except that now $e = 10$. Clearly, the predictions are less accurate with this higher embedding dimension. More generally, Fig. 7.4(b) shows ρ between predicted and actual results one time step into the future (T_p) as a function of e, for two different choices of time lag ($\tau = 1$ and $\tau = 2$). It may seem surprising that having potentially more information—more data—summarized in each e-dimensional point, and a higher-dimensional simplex of neighbours of the predictee, reduces the accuracy of the predictions; in this respect, these results differ from results given by Farmer and Sidorowich for parametric forecasting involving linear interpolation for constructing local polynomial maps [8]. This effect is caused by contamination of nearby points in the high-dimensional embedding with points whose earlier coordinates are close, but whose recent coordinates are distant. It is also possible that this empirical approach of choosing the value of e to be that which optimizes the short-term predictability may be used as a method for assessing an upper bound on the embedding dimension, and thence on the dimensionality of the attractor.

7.2.3 *Other systems*

Let us consider the time series obtained by taking points at discrete time intervals from a continuous chaotic systems such as those of the Lorenz or the Rössler model. In each case, the results for ρ are similar to those shown in Fig. 7.2(d). In more complicated cases, such as the superposition of different chaotic maps, we observe a decline in ρ versus T_p. In such cases, the signature can show a step pattern, with each step corresponding to the dominant Lyapunov exponent for each map.

What is more problematic is the comparison with ρ–T_p relationships generated by coloured noise spectra, in which there are significant short-term autocorrelations, although not long-term ones. Such autocorrelated noise can clearly lead to correlations ρ between predicted and observed values that decrease as T_p lengthens. Indeed, it seems likely that a specific pattern of autocorrelations could be hand-tailored to mimic any given relationship between ρ and T_p obtained from a finite time series. It may be that the characteristic pattern of autocorrelation in this case will in general give a flatter ρ–e relationship than those of simple chaotic time series; that is, the ρ–T_p relationships generated by autocorrelated noise may characteristically scale differently from those generated by deterministic

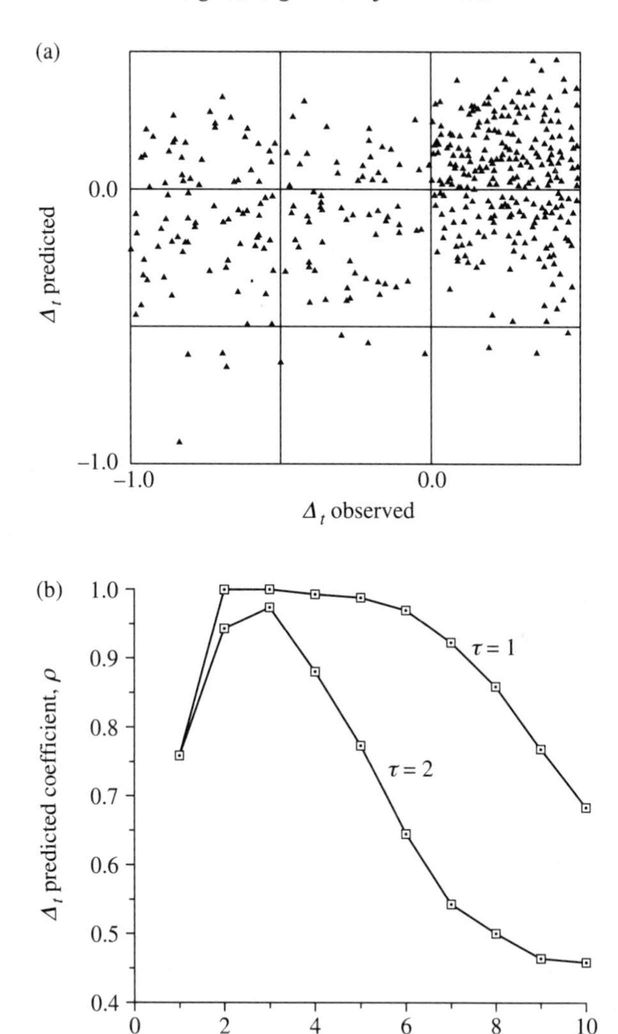

Fig. 7.4. (a) Similar to Fig. 7.2(b), this figure shows predictions one time step into the future ($T_p = 1$) versus observed values, for the second 500 points in the tent map time series of Fig. 7.2(a), with the difference that here we used an embedding dimension $e = 10$ (in constrast to $e = 3$ in Fig. 7.2(b); the lag time remains unchanged at $\tau = 1$). As discussed in the text, the accuracy of the prediction deteriorates as e gets too large ($\rho = 0.25$, $N = 500$). (b) Correlation coefficient between predicted and observed results, ρ, is shown as a function of e for predictions one time step into the future ($T_p = 1$). The relationship is shown for $\tau = 1$ and $\tau = 2$. The figure indicates how such empirical studies of the relationship between ρ and e may be used to assess the optimal e.

chaos [8]. But it seems that very long runs of data will be required to sort out such scaling laws, if indeed they exist.

Can we have confidence in applying the method to real time series arising in contexts in population biology or elsewhere? One suggestion is that the time series may tentatively be regarded as deterministically chaotic if, in addition to a decaying ρ–T_p signature, the correlation ρ between predicted and observed values obtained by the method we have described is significantly better than the corresponding correlation coefficient obtained by the best-fitting autoregressive linear predictor [9].

Other time series methods applied to experimental data begin with an estimate of the dimension of the underlying attractor. The procedure is to construct a phase space embedding for the time series and calculate the dimension of the putative attractor using a variant of the Grassberger–Procaccia algorithm [13]. In the Grassberger–Procaccia method, a correlation integral is calculated that is essentially the number of points in E-space separated by a distance less than l, and the power law behaviour l^{ν} of this correlation integral is then used to estimate the dimension D of the attractor ($D > \nu$). This dimension is presumed to give a measure of the effective number of degrees of freedom or 'active modes' of the system. An upper bound on a minimal embedding dimension is $E_{min} < 2D + 1$, where D is the dimension of the attractor [11,3]. The drawback of the Grassberger–Procaccia algorithm is that the calculations of the dimension of the attractor mainly involve only a small subset of pairs of points in the phase space. In other words, the method discards much of the information in a time series, which, because many natural time series are of limited size, can be a serious problem.

The advantage of using prediction methods is that standard statistical criteria can be used to evaluate the significance of the correlation between predicted and observed values. Prediction methods should provide a stringent test of the underlying determinism in situations of given complexity. Prediction is, after all, the *sine qua non* of determinism.

7.3 Time series from the natural world

7.3.1 *Measles in New York*

For reported cases of measles in New York City, there is a monthly time series extending from 1928 [4]. After 1963, immunization began to alter the intrinsic dynamics of this system (an extreme example of environmental changes), and so we use only the data from 1928 to 1963, giving 432 data points. It would be interesting if we were able to determine whether the dynamics reflect a noisy limit cycle [2] or low-dimensional chaos superimposed on a seasonal cycle [13–16]. This system can be tested for low-dimensional chaos using a number of methods, including the Grassberger–Procaccia algorithm [16,17,3,8], estimation of Lyapunov exponents [16], reconstruction of Poincaré return maps [13], [15], and model simulations [15], [16]. The tests prove to be inconclusive, but suggest that the measles data are indeed best described by a low-dimensional chaotic attractor.

Figure 7.5(a) shows the time series obtained by taking first differences, $x_{t+1} - x_t$, of the measles data. The first difference was taken to reduce autocorrelations and to reduce any signal associated with simple limit cycles. We then generated our predictions by using the first half of the series (216 points) to construct an ensemble of points in an e-dimensional phase space, that is, to construct a library of past patterns. The resulting information was then used to predict the remaining 216 values in the series, along the lines we have described, for each chosen value of the embedding dimension e. Fig. 7.5(b), for example, compares predicted and observed results, one time step into the future ($T_p = 1$ month), with $e = 6$. Figure 7.5(c) shows the linear correlation coefficient ρ between predicted and observed results as a function of e for $T_p = 1$. If we assume the optimal embedding dimension to be that yielding the highest correlation coefficient between prediction and observation in one time step, it is seen from Fig. 7.5(c) that $e \simeq 5$–7. This accords with previous estimates made using various other methods, and is consistent with an attractor of dimension $D = 2$–3 [13–16].

The points joined by the solid line in Fig. 7.5(d) show ρ as a function of T_p, given $e = 6$. Prediction error seems to propagate in a manner consistent with chaotic dynamics. This result, in combination with the significantly better performance of our nonlinear predictor as compared with an optimal linear autoregressive model, which consists of several optimal linear maps, agrees with the conclusion that the noisy dynamics shown in Fig. 7.5(a) are, in fact, deterministic chaos [13–16].

For data from the natural world, as distinct from artificial models, physical or biological parameters, or both, can undergo systematic changes over time. In this event, libraries of past patterns can be of dubious relevance to an altered present and even more different future. In a different context, secular trends in environmental variables can complicate an analysis of patterns of fluctuations in the abundance of bird species [18, 19]. We can gauge the extent to which secular trends might confound our forecasting methods in the following way. Rather than using the first half of the time series to compile the library of past patterns, and the second half to compute correlations between prediction and observations, we instead investigate the case in which the library and forecasts span the same period. Therefore we focus our predictions also on the first half of the series, from which the library was previously drawn. To avoid redundancy between our forecasts and the model, however, we sequentially exclude points from the library that are in the neighbourhood of each predictee, in particular, the e_τ points preceding and following each forecast. The points connected by the dashed lines in Fig. 7.5(d) show the linear correlation

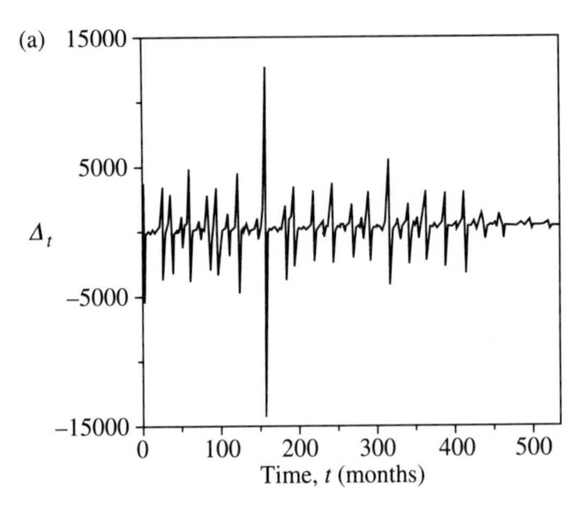
(a)
15000
5000
Δ_t
−5000
−15000
0
100
200
300
400
500
Time, t (months)

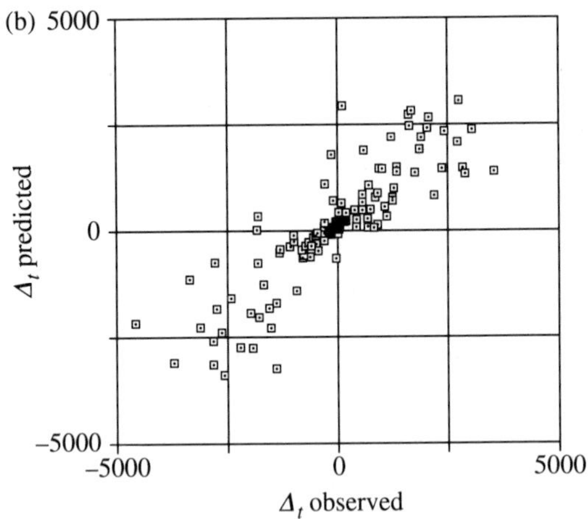
(b)
5000
0
−5000
Δ_t predicted
−5000
0
5000
Δ_t observed

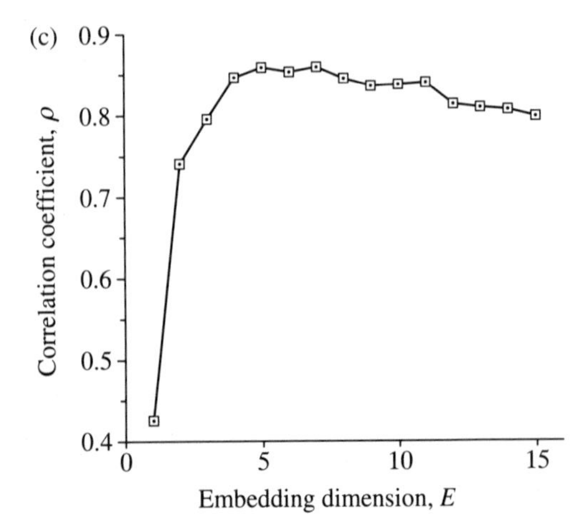
(c)
0.9
0.8
0.7
0.6
0.5
0.4
Correlation coefficient, ρ
0
5
10
15
Embedding dimension, E

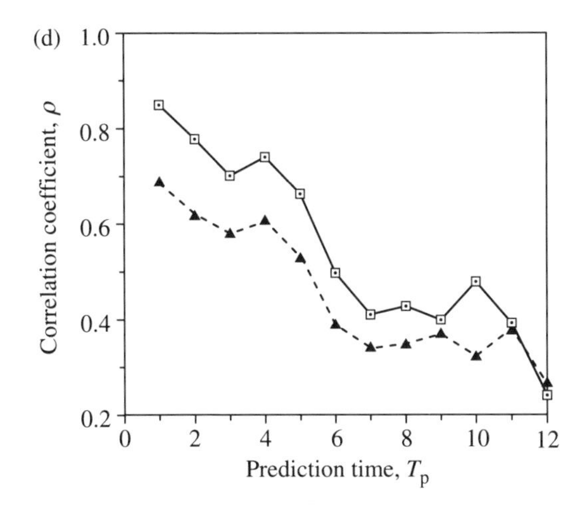

Fig. 7.5. (a) Time series generated by taking the first differences, $x_{t+1} - x_t$, of the monthly number of cases of measles reported in New York City between 1928 and 1972 (the first 432 points in the sequence shown here). After 1963, the introduction of immunization against measles had a qualitative effect on the dynamics of infection; this can be seen in the later part of the sequence illustrated here. (b) Using the methods described earlier, the first part of the series in Fig. 7.4(a) (216 points, 1928–1946) was used to construct a library of past patterns, which were then used for a basis for predicting forward from each point in the second part of the series, from 1946 to 1963. Predicted and observed values are shown here for predictions one time step into the future ($T_p = 1$), using $e = 6$ and $\tau = 1$. The correlation coefficient between predicted and observed values is $\rho = 0.85$ ($P < 10^{-5}$ for $N = 216$). For comparison, the corresponding prediction based on an autoregressive model (composed of five optimal maps; compare Fig. 7.1(b)) gives $\rho = 0.72$ (which is significantly different from $\rho = 0.85$ at the $P < 0.0005$ level). (c) As in Fig. 7.4(b), ρ between predicted and observed results, is shown as a function of e (for $T_p = 1$ and $\tau = 1$). This figure indicates an optimal embedding dimension of $e \simeq 5$–7, corresponding to a chaotic attractor with dimension 2–3. (d) Here, ρ between predicted and observed results for measles is shown as a function of T_p (for $e = 6$ and $\tau = 1$). For points connected by solid lines, the predictions are for the second half of the time series (based on a library of patterns compiled from the first half). For the points connected by the dashed lines, the forecasts and the library of patterns span the same time period (the first half of the data). The similarity between solid and dashed curves indicates that secural trends in underlying parameters do not introduce significant complications here. The overall decline in prediction accuracy with increasing time into the future is, as discussed in the text, a signature of chaotic dynamics as distinct from uncorrelated noise.

coefficient ρ versus T_p relationship which results from treating the measles data in this way, with the embedding dimension e again choosen to be 6. The reasonable agreement between these results, for which the library of patterns and the forecast span the same time period, and those of the

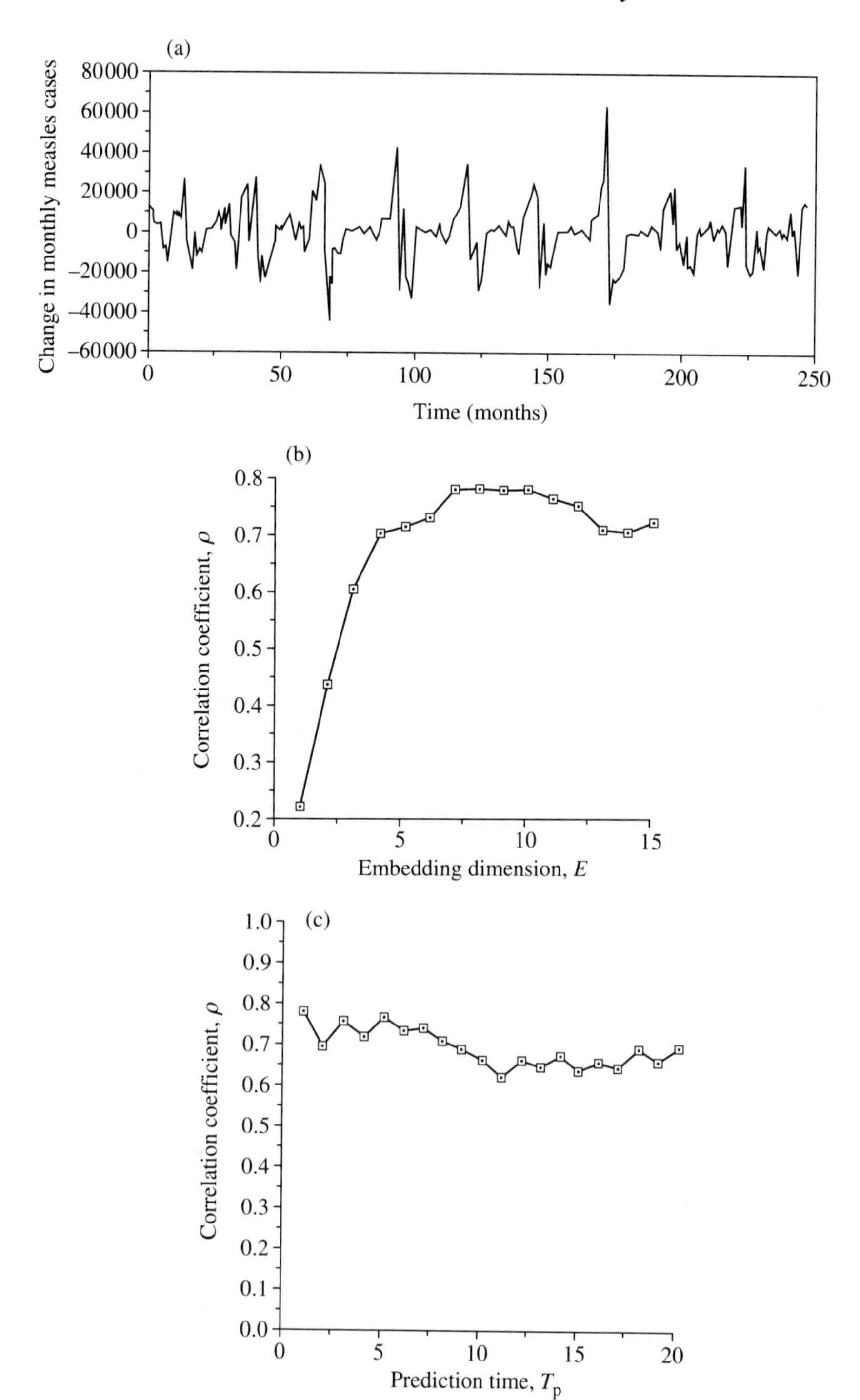

(a)
Change in monthly measles cases
80000
60000
40000
20000
0
–20000
–40000
–60000
0
50
100
150
200
250
Time (months)
(b)
Correlation coefficient, ρ
0.8
0.7
0.6
0.5
0.4
0.3
0.2
0
5
10
15
Embedding dimension, E
(c)
Correlation coefficient, ρ
1.0
0.9
0.8
0.7
0.6
0.5
0.4
0.3
0.2
0.1
0.0
0
5
10
15
20
Prediction time, T_p

simpler previous analysis (the solid line in Fig. 7.5(d)) indicates that (within the 1928–1963 time frames), secular trends in underlying parameters are qualitatively important.

7.3.2 *Measles in England and Wales*

We can compare the previous analysis of data of measles in New York with the monthly incidence in England and Wales (Fig. 7.6(c)) [20]. As before, we begin by transforming the data to first differences, partly to remove linear trends and partly to increase the density of points in phase space. The number of data points available is small ($N = 240$), roughly half the size of the New York time series, and consequently we allow the library and predictions to span the full time period. However, to avoid circularity between forecasts and the model, we sequentially exclude points from the library that are in a neighbourhood of each predictee (specifically the $(e - 1)\tau$ points preceding and following each forecast).

As shown in Fig. 7.6(b), we obtain optimal embeddings at $e = 7$–10, which are similar to the range of values found for measles in New York ($e = 5$–8). However, unlike New York, it appears that the dynamics here are not produced by low-dimensional chaos. The ρ–T_p curve (Fig. 7.6(c)) does not decay exponentially as it does for the measles incidence data for New York, but rather has a flat appearance more reminiscent of additive noise.

Thus we are faced with an apparent contradiction: why should measles in New York be chaotic while the same disease in the UK is a simple biennial cycle with additive noise? Could the contradiction be explained by differences in the population or spatial scales involved? For example, could the individual cities in the UK be chaotic but produce an emergent behaviour that appears linear when aggregated?

7.4 Spatial dynamics of measles

The sensitivity of ecological models to aggregation and scaling has been discussed in a number of different contexts [21–28].

Fig. 7.6. (a) Time series of first differences in the number of cases of measles reported in England and Wales, 1948–1966 [33]. (b) As in Figs. 7.4(b) and 7.5(c), the correlation coefficient or predictability ρ is shown as a function of embedding dimension e ($T_p = 1$ and $\tau = 2$). The figure suggests an optimal embedding dimension broadly in the range $e \simeq 7$–10. (c) Predictability ρ as a function of prediction interval of T_p for measles in England and Wales (using the optimal parameters $e = 8$ and $\tau = 2$). The library of patterns from the first half of the data was used to predict the second half. The relatively flat pattern shown here is similar to the 'addictive noise' case seen in Fig. 7.3(a). This figure suggests that the large-scale aggregate behaviour of measles across England and Wales is not chaotic (in puzzling contrast to New York City, Fig. 7.5).

We approximate measles dynamics within a single city as a chaotic logistic map ($x_{t+i} = ax_t(1-x_t)$), superimposed on a sine wave. We then investigate what happens as more of these 'sine + logistic' series are summed. In summing the series, we require the linear part (the sine waves) to be synchronized as the seasonal patterns in cities would be, but we allow the nonlinear parts to be independent to approximate spatial decoupling. In effect, this is equivalent to averaging the output from independent logistic maps, and superimposing this net output on a sine wave. To model dynamical variability in each city, the parameter a is chosen for each map uniformly in the interval (2.67, 3.67).

Figure 7.7(a) shows how the ρ–T_p signature varies with increasing aggregation. As more independent logistic maps (cities) are folded into the picture, the ρ–T_p signature becomes ever more shallow, giving much the same appearance as the linear noise case. This is corroborated by Fig. 7.7(b), where the linear predictor tends to match the nonlinear predictor more closely as more logistic series are summed. These trends are understandable in light of the following two facts. First, as more maps are superimposed, the signal becomes more complicated. High-dimensional dynamics, chaotic or otherwise, may be thought of as so complicated as to constitute 'noise'. Second, as more such series are superimposed, the amplitude of the signal should decrease as the square root of the number of independent chaotic maps; this exposes more clearly the linear parts (seasonal sine wave) of the time series which are synchronized. Thus, as more independent chaotic series are aggregated, the nonlinear part should begin to resemble noise superimposed on a sine wave.

To test the applicability of these ideas to the observed patterns in England and Wales [24], we have disaggregated the data, focusing on individual cities. The central question here is whether evidence for chaotic behaviour (not apparent in the countrywide analysis) emerges on a single-city scale. We shall focus here on a representative sample of seven large cities: Table 7.1 lists the cities, along with their population sizes and a distance matrix, while Fig. 7.8(a) shows the associated measles time series for the period 1948–1967.

Figure 7.8(b) shows the embedding analyses for each of the seven cities (again, using the full data set ($N = 240$) to compute these correlations). All of the five most populous cities, London, Birmingham, Liverpool, Manchester, and Sheffield, had optimal embeddings in a range similar to that observed for New York ($e = 5$–8; Manchester, however, also had a peak at $e = 4$), and each had a local maximum at $e = 7$. Indeed these results seem to match the embedding results for New York better than those for the pooled data for Britain. On the other hand, the two least populous and most isolated cities appeared to require higher-dimensional embeddings: for Bristol $e = 10$, and for Newcastle $e = 12$. Although one must be cautious not to overinterpret the specific figures obtained here, especially in light of

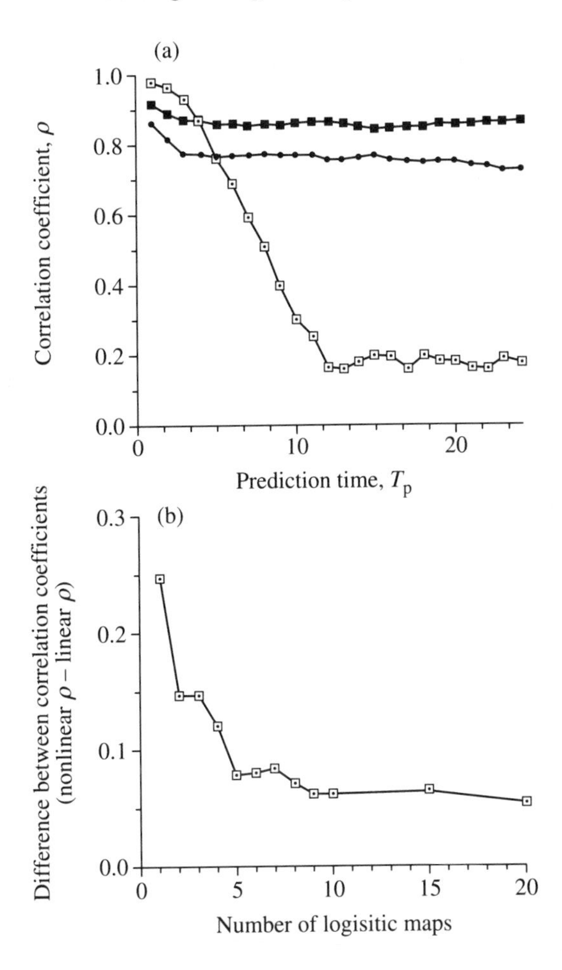

Fig. 7.7. (a) Predictability ρ as a function of prediction time T_p at different levels of aggregation. Here, we approximate measles dynamics in a single city as a chaotic logistic map superimposed on a sine curve, and investigate behaviour as more of these (sine + logistic) series are lumped (summed). The solid line with the open boxes is for one (sine + logistic) ($\tau = 1$, $e = 3$); the dashed line with closed diamonds is for 10 (sine + logistic) series summed ($\tau = 1, e = 7$); and the solid line with boxes is for 20 series summed ($\tau = 1, e = 9$). The effect of such aggregation on the dynamics is to diminish the nonlinear chaotic portion of the signal, so that the ρ–T_p signature looks increasingly like the additive noise case. (b) The difference in predictability ρ between optimal linear autoregressive methods verses our nonlinear methods is shown, as a function of the number of (sine + logistic) maps that are lumped together. The maps are as described in Fig. 7.7(a), and here $T_p = 1$ (and $e = 3, \tau = 1$). Note that the difference in ρ decreases with increasing aggregation.

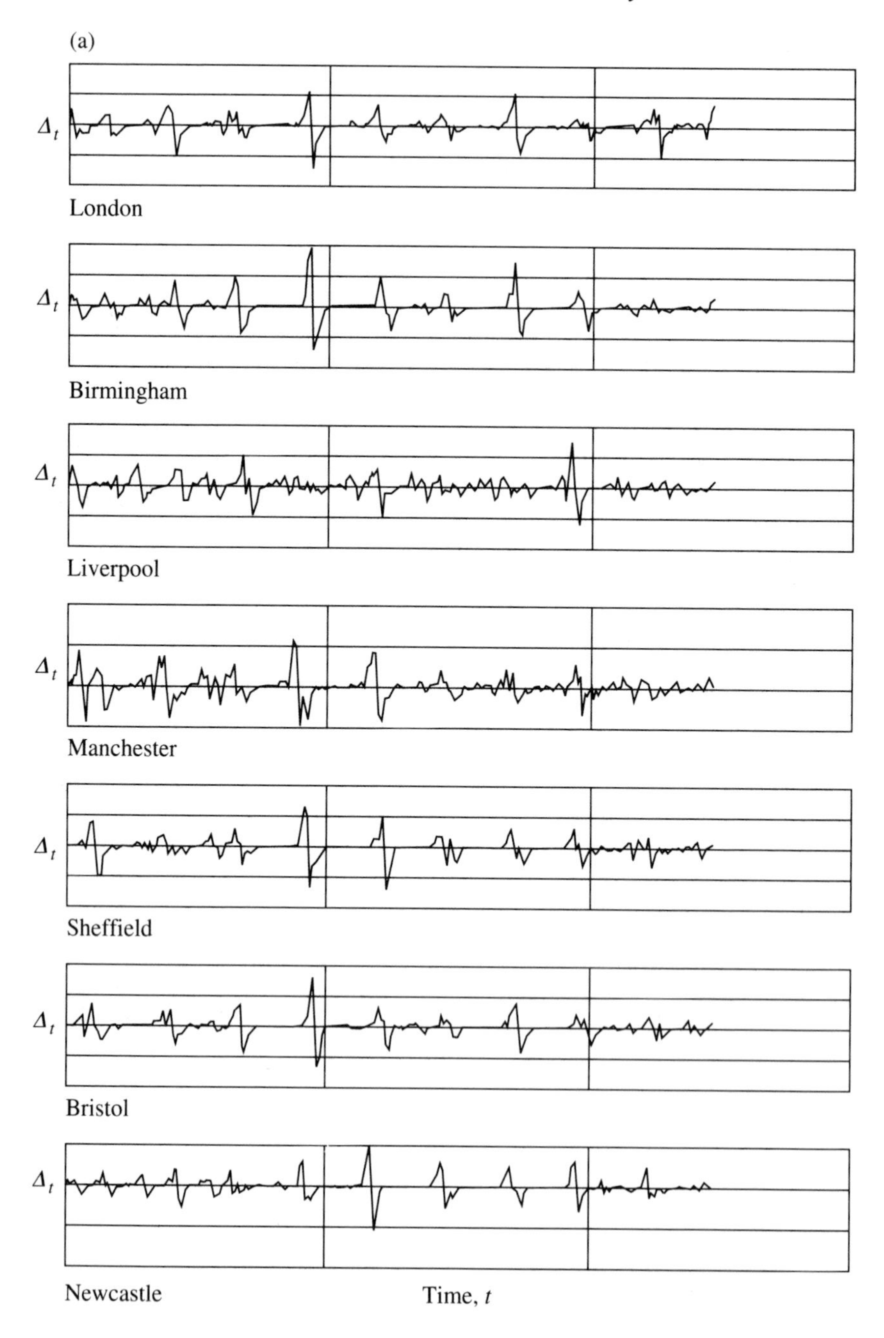
(a)
Δ_t
London
Birmingham
Liverpool
Manchester
Sheffield
Bristol
Newcastle
Time, t

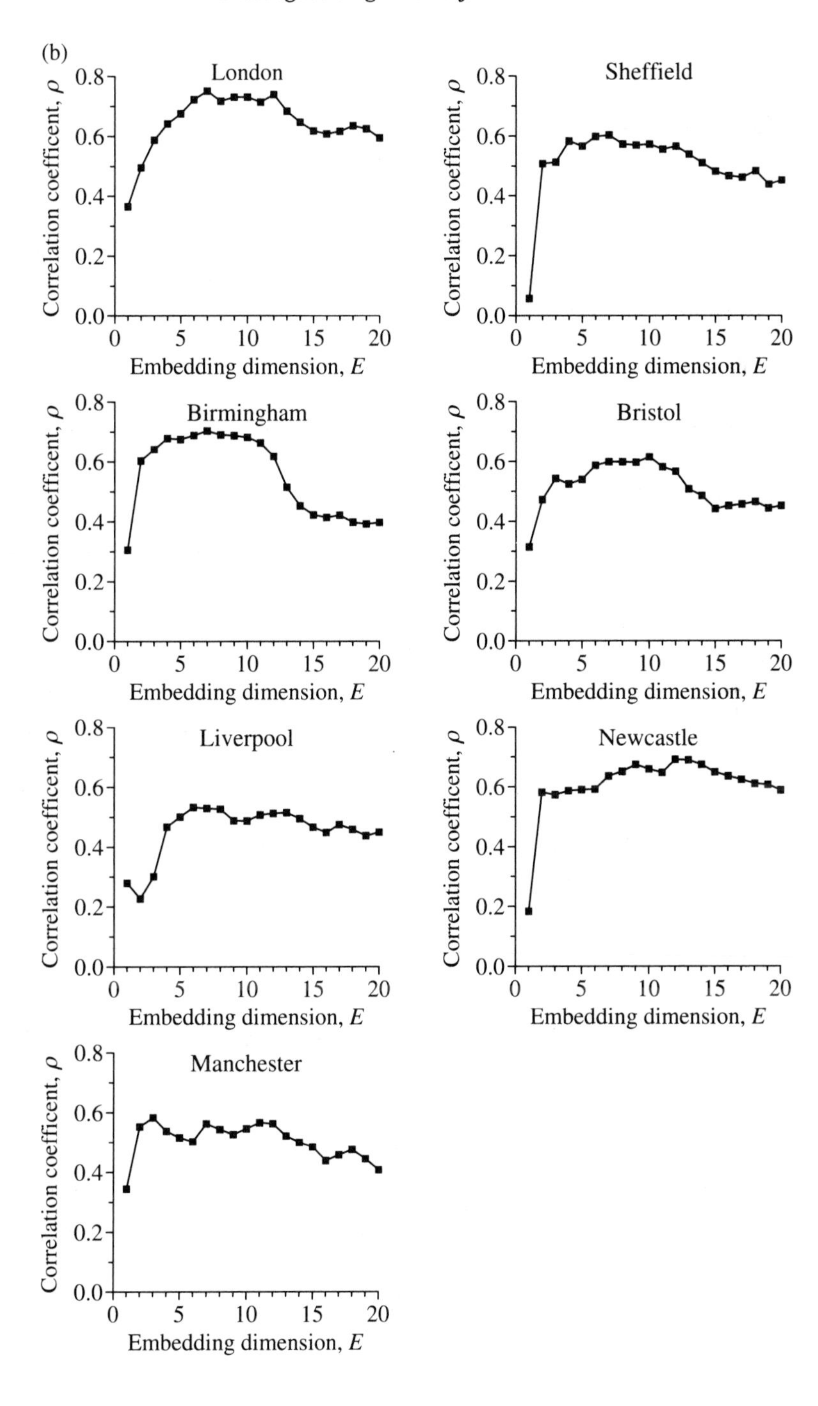
(b)
London
Sheffield
Birmingham
Bristol
Liverpool
Newcastle
Manchester
Correlation coefficent, ρ
Embedding dimension, E
0.0
0.2
0.4
0.6
0.8
0
5
10
15
20

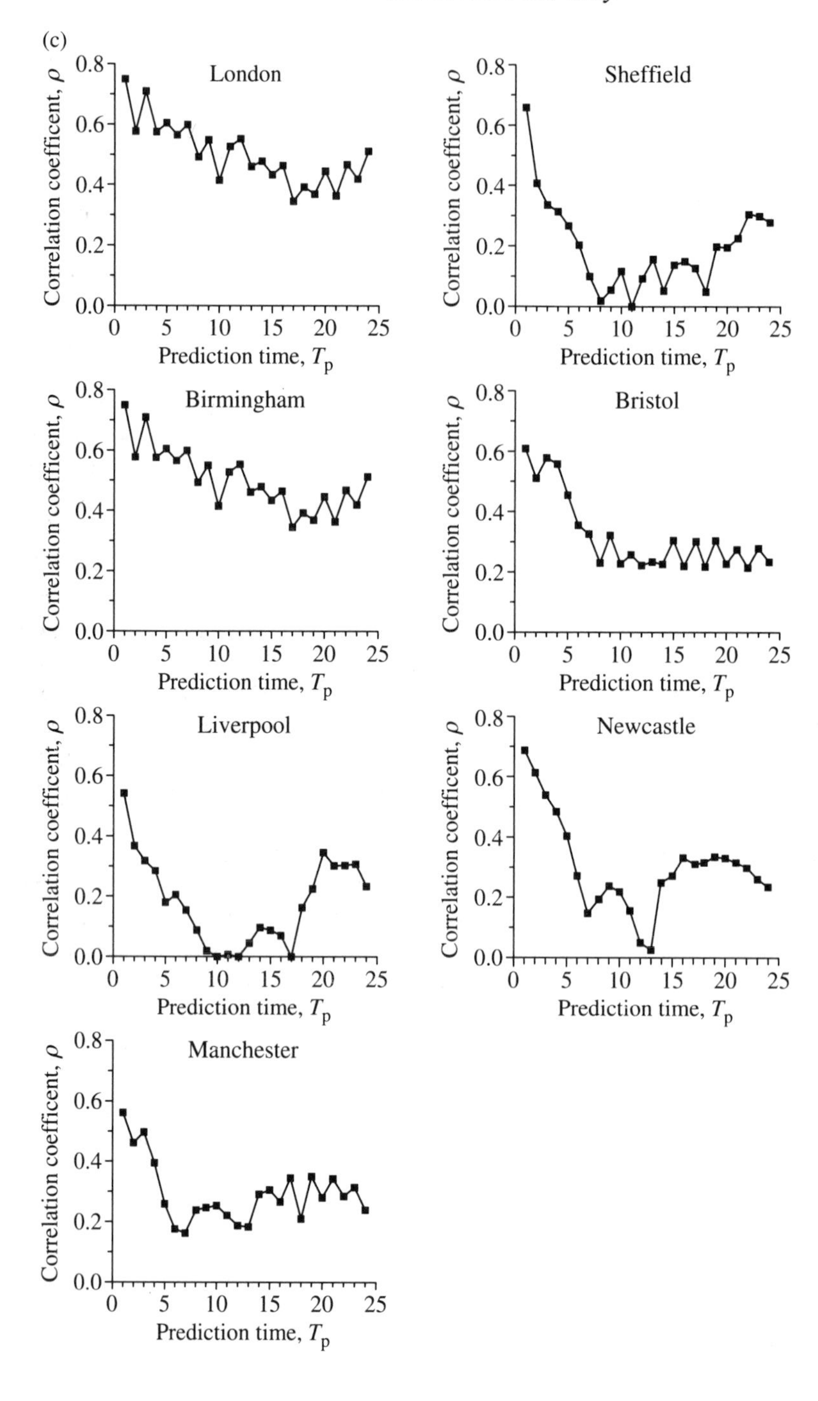
(c)
London
Sheffield
Birmingham
Bristol
Liverpool
Newcastle
Manchester
Correlation coefficent, ρ
Prediction time, T_p
0.0
0.2
0.4
0.6
0.8
0
5
10
15
20
25

Table 7.1. Demographic summary for the seven English cities used in the spatial analysis of measles.

Distance (road miles)	London	Birmingham	Liverpool	Manchester	Sheffield	Bristol	Newcastle
London	–	113	205	189	159	115	281
Birmingham	–	–	94	81	76	88	205
Liverpool	–	–	–	35	74	164	156
Manchester	–	–	–	–	38	164	132
Sheffield	–	–	–	–	–	164	128
Bristol	–	–	–	–	–		293
Population (thousands)	8282	1096	792	693	514	435	294

the low number of data points involved ($N = 240$), it is interesting to note that both Bristol and Newcastle fall well below the population threshold believed necessary for the infection to remain endemic [29].

Figure 7.8(c) shows the ρ–T_p curves for the seven cities. The results here are not flat like the ones in the aggregated data, but rather have a look very similar to the chaotic signature observed in New York, where predictability falls off steeply with increasing prediction time. Thus it appears that the scale considerations may help to resolve the apparent contradiction between the lumped analysis for measles in England and Wales, and the earlier analysis for New York City measles. In this regard, it is interesting that London, the most populous and geographically the largest of the British cities by almost an order of magnitude, appears to show the gradual decline in its ρ–T_p curve (that is, the ρ–T_p curve for London has a character intermediate between those for other UK cities and that for England and Wales as a whole).

Fig. 7.8. (a) Time series of first differences in the reported monthly cases of measles in seven major British cities between 1948 and 1966 (arranged in order of city population size). (b) Predictability ρ as a function of embedding dimension e for measles from seven major British cities. Clearest results were obtained using $\tau = 1$ for Liverpool, Manchester, Sheffield, and Newcastle, and $\tau = 2$ for London, Birmingham, and Bristol. (c) Predictability ρ as a function of prediction interval T_p for measles from seven major British cities. The following parameters were used: London $\tau = 2$, $e = 7$; Birmingham $\tau = 2$, $e = 7$; Liverpool $\tau = 1$, $e = 7$; Manchester $\tau = 1$, $e = 7$; Sheffield $\tau = 1$, $e = 6$; Bristol $\tau = 2$, $e = 10$; Newcastle $\tau = 1$, $e = 12$. With the possible exception of London, all of the above cities show the characteristic decline in predictability with increasing prediction interval associated with chaotic dynamics, as seen in New York City.

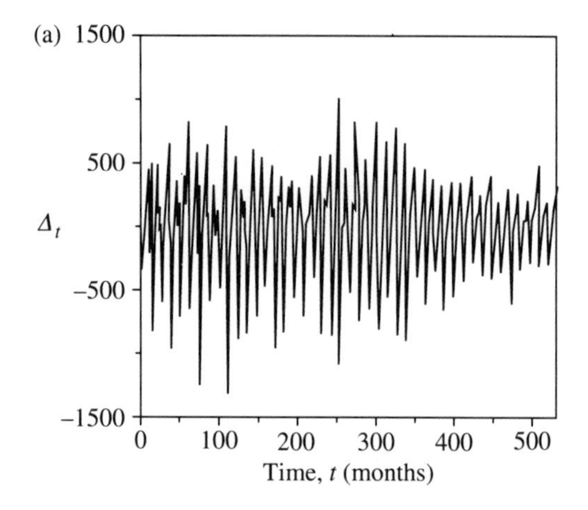
(a)
1500
500
Δ_t
−500
−1500
0
100
200
300
400
500
Time, t (months)

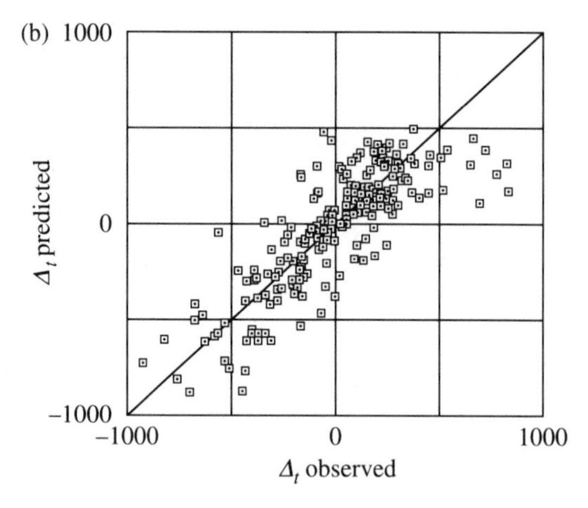
(b)
1000
0
−1000
Δ_t predicted
−1000
0
1000
Δ_t observed

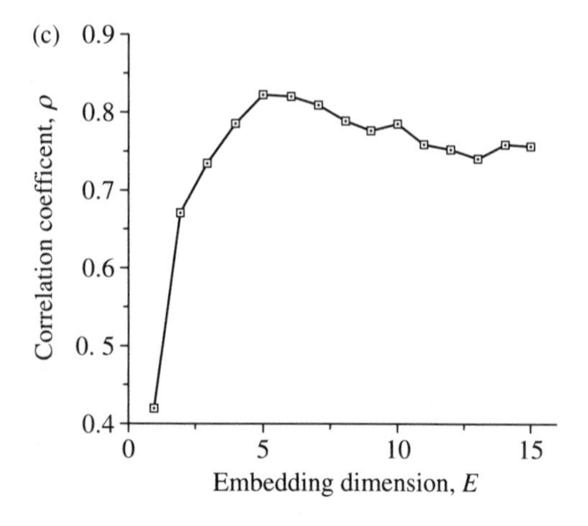
(c)
0.9
0.8
0.7
0.6
0. 5
0.4
Correlation coefficent, ρ
0
5
10
15
Embedding dimension, E

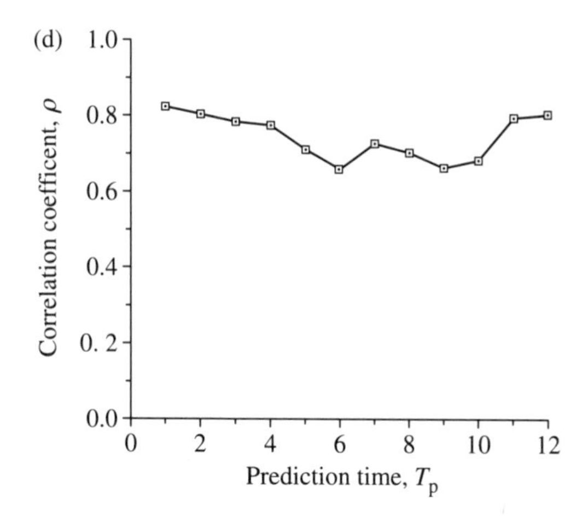

Fig. 7.9. (a) As for Fig. 7.5(a), except the time series comes from taking first differences of monthly numbers of reported cases of chickenpox in New York City from 1928 to 1972. (b) As in Fig. 7.5(b), predicted and observed numbers of cases of chickenpox are compared, the predictions being one time step into the future, $T_p = 1$ (here, $e = 5$ and $\tau = 1$). The correlation coefficient between predicted and observed values is $\rho = 0.82$; an autoregressive linear model alternatively gives predictions which have $\rho = 0.84$. In contrast to Fig. 7.3.5(b) for measles, here there is no significant difference between our prediction technique and standard linear autoregression methods. (c) As in Figs. 7.4(b) and 7.5(c), ρ between predicted and observed results is shown as a function of e (for $T_p = 1$ and $\tau = 1$). This figure suggests an optimal embedding dimension of around $e \simeq 5$. (d) Compare with Fig. 7.5(d); ρ between predicted and observed values, as a function of T_p (with $e = 5$ and $\tau = 1$), is shown. Here, the lack of dependence of ρ on T_p, which is in marked contrast with the pattern for measles in Fig. 7.5(d), indicates pure additive noise (superimposed on a basic seasonal cycle).

In conclusion, the quantitative difference between the patterns seen for the incidence of measles aggregated over England and Wales, versus those seen in individual cities, provides a striking illustration of how the scale on which we collect and analyse data can affect our interpretation. Sufficiently aggregated, the England and Wales data for measles suggest a dynamical pattern of approximately a two-year cycle with additive noise. Disaggregating city-by-city, a more detailed pattern of chaotic dynamics emerges.

7.4.1 *Chickenpox*

If we repeat the process just described for measles, but now for monthly records of cases of chickenpox in New York City from 1949 to 1972 [21], the results can be seen in Fig. 7.9. Figure 7.9(a) shows the time series of differences, $x_{t+1} - x_t$. The 532 points in Fig. 7.9(a) are divided into two halves, with the first half used to construct the library on which the predictions are made for the second 266 points. The predictions are compared with the actual data points, as shown for predictions one time step ahead in Fig. 7.9(b), with the prediction time $T_p = 1$ month and $e = 5$. Figure 7.9(c) shows that the optimum value of the embedding dimension is about 5–6. By contrast with Fig. 7.6(c) for measles, Fig. 7.9(d) for ρ between predicted and observed results for chickenpox shows no dependence on T_p: one does as well predicting the incidence next year as next month. Moreover, the optimal linear autoregressive model performs as well as our nonlinear predictor. We take this to indicate that chickenpox has a strong annual cycle, as does measles, with the fluctuations being additive noise; this is in contrast to measles, for which fluctuations derive mainly from the dynamics.

The contrast between measles and chickenpox can be explained on biological grounds [30]. Measles has a fairly high basic reproductive rate ($R_0 = 10$–20), and, after a brief interval of infectiousness, those who recover are immune and uninfectious for life; these conditions tend to produce long-lasting 'interepidemic' oscillations, with a period of about two years, even in the simplest models [31]. This, in combination with seasonal patterns, makes it plausible that measles has complex dynamics. Chickenpox is less 'highly reproductive' and may recrudesce as shingles in later life; this makes for an infection less prone to show periodicities other than basic seasonal ones associated with schools opening and closing, and therefore suggests seasonal cycles with additive noise. Whatever the underlying biological explanation, the patterns in Fig. 7.6(c) and 7.9(d) differ in much the same way as those illustrated in Fig. 7.3(a) for artificially generated time series of Fig. 7.3(b, c).

7.5 Marine plankton

Another biological time series is provided by Allen's weekly record of marine planktonic diatoms gathered at Scripps Pier, San Diego, between 1920 and 1939; the number of data points is $N = 830$. With the exception of the work of Tont [5], this collection of information has been little analysed, and not at all in the light of contemporary notions about nonlinear dynamics. The data comprises weekly totals of numbers of individuals of diatom species, tallied in daily seawater samples over ~20 years. As for our analysis of the measles and chickenpox data above, we do not 'smooth' the diatom data in any of the usual ways, although we take first differences for the reasons stated earlier. The resulting time series is shown in Fig. 7.10(a).

The results of using the first half of the diatom series to predict the second half are shown in the usual way in Fig. 7.10(b). Figure 7.10(c) shows the correlation coefficient ρ between predicted and observed results looking one step ahead, $T_p = 1$, as a function of the embedding dimension e. The optimum embedding dimension seems to be about 3. The consistent decay in predictive power as one extrapolates further into the future is consistent with the dynamics of the diatom population being partly governed by a chaotic attractor. We note, however, that deterministic chaos at best accounts for about 50% of the variance, with the rest presumably deriving from additive noise; the relatively low dimension of the attracting set for diatoms compared with measles makes it plausible that the noisier fit of predicted fluctuations in diatoms, versus the predicted monthly fluctuations in measles, reflects a much higher sampling variance for diatom than for reported measles cases.

7.6 Predator–prey systems and infectious diseases

Let us see how the construction of model predator–prey systems can suggest a reason for dynamical chaos occurring in natural populations. The nonlinearities in mathematical models for species interacting as competitors, mutualists, or prey and predator are even more diverse than for single populations (as we have seen in Chapter 6). For instance, the simplest model for systems of three species competing intransitively (such that A beats B, B beats C, and C beats A, as in the game 'scissors, paper, stone') exhibits irregular cycles whose periods lengthen progressively with time [32]. In addition, it is more difficult to unravel the details and to measure parameters for multispecies associations than for single populations (where the task is hard enough). As a consequence, there are relatively few studies interweaving theoretical and experimental work on interacting species in the laboratory, and even fewer in the field.

Interactions between human and other animal hosts and their viral, bacterial, and protozoan pathogens ('microparasites' *sensu* [30]) are, however, sometimes simple enough to permit the nonlinear dynamical behaviour to be compared with public health and other data. Some host–pathogen systems will now be discussed in this light.

We begin with something that is more a metaphor than a realistic study. Consider a plant or animal population with discrete, nonoverlapping generations, whose density is regulated by a lethal pathogen which spreads in epidemic fashion through each generation before the reproductive age is attained. The host population N_{t+1} in generation $t + 1$ will then be related to that N_t in generation t by the equation

$$N_{t+1} = \lambda N_t[1 - I(N_t)]. \tag{7.4}$$

Here the pathogen is assumed to be the only regulatory factor, so that the intrinsic growth rate λ per head for the disease-free population is constant. The fraction I of the population which is infected—and thus killed—by the

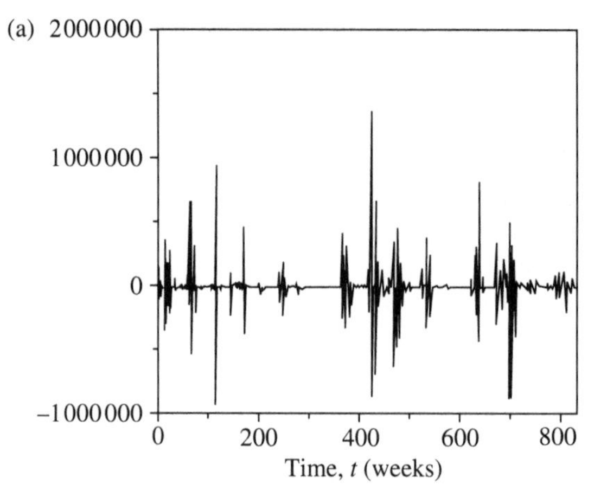
(a)
2000000
1000000
0
−1000000
0
200
400
600
800
Time, t (weeks)

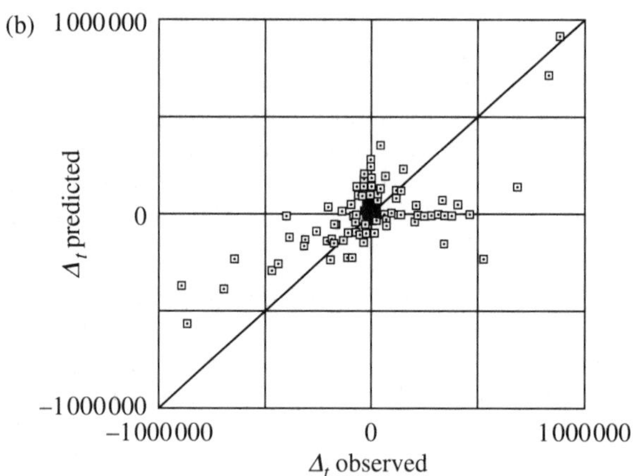
(b)
1000000
0
−1000000
−1000000
0
1000000
Δt predicted
Δt observed

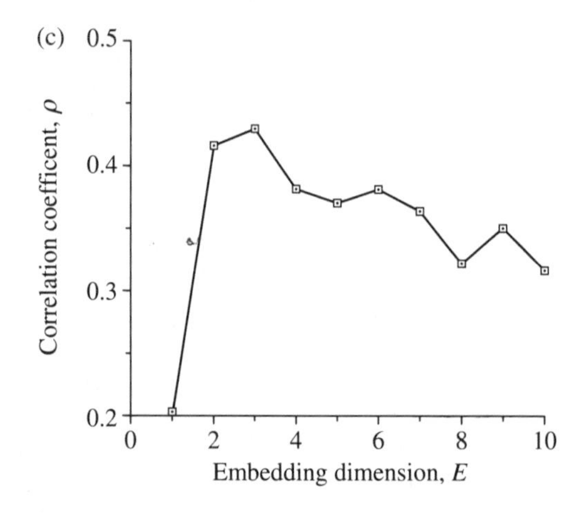
(c)
0.5
0.4
0.3
0.2
0
2
4
6
8
10
Correlation coefficent, ρ
Embedding dimension, E

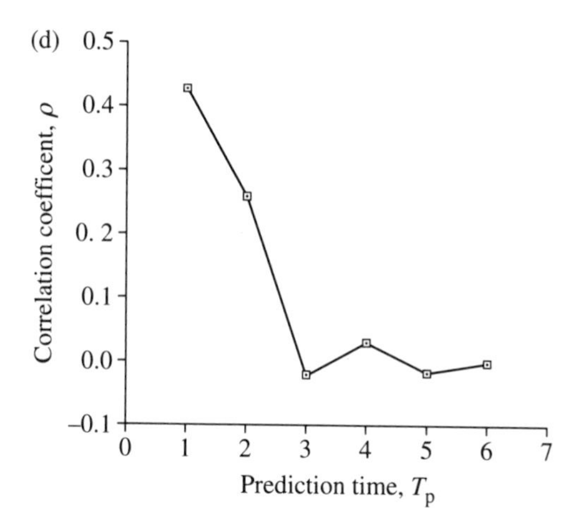

Fig. 7.10. (a) Time series of first differences, $x_{t+1} - x_t$, of the weekly numbers of diatoms in seawater samples taken at Scripps Pier, Dan Diego, from 1929 to 1939 ($N = 830$). (b) Using the first half of the time series in (a) to construct a library of patterns, we use the simplex projection method described in the text to predict one week (T_p) into the future form each point in the second half of the series ($N = 415$); here $e = 3$ and $\tau = 1$. The correlation coefficient between predicted and observed values is $\rho = 0.42$ ($P < 10^{-4}$ for $N = 415$); the best autoregressive linear predictions (composed of three optimal linear maps) give $\rho = 0.13$, which is significantly less than the nonlinear result ($P < 0.0005$). (c) As in Fig. 7.5(c) and 7.6(c), ρ between predicted and observed values is shown as a function of the choice of e for predictions two time steps into the future ($T_p = 2$ and $\tau = 1$). This figure indicates an optimal e of about 3, consistent with an attractor of dimension about 2. (d) As in Figs. 7.5(d) and 7.6(d), ρ is shown as a function of T_p (for $e = 3$ and $\tau = 1$). Here, the correlation coefficient decreases with increasing prediction interval, in the manner characteristic of chaotic dynamics generated by a low-dimensional attractor. That ρ is about 50% at best, however, indicates that roughly half the variance in the time series comes from additive noise. The dynamics of this system therefore seems to be intermediate between those of measles (for which Fig. 7.5(d) indicates deterministically chaotic dynamics) and chickenpox (for which Fig. 7.6(d) indicates purely additive noise superimposed on a seasonal cycle).

time the epidemic has run its course depends on the density of the host population according to the standard Kermack–McKendrick [33] relation:

$$I = 1 - \exp[-IN/N_T]. \tag{7.5}$$

Here N_T is a 'threshold' host density, which is determined by epidemiological parameters characterizing the transmission and virulence of the pathogen. If N is below the threshold density, $N < N_T$, the pathogen has a reproductive rate below unity and the epidemic cannot spread; equation (7.4) has only the trivial solution $I = 0$. That is, for $N_t < N_T$, equation

(7.4) gives simple growth at the rate λ, that is, $N_{t+1} = \lambda N_t$. For $N > N_T$, equation (7.4) has a nontrivial solution, $I \neq 0$; the fraction $1 - I$ escaping infection becomes smaller and smaller as N increases above N_T.

Equation (7.4), in combination with (7.5), thus represents a one-hump map, relating N_{t+1} to N_t, of exactly the kind discussed in Chapter 5. Acquaintance with the one-humped maps discussed earlier might suggest this system will exhibit a stable point, then a cascade of period doublings, and finally a regime of apparently chaotic behaviour, as λ increases. In fact, the system defined by equations (7.4) and (7.5) has no stable points and no stable cycles for any values of $\lambda > 1$ (and for $\lambda < 1$ the host population simply declines to extinction). All population trajectories are asymptotically aperiodic. This purely deterministic model is thus 'completely chaotic', in the sense that it generates smooth distribution functions or 'invariant measures', which describe the probability that the population will have any specific value.

Figures 7.11 and 7.12 illustrate this dynamical behaviour. Figure 7.11 is obtained by plotting long runs of iterations of equation (7.4) for specific values of λ, and it gives some idea of the probability distributions for host

Fig. 7.11. Plot of population values (on a logarithmic scale, $\ln(N/N_T)$) generated by iterating the map, many times, for each of a sequence of λ values. The diagram gives an impression of the probability distribution of population values generated by this purely deterministic difference equation.

densities. For relatively small values of λ ($\lambda > 1$), the host population alternates between a band of relatively high values and a band of relatively low values; these bands are narrow for λ close to unity (and indeed there are four bands, corresponding to trajectories looking rather like a 4-point cycle, for λ less than 1.68). For $\lambda > 2.91$, the two bands coalesce, and the population can take any value between the upper and lower bounds. Figure 7.12 illustrates these properties by displaying three specific trajectories: for $\lambda = 1.3$, the trajectory, although in fact chaotic, is close to a 2-point cycle; for $\lambda = 2$, the aperiodicity is marked, although the population density still clearly alternates up and down; and, for $\lambda = 4$, the trajectory is quite irregular.

It is interesting to speculate what might have happened if this particular model had been studied in the 1930s, as it could well have been. When Moran and Ricker studied the map (7.4) in the 1950s, they did find cycles and chaotic dynamics in their numerical investigations (carried out on mechanical calculators), but they were looking for steady solutions and, having found them, they went no further. However, the model defined by (7.4) and (7.5) has *only* chaotic solutions, with no stable points and no stable cycles. Deterministic chaos might have forced itself on our attention long ago if only a population biologist had thought to look at such a simple and natural model.

In human populations, the demographic and epidemiological processes (such as birth and infection) tend to occur continuously in time. Studies

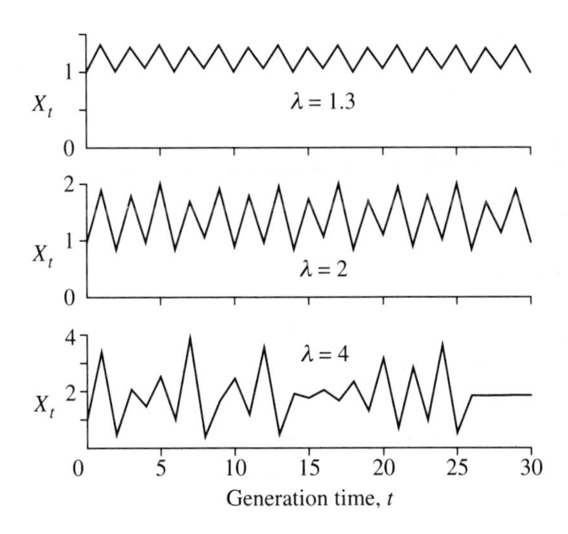

Fig. 7.12. Variation in the rescaled population, $X = N/N_T$, with time t as described by equation (7.4). As discussed in the text, these three population trajectories illustrate the character of the chaotic behaviour exhibited by this equation.

of human host–pathogen interactions thus usually employ differential equations.

These differential equations for host–pathogen interactions are essentially special cases of prey–predator equations of the kind first studied by Lotka and Volterra. It is thus not surprising that they tend to show oscillations whose periods obey the general Lotka–Volterra relation:

$$T \simeq 2\pi(A\tau)^{1/2}. \tag{7.6}$$

Here, A is the average age of hosts at infection (the characteristic lifetime of susceptible 'prey') and τ is the duration of the infection (the characteristic lifetime of the 'predatory' infection). Various kinds of realistic complications can, in effect, 'pump' this propensity to oscillation, so that cycles with period T of equation (7.6) are perpetuated indefinitely. Such mechanisms include the demographic stochasticity associated with fluctuations in host births and deaths [29], incubation periods [34], and seasonality in transmission rates [4, 35, 36]. In particular, annual variation in transmission tends to interact with the intrinsic period T of equation (7.6) in such a way as to enhance subharmonics of the annual cycle that resonate with T; the outcome can be regular or chaotic fluctuations in the amplitude of the annual cycle, with peaks roughly every T years.

Public health data for various infections of childhood in developed countries do show rough cycles in the annual incidence, with periods that are characteristically different for different infections (see Table 7.2). Figure 7.13 departs from the rather impressionistic analysis of much earlier work, showing the frequency spectrum for the time series generated by weekly notifications of measles in England and Wales from 1948 to 1968. In Fig. 7.13 the minor peak at a frequency 1 per year corresponds to an annual periodicity in measles incidence, which is probably associated with the annual patterns in school openings. The more pronounced peak at a frequency of 0.5 per year in Fig. 7.13 corresponds to the much-discussed two-

Table 7.2. Estimates of the average duration of latent plus infectious interval (τ), average age at infection (A), predicted interepidemic period (T from eqn (7.6)), and observed interepidemic period, for childhood infections in England and Wales and (more roughly) for smallpox in India [37, 38].

Infection	Latent plus infectious interval, τ (days)	Average age at infection, A (years)	Theory	Observed
Measles	12	4–5	2.3–2.5	2
Pertussis	25	4–5	3.3–3.6	3
Mumps	19	6–7	3.5–3.8	3
Smallpox	2–3 weeks	teens	4.0–6.0	5

year cycle in measles incidence, which shows up clearly as the main feature in the frequency spectrum.

As indicated in Table 7.2, rough theoretical estimates based on equation (7.6) are in good agreement with the observed 'interepidemic' periods for measles, pertussis, mumps, and (more approximately) smallpox. Note that this tentative explanation of the observed periodicities rests on the assumption that some mechanism pumps the intrinsic cycle of period T, and is independent of exactly what the mechanism is.

The basic propensity for host–pathogen associations to oscillate can manifest itself in other ways, which may have practical implications. Consider an immunization programme that has the long-term aim of reducing, or even eradicating, the incidence of particular infection. Such a programme represents a perturbation to the host–pathogen system. The system is unlikely to move smoothly and monotonically to the new final state, but rather is likely to exhibit pronounced oscillations en route to this final state.

Figure 7.14 illustrates this phenomenon. The y-axis shows the number of cases of congenital rubella syndrome (CRS)—taken to be proportional to the number of women contracting rubella (German measles) in the first trimester of pregnancy—as a function of the number of years after initiating a programme of immunizing either 20% or 80% of all two-year-old boys and girls; the number of case of CRS is shown as a ratio to the annual average number before any immunization. At 80% coverage, there are violent oscillations, with an approximate ten-year period, in the incidence of CRS. These oscillations damp to a lower level of incidence on a time scale too long to be seen in Fig. 7.14. Figure 7.14 represents a ludicrous oversimplification of the real immunization programme in the USA, being based on an excessively simple model, and neglecting both immunization of adult women as part of postnatal care and other realistic complications.

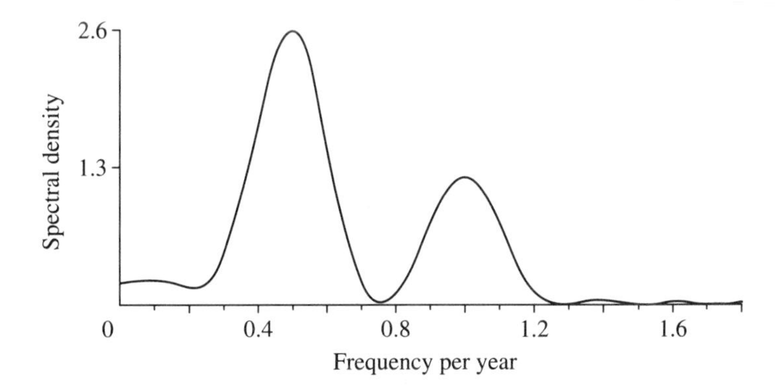

Fig. 7.13. Frequency spectrum constructed from the weekly measles reports for England and Wales, 1948–1968, showing the pronounced peak corresponding to two-year cycles.

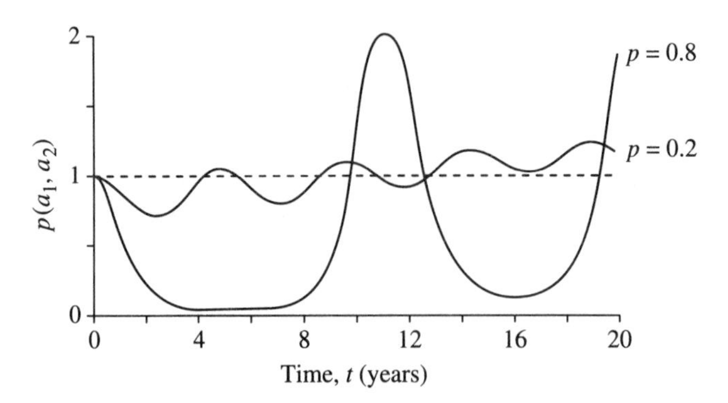

Fig. 7.14. Number of cases of CRS as a function of time, after instituting a programme of vaccinating 20 per cent or 80 per cent of two-year-old children in year 0, as predicted by a simple model [38].

The figure does, however, capture the major feature of the actual incidence of CRS in the USA, as shown in Fig. 7.15. The data in Fig. 7.15 show that the incidence of CRS at first fell after immunization was begun in 1970, but that it rose back almost to the pre-immunization level about ten years later in 1979. Figure 7.15 also shows the total number of cases of rubella,

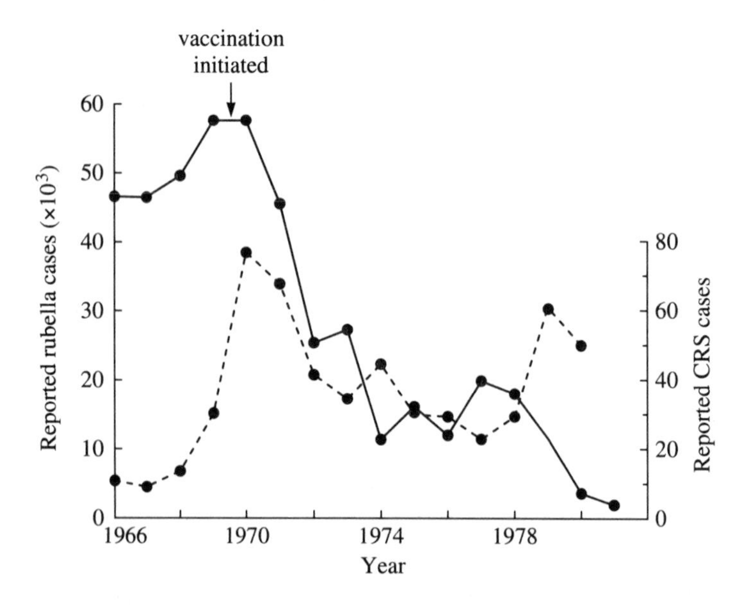

Fig. 7.15. Reported cases of rubella (solid line) and CRS (dashed line) in the USA between 1966 and 1981. The apparent rise in the pre-immunization years, before 1970, may be associated with under-reporting.

as distinct from cases in pregnant women; for the total incidence, the oscillatory pattern is much less marked in both theory and practice.

In the absence of the theoretical understanding developed above, the upswing in the incidence of CRS shown in Fig. 7.15 could be mistaken as deriving from a failure in vaccine production or some other extrinsic cause, rather than being recognized as intrinsic to the dynamics of the system.

References

1. Tong, H. and Lim, K. S. (1980). Threshold autoregression, limit cycles and cyclical data. *J. R. Stat. Soc.*, **42**, 245–92.
2. Packard, N. H., Crutchfield, J. P., Farmer, J. D., and Shaw, R. S. (1980). Geometry from a time series. *Phys. Rev. Lett.*, **45**, 712–15.
3. Takens, F. (1980). Detecting strange attractors in turbulence. In *Dynamical systems and turbulence*, Warwick, Lecture Notes in Math. 898 (eds. D. A. Rand and L.-S. Young), pp. 366–81. Springer, Berlin.
4. London, W. P. and Yorke, J. A. (1973). Recurrent outbreaks of measles, chickenpox, and mumps. I. Seasonal variation in contact rates. *Am. J. Epidemiol.*, **98**, 453–68.
5. Tont, S. A. (1981). Temporal variations in diatom abundance off Southern California in relation to surface temperature, air temperature and sea level. *J. Mar. Res.*, **39**, 191–201.
6. Lorenz, E. N.(1969). Atmospheric predictability as revealed by naturally occurring analogues. *J. Atmos. Sci.*, **26**, 636–46.
7. Sugihara, G. and May, R. M. (1990). Nonlinear forecasting as a way of distinguishing chaos from measurement error in time series. *Nature*, **344**, 734–41.
8. Farmer, J. D. and Sidorowich, J. J. (1989). Exploiting chaos to predict the future and reduce noise. In *Evolution, learning and cognition* (ed. Y. C. Lee). World Scientific, Singapore.
9. Casdagli, M. (1989). Nonlinear prediction of chaotic time series. *Physica D*, **35**, 335–56.
10. Press, W. H., Flannery, B. P., Teulosky, S. A., and Vetterling, W. T. *Numerical recipes in C*. Cambridge University Press.
11. Grassberger, P. and Procaccia, I. (1983). Characterization of strange attractors. *Phys. Rev. Lett.*, **50**, 346–769.
12. Schwartz, I. (1985). Predictability in epidemic models. *J. Math. Biol.*, **21**, 347–61.
13. Schaffer, W. M. and Kot, M. (1985). Nearly one dimensional dynamics in an epidemic. *J. Theor. Biol.*, **112**, 403–7.
14. Schaffer, W. M. and Kot, M. (1986). Differential systems in ecology and epidemiology. In *Chaos: an introduction* (ed. A. V. Holden). Princeton University Press.
15. Schaffer, W. M., Ellner, S., and Kot, M. J. (1986). The effect of noise on dynamical measures of ecological systems. *J. Math. Biol.*, **24**, 479–523.
16. Schaffer, W. M., Olsen, L. F., Truty, G. L., Fulmer, S. L., and Graser, D. J. (1988). In *From chemical to biological organisation* (eds. M. Markus, S. C. Muller, and G. Nicolis) Springer, New York.

17. Yule, G. U. (1927). On a method of investigating periodicities in disturbed series with special reference to Walker's sunspot numbers. *Phil. Trans. R. Soc. A*, **266**, 267–78.
18. Pimm, S. L. and Redfearn, A. (1988). The variability of population densities. *Nature*, **334**, 613–14.
19. Lawton, J. H. (1988). More time means more variation. *Nature*, **334**, 563–5.
20. Data extracted from the Registrar General's Weekly Returns, for the period 1948 to 1967.
21. Cohen, J. E. (1979). Long-run growth rates of discrete multiplicative processes in Markovian environments. *J. Math. Anal. Applic.*, **69**, 243–5.
22. Livdahl, T. and Sugihara, G. (1984). Nonlinear interactions of populations and the importance of estimating per capita rates of change. *J. Anim. Ecol.*, **53**, 573–80.
23. Sugihara, G. (rapporteur). (1984). Ecosystems dynamics. In *Exploitation of marine communities* (ed. R. M. May), pp. 131–53. Springer, New York.
24. Sugihara, G., Grenfell G., and May R. M. (1990). Distinguishing error from chaos in ecological time series. *Phil. Trans. R. Soc. Lond. B*, **330**, 235–51.
25. Ives, A. R. and May, R. M. (1985). Competition within and between species in a patchy environment: relations between microscopic and macroscopic models. *J. Theor. Biol.*, **115**, 65–92.
26. Allen, T. F. H. and Starr, T. B. (1985). *Hierarchy*, Chicago University Press.
27. O'Neill, R. V. (1986). *A hierarchical concept of ecosystems*. Princeton University Press.
28. Sugihara, G., Schoenly, K., and Trombla, A. (1989). Scale invariance in food web properties. *Science*, **245**, 48–52.
29. Bartlett, M. S. (1957). Measles periodicity and community size. *J. R. Stat. Soc. A*, **120**, 48–70.
30. Anderson, R. M. and May R. M. (1979). Population biology of infectious diseases. *Nature*, **280**, 361–7, 455–61.
31. Anderson, R. M., Grenfell, B. T., and May, R. M. (1984). Oscillating fluctuations in the incidence of infectious disease and the impact of vaccination: time series analysis. *J. Hyg.*, **93**, 587–608.
32. May, R. M. and Leonard, W. J. (1975). Nonlinear aspects of competition between three species. *SIAM J. Appl. Math.*, **29**, 243–53.
33. Kermak, W. O. and McKendrick, A. G. (1927). A contribution to the mathematical theory of epidemics. *Proc. R. Soc. Lond. A*, **115**, 700–21.
34. Hethcote, H. W., Stech, H. W., and Van den Driessche, P. (1981). Nonlinear oscillations in epidemic models. *SIAM J. Appl. Math.*, **40**, 1–9.
35. Grossman, Z. (1980). Oscillatory phenomena in a model of infectious disease. *Theor. Popul. Biol.*, **18**, 204–43.
36. Aron, J. L. and Schwartz, I. B. (1984). Seasonality and period doubling: bifurcations in an epidemic model. *J. Theor. Biol.*, **110**, 665–79.
37. Anderson, R. M. and May R. M. (1985). Vaccination and herd immunity to infectious disease. *Nature (London)*, **318**, 323–29.
38. Anderson, R. M. and May R. M. (1985). Age related changes in the rate of disease transmission: implications for the design of vaccination programmes. *J. Hyg.*, **94**, 364–436.

8
The chaos of disease response and competition

David Holton and Robert M. May

Leon Glass and Michael Mackey at McGill University in Montreal were among the first to explore the possibility that many medical problems may have their roots in some underlying dynamical effect: the so-called *dynamical diseases*. Mathematical models suggest that, when physiological parameters (such as CO_2 transport rates in the blood stream) are changed, processes that normally are rhythmical may be replaced by erratic or chaotic fluctuations. For instance, in some blood diseases the numbers of blood cells show large oscillations that are not normally present. Glass and Mackey demonstrated that simple, but realistic, mathematical models for controlling blood cell production display similar periodic and chaotic oscillations as a particular parameter is varied. The changes in the parameters have themselves a physiological interpretation. Clifford Gurney of the University of Chicago has performed successful experiments, based on Mackey and Glass's models, which produce oscillations in the numbers of blood cells.

Breakdowns in cardiac rhythms are obvious candidates for a 'dynamical disease'. Perhaps the best studies of the dynamics of heartbeats comes from Petri dishes rather than humans. Glass and co-workers [1–3] showed that a cluster of heart cells from chick embryos will beat spontaneously with an innate and regular rhythm. Applying a strong electric field to this cell aggregate resets the phase of the heartbeat, that is, the next beat will be earlier or later than normal. Introducing a periodic series of such electrical impulses means that the heart is pushed by two forces of different periods: one with the heart cell's intrinsic period and the other with the rhythm of the electrical shock. The ensuing heartbeat depends on the relationship between the the two periods and amplitude of the forcing.

In some cases, the heart cells resonate with some harmonic of the stimulus, beating once for each jolt, or twice, or perhaps 36 times for 59 jolts! In other instances, the cells fire apparently at random, giving irregular or chaotic patterns. Glass and his colleagues interpret the dynamics of these periodic or chaotic patterns in terms of the complex bifurcations that result from the interplay between inner physiological rhythms of heart cells and the frequency of the forcing electrical stimuli. These experiments show that you can induce and study chaos in an artificial system that mimics behaviour seen in cardiac processes. Applying these to cardiac arrhythmias, or to electrocardiograms before and after heart attacks, is, however, still at an early stage.

8.1 Dynamical diseases

Neurophysiology [4] also offers a wide range of phenomena that are candidates for 'dynamical diseases', or abnormal oscillations and complex rhythms posing clinical problems. Sometimes, there is a marked oscillation in a neurological control system that does not normally have a rhythm. Examples are ankle tremor in patents with corticospinal tract disease, various movement disorders (Parkinson's tremors, for instance), and abnormal paroxysmal oscillations in the discharge of neurons that occur in many seizures.

Alternatively, there can be qualitative changes in the oscillations within an already rhythmic process, as in abnormality in waking, altered sleepwake cycles, or rapidly cycling manic depression. Yet again, clinical events may recur in seemingly random fashion, as in seizures in adult epileptics. Neural processes are, however, so complex that it is not easy to see how models for these dynamical diseases—if, indeed, they exist—can be developed, tested, and understood.

One approach, taken by Paul Rapp [5] at the College of Medicine of the State of Pennsylvania, rests on analysing the dynamical complexity of electroencephalograms (EEGs) which recorded patterns of brain activity of human subjects as they performed various tasks. Rapp found that the complexity of the patterns changed in response to change in intellectual effort. One study, for example, asked subjects to count backwards from 700 in steps of 7. Rapp characterized the changing complexity of the resulting EEG patterns using what is becoming a standard method for analysing chaotic rhythmic processes. He computed the fractal dimension of the time series, and found the dimension rose from its background value of around 2.3 to around 2.9 during the tests. He infers that the higher-dimensional, more complicated EEG patterns correspond to a more alert state.

If we want to understand the dynamics of neurophysiological processes more clearly, we need a simpler system that we can control more precisely, such as the light reflex of the pupil of the eye. This reflex is a neural control mechanism with a delayed negative feedback, which regulates the amount of light reaching the retina by changing the area of the pupil. One can demonstrate the phenomena informally by playing with a torch in front of a mirror.

The time delay in the pupil's response, or 'pupil latency', is around 0.3 seconds. As the gain and/or delay in the feedback loop increases, the pupil light-reflex becomes unstable and starts to oscillate periodically. In an experiment by Longtin *et al.* [6] (also at McGill), who designed the electronically constructed circuit to mimic 'mixed' feedback, the pupil reflex became unstable and produced aperiodic oscillations. This is conjectured to be as a result of the interaction of complex, possibly chaotic, dynamics and neural noise.

8.2 Application to laboratory and field populations

The ideal of a population being subjected to a purely homogeneous and deterministic environment is perhaps an artificial one, likely only to be found in the laboratory. A number of such experiments do exist that seem to bear out the deterministic approach. For example, studies by Utida and by Fuji of different strains of stored-product beetles show population densities settling monotonically to a steady level, or damped oscillations, or stable cycles alternating between high and low population values, as the intrinsic growth rate r increases. Murdoch and McCauley [7] have reviewed a collection of their own and other peoples' laboratory studies of *Daphnia* populations, finding that qualitative changes in dynamical behaviour are

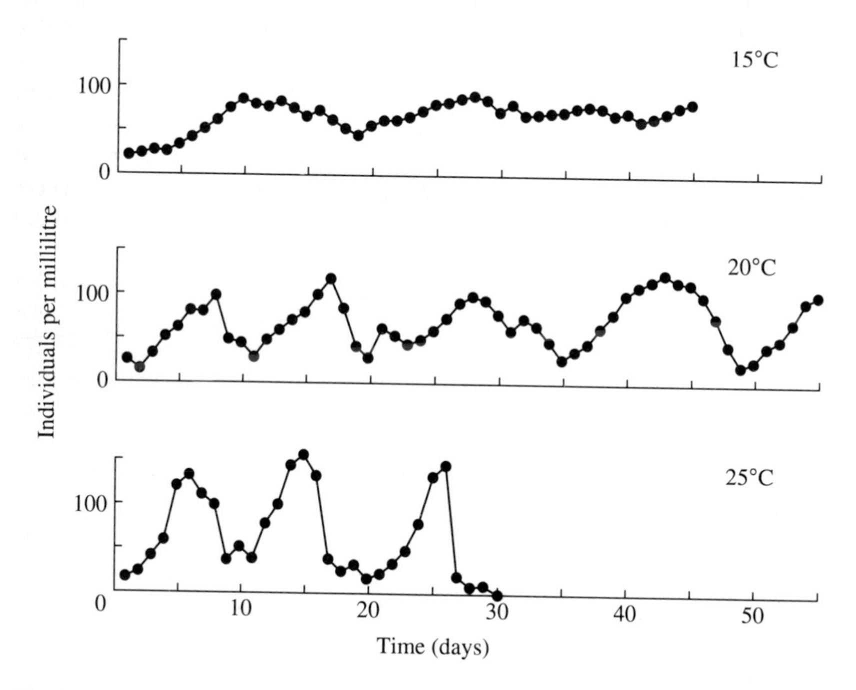

Fig. 8.1. The dynamical behaviour of laboratory populations of rotifers is shown as a function of time, for three different temperatures. After Halbach (1979).

indeed produced by what amounts to changes in the steepness of the map relating N_t to N_{t+1}. Figure 8.1 shows the dynamical behaviour of laboratory rotifers, as a function of time, at three different laboratory temperatures. As temperatures rise, the density-dependent propensity to 'boom and bust' becomes more pronounced and results in the observed transition from a relatively steady state at 15°C to oscillations that are too severe to be sustained at 25°C (this system may be accurately modelled by using differential equations with time lags). A series of similar experiments, with similar results, was performed by Goulden and Hornig [8].

A typical run from Nicholson's [9] classic series of laboratory studies of Australian blowflies under constant conditions is shown in Fig. 8.2. He consistently found the repeated, if ragged, cycles depicted in Fig. 8.2, with the total number of individuals (the y-axis) scaling linearly with the constant amount of food supply. The smooth curve shows that a reasonable fit to these data can be obtained by the naive use of a time-delayed logistic equation, in which the time lag in regulatory effects corresponds roughly to the egg-to-adult development time (the theoretical curve is for a nine-day lag, to be compared with the actual figure of 11–14 days). Nisbet and Gurney [10] and Brillinger *et al.* [11] have given more accurate analyses of these data. In the age-structured model of Brillinger *et al.*, the blowflies are found to be exhibiting chaotic dynamical behaviour: the chaotic behaviour has almost periodic structure most of the time, but occasionally has episodes in which the dynamics appear altogether random. Interestingly, Nicholson occasionally found his regular cycles 'breaking down' into more random patterns, at which point he terminated the run (believing something had gone wrong with the genetic make-up of his population). It would be interesting to repeat Nicholson's experiment, to see if the phenomenon suggested by Brillinger *et al.*'s model does occur, that is, to see whether,

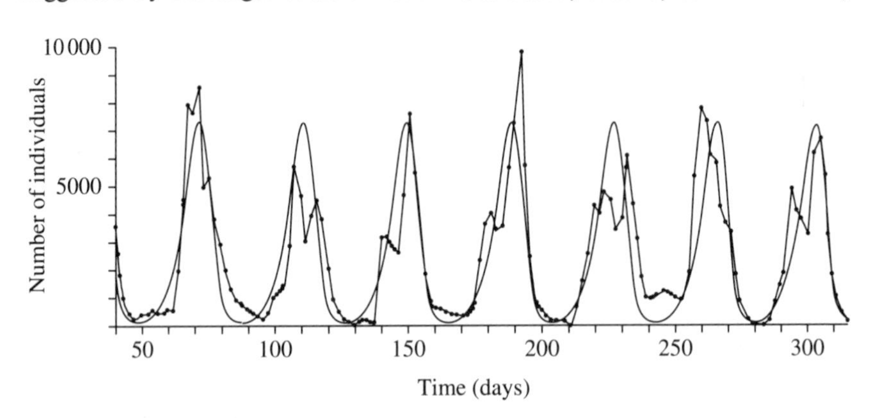

Fig. 8.2. The data for Nicholson's [9] studies of sheep blowflies in a constant laboratory environment may be roughly described by a limit cycle, generated by the time-delayed logistic. After [12].

following such an interval of breakdown, the system reverts after a time to the familiar almost-periodic structure if the experiment is kept running.

But of course, in the natural world, environmental stochasticity, heterogeneity, and interactions with other species are all likely to be important, which tends to preclude any direct confrontation between data and simple theory. Thus, for example, the explanation given by May [12] of 'wildlife's four-year cycle' as arising naturally for populations with high intrinsic growth rates in strongly seasonal environments (where regulatory effects are likely to operate with one-year lags) has been reasonably criticized by Stenseth [13] as being too simple to be sensible. An interesting recent development is by Schaffer [14] and Schaffer and Kot [15]. By using techniques developed for the study of chaotic physical systems, they have suggested that the apparently complicated behaviour of some populations (such as lynx in Canada) may be described heuristically as generated by relatively low-order (one-dimensional) nonlinear maps.

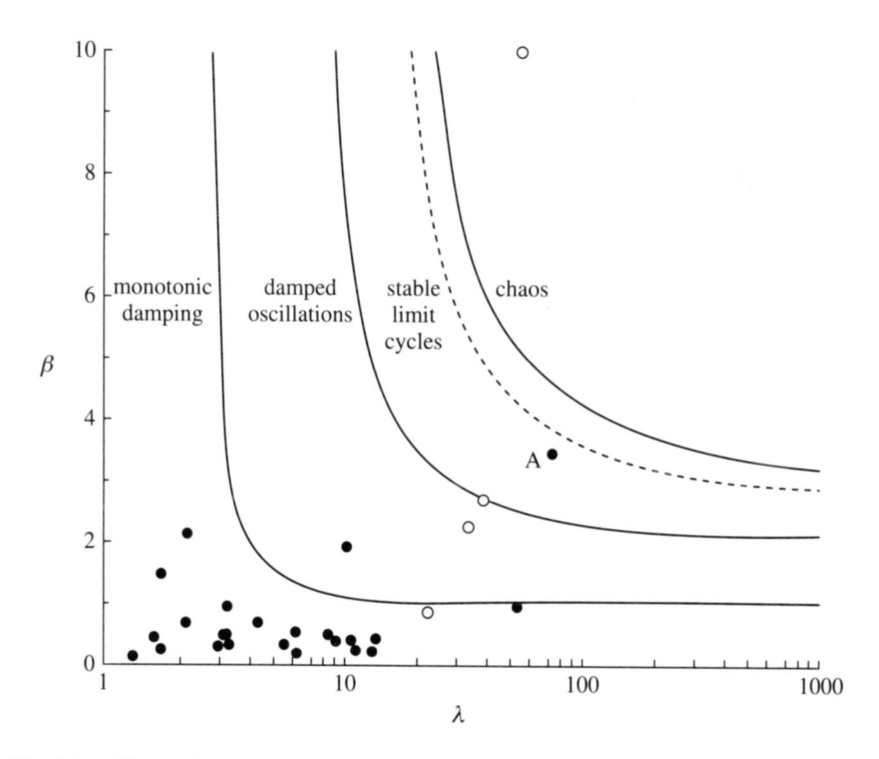

Fig. 8.3. The various regimes of dynamical behaviour of equation (8.1) are shown as functions of the population parameters λ and β (the broken line indicates where period-2 cycles give way to higher, period-2^n, cycles). The solid circles come from analyses of life table data on field populations, and the open circles from laboratory populations. After [16], where details are given.

Hassell *et al.* [16] analysed information on 28 populations of seasonally breeding insects with non-overlapping generations. In each case, the data were fitted—sometimes in a procrustean way—to the difference equation

$$N_{t+1} = \lambda N_t (1 + aN_t)^{-\beta}. \quad (8.1)$$

The intrinsic growth rate parameter λ was first estimated independently, and the census data then used to estimate the scaling parameter a and the density-dependence parameter β. Figure 8.3 shows the theoretical domains of stability behaviour for this difference equation: depending on the values of λ and β, there is a stable cycle in which the population alternates up and down, or chaos; the parameter a does not affect the stability properties. The points show the parameter values for the actual populations, with solid circles denoting field populations and open circles laboratory ones (whose freedom from many natural causes of mortality may well encourage exaggeratedly nonlinear behaviour). One should, however, not make too much of the relative steadiness of most field populations suggested by Fig. 8.3. For one thing, there is the bias that long runs of data are more likely available for steady populations than for those that bounce around. For another, the assumption that a population's interactions with other species can be subsumed in passive parameters, such as λ and β, does violence to the multispecies reality.

A more comprehensive review of these laboratory and field studies, along with others, is given by May [12].

8.3 From individual behaviour to population dynamics

The parameters characterizing the dynamics (the λ, a, and β of equation (8.1), for example) have been regarded as population-level quantities, to be determined empirically from census data. The population-level parameters are, however, ultimately determined by the foraging, egg-laying, and other types of behaviour of the individuals constituting the population. Such population-level implications of the behaviour of individuals [4] are usually of little interest to behavioural ecologists, who tend to be more concerned with how the behaviour of animals, as individuals or groups, may have evolved.

Consider a habitat comprising of n plants which are food for an insect species such as a caterpillar or sawfly. Assume that there are N_t adult insects in generation t and that they disperse themselves among plants such that the final distribution of adults obeys the probability distribution $p(i; m_t)$; here p is the probability of finding i egg-laying adults in a given patch, given that the average number of adults per patch in generation t is $m_t = N_t/n$. Suppose further that each adult lays a single clutch of F

eggs, and that the probability of an egg surviving to produce a mature adult for dispersal in the next generation $t + 1$ has the density-dependent form $\exp(-dj)$, where j is the number of eggs in the patch and d characterizes the strength of interactive effects. The relation between the total number of adults in successive generations is then obtained by summing over the average outcomes in each of the n patches, and is [18]

$$N_{t+1} = nF\left(\sum_i ie^{-iFd}p(i; N_t/n)\right). \tag{8.2}$$

We now take the adults to be distributed according to the negative binomial distribution, with 'clumping parameter' k. That is, the distribution has a variance/mean ratio equal to $1 + m/k$: for large k the adults are essentially distributed independently randomly; for small k there is marked aggregation, with most of the adults clumped in relatively few patches. For this distribution, equation (8.2) simplifies to the form

$$N_{t+1} = \lambda N_t(1 + aN_t)^{-\beta}. \tag{8.3}$$

Here, the population parameters λ, a, and β are expressed in terms of the parameters characterizing individual behaviour: $\lambda = F\exp(-dF)$, $a = [1 - \exp(-dF)]/nk$, and $\beta = k + 1$. In this example, the expression relating the overall population density in successive generations (equation (8.3)) is exactly the same as that used by Hassell *et al.* [16] to analyse the population data in Fig. 8.3. There is, however, an important difference. Whereas the parameters in equation (8.1) earlier were phenomenological entities, to be determined empirically from the population data, the parameters in equation (8.3) can be expressed in terms of quantities characterizing the behaviour of individuals. The parameter β is thus expressed in terms of the clumping parameter k as $\beta = k + 1$. Hence any change in foraging behaviour that affects the distribution of adult individuals among plants—that is, which affects the clumping parameter k—may be translated into a change in the stability properties of the population.

Once we have such a model determining the population dynamics from the behaviour of the individuals, we can impose environmental noise on various aspects of individual behaviour and see how the system responds. It turns out that noise introduced in this way may render the density-dependent effects undetectable by conventional methods, such as 'k-factor analysis', which focus on differences between overall population levels in successive generations. The reason is simple. In these models, the regulatory effects involve both overall population magnitudes and density-dependent differences among patches within any one generation. In some circumstances, the noise is such that the intragenerational regulatory effects outweigh the intergenerational differences in the overall population density, so that the k-factor analysis is unhelpful. This point may be clarified by example.

Hassell *et al.* [19] have analysed data for the population dynamics of the whitefly (*Aleurotrachelus jelinekii*) on three distinct viburnum bushes at Silwood park. For the eggs, the four larval instars, and the emergent adults, there are data both for total populations and for subpopulations on individual leaves (the patches of the system), over the period 1962–1984. When analysed by *k*-factor analysis applied to overall population densities, these data show no convincing evidence of density dependence of any kind [20]. A more recent analysis uses the population data for individually marked leaves to construct a model of the form

$$E_{t+1} = nF\left(\sum_{i=0}^{\infty} i(ai^{-b})p(i;m_t)\right). \tag{8.4}$$

Here, E_t is the total number of eggs on a bush in generation t; n is the number of leaves (patches) on a bush (which can change over time); F is the effective fecundity per adult; $p(i;m)$ is the probability of having i eggs on a leaf (with m the mean number of eggs per leaf, $m = F(\text{total adults})/n$); and the expression ai^{-b} describes the density-dependent survival probability of an egg on a leaf containing a total of i eggs (with a and b assessed from the time history of individual leaves). Hassell *et al.* [16] take the egg distribution $p(i)$ to be the negative binomial, with a clumping parameter

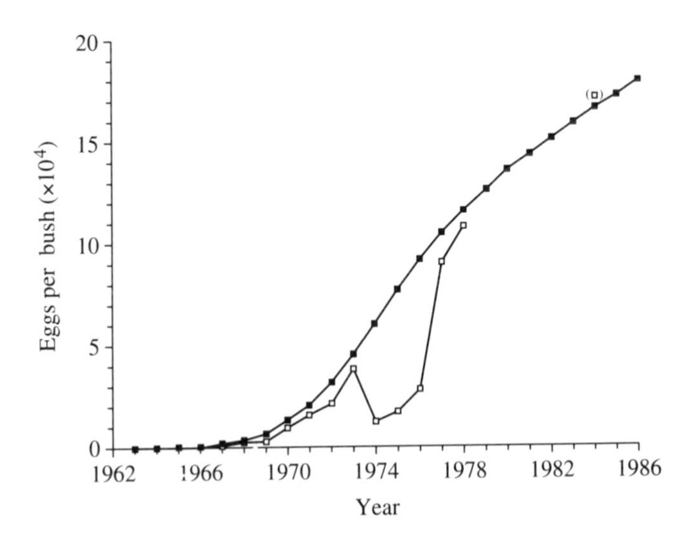

Fig. 8.4. Data for the total number of whitefly eggs on viburnum 'bush B' at Silwood Park, 1963–1978 and 1984. The solid curve is the theoretical expectation according to the model of equation, with the parameter values determined from the data (for individual leaves and for the bush as a whole) from 1963 to 1978. The 1984 census point is enclosed in brackets to emphasize that it was not used in constructing the theoretical result, and thus provides an independent check. After [19].

k that obeys the observed trend to higher values at higher population densities.

Figure 8.4 shows the fit between the model of equation (8.4) and the data for whitefly eggs on 'bush B' at Silkwood Park. The parameters in the model were determined from the earlier data (1963–1978) so that the census point in 1984 represents an independent check on the predictions of the model.

In a purely deterministic world, there would be no problem in identifying these purely density-dependent effects. Although the primary such effect comes from differences among leaves within any one generation, its influence on overall population levels between generations would show up. This is illustrated by the solid lines in Figs. 8.5 and 8.6, which show the application of conventional k-factor analysis for investigating changes in the total egg-to-adult mortality as a function of initial egg density: the solid lines illustrate the density-dependent relations corresponding to equation (8.3) with the parameters for the theoretical line in Fig. 8.4.

Hassell *et al.* [19] emphasize, however, that the clear-cut results in Figs. 8.5 and 8.6 can be obscured by certain kinds of noise in the parameters

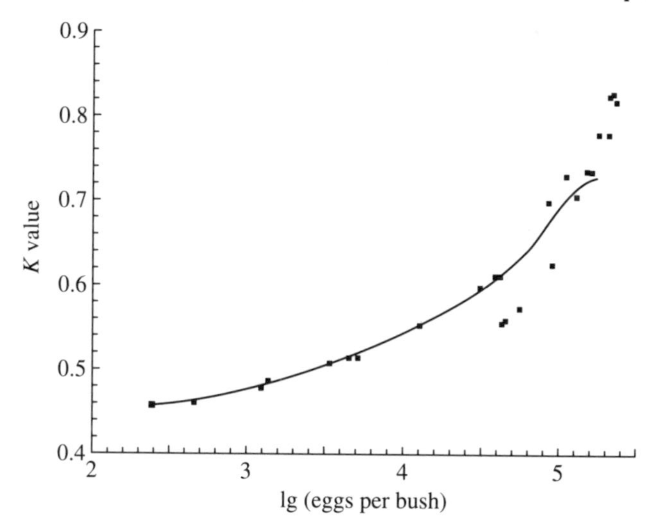

Fig. 8.5. The solid line shows the density-dependent relation between 'k value' (essentially, egg-to-adult mortality) and the total number of eggs per bush (plotted logarithmically) in the absence of any stochastic fluctuations in any of the underlying parameters that characterize the system at the level of individual behaviour; that is, the solid line corresponds to parameter values that give rise to the theoretical curve in Fig. 8.4. The points scattered about this curve show what happens when the per capita clutch size F exhibits random variation that is distributed normally with mean and standard deviation corresponding to those actually observed for bush B ($F = 5.6 \pm 3.1$). In this case, the noise does not obscure the density-dependent relation. After [19].

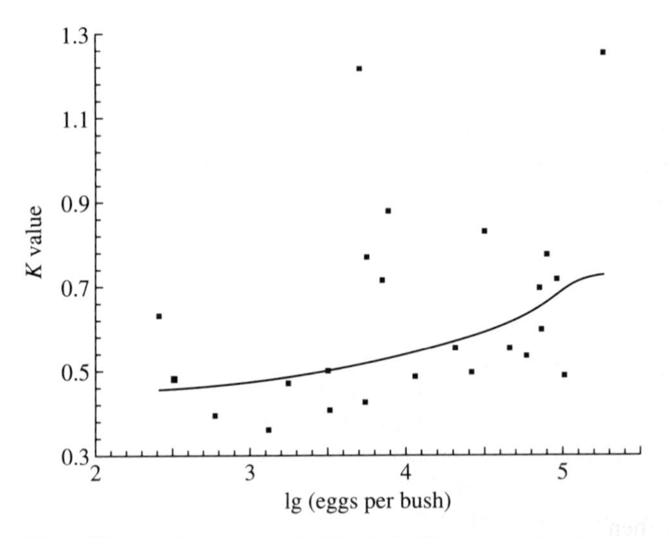

Fig. 8.6. The solid curve is exactly as in Fig. 8.5. The scattered points are generated by letting the leaf-to-leaf 'density-independent' parameter a of equation (8.4) vary according to a normal distribution with mean and standard deviation corresponding to those actually observed in some of the date ($a = 0.51 \pm 0.17$). Such noise at the level of individual behaviour in a patch now obscures the overall density dependence, rendering it undetectable by analysis that uses only the total populations in successive generations. After [19].

characterizing individual whitefly behaviour. As shown in Fig. 8.5, variability in the whitefly fecundity F has little effect; the signal is clearly discernible against noise. In contrast, Fig. 8.6 shows effects of modest levels of variability in the parameter a of the formula ai^{-b} characterizing density dependence within a patch; here the noise renders the signal undetectable by conventional k-factor techniques. The underlying density dependence in Fig. 8.6 emerges clearly, however, when the analysis focuses on intragenerational differences among patches. Hassell *et al.* [19] suggest that the features revealed in Fig. 8.6 explain why density dependence was not detected in earlier analyses of the whitefly data. Their main conclusion is, however, a more general one: 'The large body of existing life table data is not a reliable source for revealing the causes of regulation of natural populations. Within-generation heterogeneity can be a very powerful cause of population stability . . . and there is the risk of overlooking this when using data on mean populations per generation.'

8.4 Population genetics and chaotic polymorphisms

However complex the dynamics in simple population models, things can get messier when population genetics is combined with population biology in models where fitness functions are frequency or density dependent.

The simplest such models that we have already seen in Chapter 6 deal with a single diallelic locus, with p_t and q_t the relative proportions of the two alleles A and a, respectively, in generation t. In a diploid population with random mating, the frequency p_{t+1} of A in generation $t+1$ is related to that in generation t by the first-order difference equation

$$p_{t+1} = (p_t^2 W_{AA} + p_t q_t W_{aA})/(p_t^2 W_{AA} + 2p_t q_t W_{Aa} + q_t^2 W_{aa}). \qquad (8.5)$$

The quantities W_{ij} represent, as before, the fitness or relative reproductive success of the three genotypes. If the fitnesses themselves depend on the gene frequency, p_t, as can happen in a variety of biologically reasonable situations, the resulting map on the unit interval can be nonlinear.

In particular, one can assume that each of the genotypes is susceptible to a particular pathogen (to which the other two are resistant) which spreads in epidemic fashion. The fitness of genotype ij (ij = AA, Aa, aa) in generation t is then

$$W_{ij} = \lambda_{ij}[1 - \gamma_{ij} I(N_{ij})]. \qquad (8.6)$$

Here, λ_{AA} is the fitness, or relative reproductive success, of genotype AA in the absence of disease; γ_{AA} is the proportion of those infected who die; and $N_{AA} = N_t p_t^2$, with N_t the total density of the population in generation t. Similar definitions apply to the corresponding quantities for the genotypes Aa and aa. In each case, the Kermack–McKendrick [21] relation gives the implicit expression for I:

$$1 - I_{ij} = \exp[-I_{ij} N_{ij}/N_T]. \qquad (8.7)$$

Here, as before, N_T is the threshold density for transmission of the infection.

Two cases can now be distinguished. If overall population density in each generation is held constant by other ecological constraints ($N_t = K$), the proportion of each of the three genotypes to be infected—and thence, via (8.6), the fitness functions W_{ii}—depend only on the gene frequency p_t. For such frequency-dependent selection, gene frequencies in successive generations obey (8.5) with frequency dependent W_{ij} from (8.6) and (8.7). The map has fixed points at (0, 0) and (1, 1), but in general also has an interior fixed point by virtue of the propensity of p_t to increase from low values and decrease from high values (because disease spreads less effectively among rare genotypes and more effectively among common ones). The result can be a stable polymorphism, but for plausible values of the epidemiological parameters there can alternatively be cyclic or chaotic fluctuations in gene frequency.

If overall population density is itself regulated by the different diseases afflicting the three different genotypes, then both total population density and gene frequency can fluctuate cyclically or chaotically. Indeed, if all three diseases are lethal ($\gamma_{ij} = 1$ in all cases), it follows that the total population, and consequently the relative proportions of the genes A and a, have chaotic dynamics [22].

These results have intrinsic mathematical interest, but their significance is that—like the results outlined in the previous sections—they suggest we should think again about empirical aspects of biological studies. Whether chaotic fluctuations in gene frequencies are observed in natural populations is not known, because most studies have not reckoned with the possibility that gene frequencies may be continually changing, driven by their own dynamics [22].

8.5 Dynamics from a model of the dynamical interaction between HIV and the immune system

8.5.1 *Background*

The early models for the transmission dynamics of HIV/AIDS have followed the conventional epidemiological assumption that infected individuals exhibit some constant level of infectiousness, beginning soon after acquiring HIV, and lasting until the patient develops AIDS (and dies soon after).

The assumption of a constant level of infectiousness is contrary to existing evidence. The patterns of infectiousness appear to vary significantly through the incubation time. The average incubation period of AIDS appears to be about 7–8 years [23], but over this period marked fluctuations of HIV antigen-specific antibodies among patients are found.

We sketch here one possible 'first step' for a model of the dynamics of the interaction between HIV and the immune system. In particular, this model can explain the variability in incubation times between infection with HIV and the clinical manifestation of the full-blown AIDS disease.

8.5.2 *A model of the interaction of HIV and the immune system [24]*

Suppose we consider a population of free HIV virus, $V(t)$, at time t, and a population of T lymphocytes that are subdivided into populations of nonactivated cells, $P(t)$, activated but uninfected cells, $X(t)$, and infected cells, $Y(t)$. The nonactivated cells are assumed to be recruited at a constant rate Λ and are removed either by mortality at a rate μ or by activation at a net rate γPV, where γ is the probability at activation through contact. The population of activated but unaffected cells grows as a result of recruitment from the immature cells at the rate γPV and the growth through clonal expansion at a net per capita rate r. The population $X(t)$ also suffers losses, at a net rate $-\beta XV$ (where β is the probability of infection via contact), as activated cells are infected with HIV. The population $Y(t)$ of infected cells is assumed be unable to proliferate, so that this population grows through recruitment from the activated but uninfected population via infection by contact with the virus at the rate βXV, and suffers deaths as a result of infection with the HIV virus at a per capita rate α.

The population $V(t)$ of free virus is assumed to increase as a result of infected cells releasing a large number of viral particles (λ per cell) when they die, at a rate αY; this produces free virus at the overall rate $\lambda\alpha Y$. The viral population $V(t)$ is assumed to decrease as a result of death at a per capita rate b, of absorption by activated cells at a net rate $\delta(X + Y)V$, and of the activities of uninfected lymphocytes in stimulating antibody and cell-mediated viral destruction at a net rate σXV.

These assumptions yield the following set of nonlinear differential equations for $P(t)$, $X(t)$, $Y(t)$, and $V(t)$:

$$\mathrm{d}P/\mathrm{d}t = \Lambda - \mu P - \gamma PV, \tag{8.8}$$

$$\mathrm{d}X/\mathrm{d}t = \gamma PV + rX - \beta XV, \tag{8.9}$$

$$\mathrm{d}Y/\mathrm{d}t = \beta XV - \alpha Y, \tag{8.10}$$

$$\mathrm{d}V/\mathrm{d}t = \lambda\alpha Y - bV - \delta(X + Y)V - \sigma XV. \tag{8.11}$$

This simplified model embodies a current opinion that the virus infects only those cells that have been activated by previous contact with the virus (although alternative views do exist [25]). The model assumes that the virus is cytopathic, and that it is killed by the action of the immune system, which we represent by T cells.

8.5.3 *Dynamical properties of the model*

In the absence of HIV ($V = 0$), the system settles to an obvious stable state with no activated cells ($X = Y = 0$) and the immature cells at an equilibrium value $P = \Lambda/\mu$.

When V is nonzero (HIV is introduced), three qualitatively different patterns of dynamical behaviour occur. It is helpful first to define the parameter R as the net reproductive rate of the virus within the human host

$$R = \beta\lambda/(\delta + \sigma). \tag{8.12}$$

If R has a value $R < 1$, viral 'reproduction' is effectively below threshold, and the virus is unable to establish itself within the host; V and Y tend to zero, while the population of activated but uninfected cells increases ($X \to \infty$), owing to clonal expansion. If we crudely describe the action of T suppressor cells by the subtraction of a regulatory term $-dX^2$ in equation (8.9), then the activated cells will establish an equilibrium at $\bar{X} = r/d$.

If $R > 1$, the virus can establish itself. Two qualitatively different possibilities can now be distinguished in the limit $b \rightarrow 0$. For

$$r > \beta\alpha\lambda(1 - 1/R)/\delta, \tag{8.13}$$

the virus can establish itself, but its effects are not sufficient to halt the growth of the population of activated cells. When the effects of regulation by T suppressor cells are included, the system tends to settle to, or to oscillate around, a state of constant abundance of activated cells, both infected and uninfected.

Conversely, when the inequality in equation (8.13) is reversed, the virus not only persists but can regulate the activated but uninfected cell population; the system tends to exhibit an initial burst of high viraemia, and then oscillate around a state where the abundance of T cells is regulated by the virus, even though viraemia may be low. Sustained oscillations are observed when r and $\beta\alpha\lambda$ $(1 - 1/R)/\delta$ are comparable in magnitude.

More generally, the phase space is messier if b is chosen sufficiently large that bV exceeds, or is comparable in magnitude with, $\delta(X + Y)V$ and σXV (see e.g. Figs. 8.7–8.9).

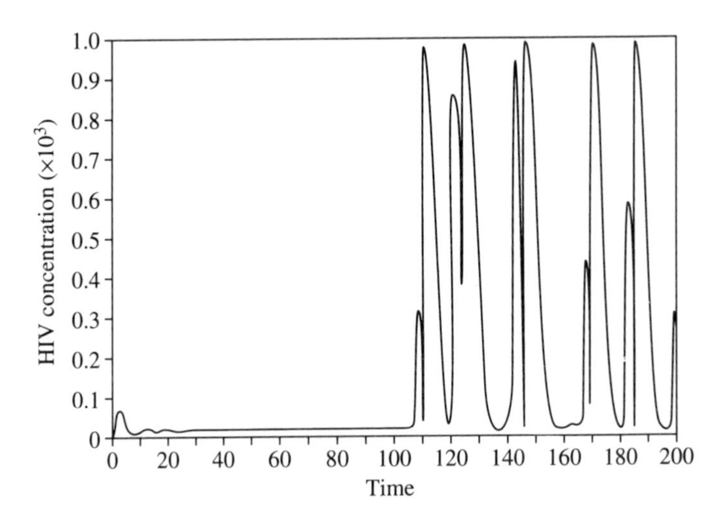

Fig. 8.7. This shows the concentration of free HIV virus in the blood as a function of time, as described by equations (8.8)–(8.11) and (8.14). We see an initial peak in the antigenaemia, which drops to a steady low level; at $t = 100$ the opportunistic infection is introduced, and its interaction with the HIV-activated immune system produces apparently chaotic fluctuations in HIV antigenaemia thereafter. Specifically, this example is for the above system of equations, with the parameters having the broadly representative values $\Lambda = 1$, $\mu = 0.1$, $\gamma = 0.01$, $r = 1$, $d = 0.001$, $\lambda\alpha = 10$, $b = 1$, $\delta = 0.01$, $\sigma = 0.1$, $k = h = 0.01$, $c = 1$, $\beta = 0.1$, and $\alpha = 2$ (all rates are in units of week^{-1} although, from the point of view of the qualitative dynamics of the system, the units may be taken to be arbitrary).

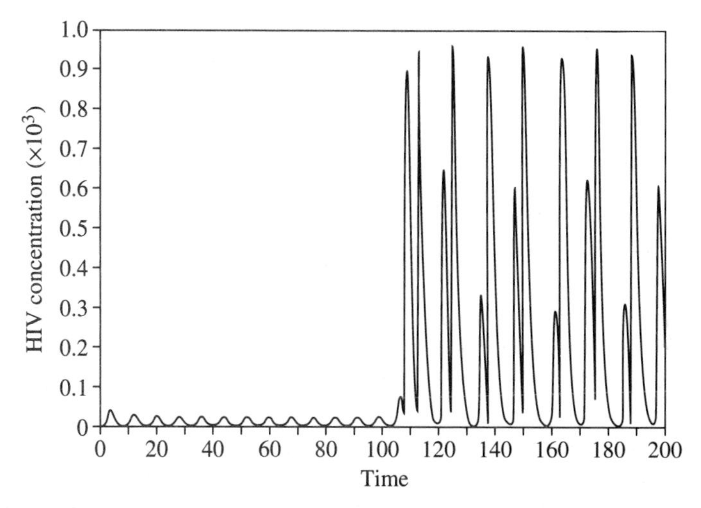

Fig. 8.8. As for Fig. 8.7, except here the parameter β has the value $\beta = 0.03$. The patterns in HIV antigenaemia are broadly similar to those of Fig. 8.7.

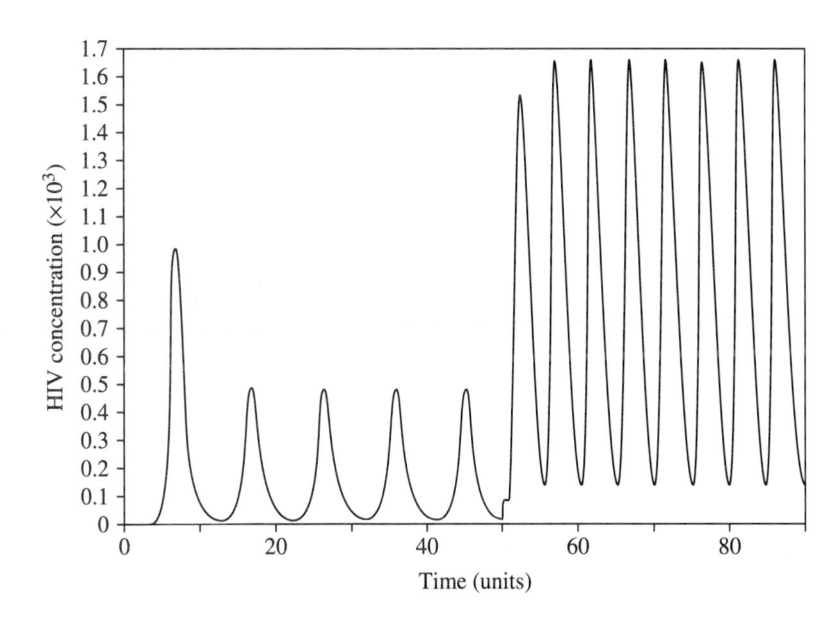

Fig. 8.9. As for Fig. 8.7, except here the parameters β and α have the values $\beta = 0.03$ and $\alpha = 1$. In this case, the opportunistic infection is introduced at $t = 50$ (rather than at $t = 100$).

Suppose we now introduce a second infection which replicates within the host and which stimulates these same T cells to proliferate by clonal expansion. The population of such a subsequent infectious agent is represented by $I(t)$. The action of this population is described by adding an extra term of the form kIX to equation (8.9), to describe growth via contact:

$$\mathrm{d}X/\mathrm{d}t = \gamma PV + rX + kIV - \beta XV. \tag{8.14}$$

The equation describing changes in $I(t)$ is

$$\mathrm{d}I/\mathrm{d}t = cI - hIX. \tag{8.15}$$

Here, c is the per capita reproductive rate of the opportunistic infection, and the term hIX represents the killing of the infectious agent by antibody or cell-mediated action.

In the absence of any HIV infection, the activated T cells always eliminate the opportunistic infection. When the host individual has already experienced HIV infection, the subsequent opportunistic infection can result in oscillating fluctuations in the abundance of HIV and the opportunistic infectious agent, or in chaotic dynamical behaviour. Numerical simulation of the behaviour of the above model for some typical parameter choices is shown in Figs. 8.7–8.9.

In conclusion, this simple model (created by the cytopathic influence of HIV on the lymphocyte population) can give rise to oscillatory or chaotic fluctuations in the abundance of HIV, in infected and uninfected lymphocyte populations, and in populations of the opportunistic infectious agent. The model, although a very crude caricature of the interaction between HIV and the immune system, suggests that a marked rise in antigenaemia is triggered by the arrival, or activation, of an opportunistic infection that stimulates further proliferation of the T4 lymphocytes. Second, our model suggests a possible interpretation of some of the relations among quantities that characterize the progression of disease in patients infected with HIV. The potential for highly chaotic dynamics may help to explain the great variability from patient to patient.

Finally, we must emphasize that the models investigated here are oversimplified caricatures. However, it is hoped that these caricatures can be systematically refined to reveal behaviour closer to reality.

References

1. Glass, L., Shrier, A., and Belair, J. (1986). Chaotic cardiac rhythms. In: *Chaos* (ed. V. I. Holden), Chap. 11.
2. Glass, L., Guevara, M. R., Shrier, A., and Perez, R. (1983). Bifurcation and chaos in a periodically stimulated cardiac oscillator. *Physica* D, **7**, 89–101.
3. Guevara, M. R., Glass, L., and Shrier, A (1981). Phase-locking, period-

doubling bifurcations and irregular dynamics in periodically stimulated cardiac cells. *Science*, **214**, 1350–53.
4. Freeman, W. J. (1991). The physiology of perception. *Scientific American*, February.
5. Rapp, P. (1985). *Nonlinear oscillations in biology and chemistry*. Springer, Berlin.
6. Longtin, A., Milton, J, *et al.* (1989). *J. Theor. Biol.*, **138**, 129.
7. Murdoch, W. W. and McCauley, E. (1985). Three distinct types of dynamic behaviour shown by a single plankton system. *Nature*, **316**, 628–30.
8. Goulden C. E. and Hornig, L. L. (1980). Population oscillations and energy reserves in planktonic cladocera and their consequences to competition. *Proc. Natl. Acad. Sci. USA*, **77**, 1716–20.
9. Nicholson, A. J. (1954). An outline of the dynamics of animal populations. *Austral. J. Zool.*, **2**, 9–65.
10. Nisbet, R. M. and Gurney, W. S. C. (1982). *Modelling fluctuating populations*. Wiley, New York.
11. Brillinger, D. R., Guckenheimer, J., Guttorp, P., and Oster, G. F. (1980). Empirical modeling of population time series data: The case of age and density dependent vital rates. In: *Some mathematical questions in biology*, Vol. 13 (ed. G. F. Oster), pp. 65–90. AMS, Providence, RI.
12. May, R. M. (ed.) (1981). *Theoretical ecology: principles and applications*, 2nd edn, Chap. 2. Blackwell, Oxford, and Sinauer, Sunderland, MA.
13. Stenseth, N. C. (1985). Why mathematical models in evolutionary ecology? In: *Trends in ecoloaical research for the 1980s* (ed. J. H. Cooley and F. B. Golley), pp. 239–87. Plenum, New York.
14. Schaffer, W. M. (1984). Stretching and folding in lynx fur returns: Evidence for a strange attractor in nature? *Am. Nat.*, **124**, 798–820.
15. Schaffer, W. M. and Kot, M. (1986). Differential systems in ecology and epidemiology. In: *Chaos* (ed. A. V. Holden). Princeton University Press.
16. Hassell, M. P., Lawton, J. H., and May R. M. (1976). Patterns of dynamical behaviour in single-species populations. *J. Anim. Ecol.*, **45**, 471–86.
17 Hassell, M. P. and May, R. M. (1985). From individual behaviour to population dynamics. In: *Behavioural ecology* (ed. R. Sibley and R. Smith), pp. 3–32.
18. dejong, G. (1979). The influence of the distribution of juveniles over patches of food on the dynamics of a population. *Neth. J. Zool.*, **29**, 33–51.
19. Hassell, M. P., Southwood T. R. E., and Reader, P. M. (1986). The dynamics of the viburnum whitefly (*Aleurotrachelus jelinekii* (Frauenf.)): A case study on population regulation. *J. Anim. Ecol.* (In press.)
20. Southwood, T. R. E. and Reader, P. M. (1976). Population census data on key factor analysis for the viburnum whitefly, *Aleurotrachelus jelinekii* (Frauenf.)), on three bushes. *J. Anim. Ecol.*, **45**, 313–25.
21. Kermack, W. O. and McKendrick, A. G. (1927). A contribution to the mathematical theory of epidemics. *Proc. R. Soc. Lond.* A, **115**, 700–721.
22. May, R. M. and Anderson, R. M. (1983). Epidemiology and genetics in the coevolution of parasites and hosts. *Proc. R. Soc. Lond.* B, **219**, 281–313.
23. Medley, G. F., Anderson, R. M., Cox, D. R., and Billard, K. The distribution

of the incubation period of AIDS. *Proc R. Soc. Lond.* B, **233**, 367–77.

24. Anderson, R. M. and May, R. M. (1990). Complex dynamical behaviour in the interaction between HIV and the immune system. In: *Cell to cell signalling: From experiment to theoretical models* (ed. A. Goldbeter). Academic Press, New York.
25. Barnes, D. M. (1988). AIDS virus coat activates T cells. *Science*, **242**, 515.

9

Chaos and fractal basin boundaries in engineering

J. M. T. Thompson

The aim of this chapter is to sketch some of the applications of chaos, and more generally the global geometrical techniques of nonlinear dynamics, to practical problems in engineering. This is done in an informal, pictorial way, with a minimum of the underlying theory, which can be found elsewhere [1, 2]. The material is based on recent research work at University College London directed towards current problems of marine technology encountered by the North Sea oil industry. This research has emphasized the importance of chaotic transients and the associated fractal basin boundaries, which are more important in an engineering context than the more familiar steady-state chaotic attractors.

9.1 Poincaré sampling of driven oscillators

The only bit of theory that we shall need is given in Fig. 9.1, which shows the three-dimensional phase space of a general one-degree-of-freedom nonlinear oscillator under periodic forcing, spanned by the displacement, velocity, and time. Such oscillators are of continuous concern to mechanical and electrical engineers.

The phase space is full of noncrossing trajectories spiralling around the time axis. Each T-interval is identical in form due to the periodicity of the forcing, allowing the convenient toroidal representation [1]. Sampling the displacement x and the velocity y, stroboscopically at time interval T, gives us the Poincaré mapping shown, which summarizes all information about the three-dimensional flow. Computer time integrations over one forcing cycle can be used to generate numerically the functions G, and H, which are used in our path- and bifurcation-following algorithms and stability tests.

For a dissipative system such as this, there will be recognizable transient motions, settling onto one or more attracting steady states such as the subharmonic of order $n = 2$ illustrated. This $n = 2$ solution is identifiable in the Poincaré section as two repeating dots, A, B, A, ...

The attractors can be harmonic, subharmonic, or even chaotic, and each

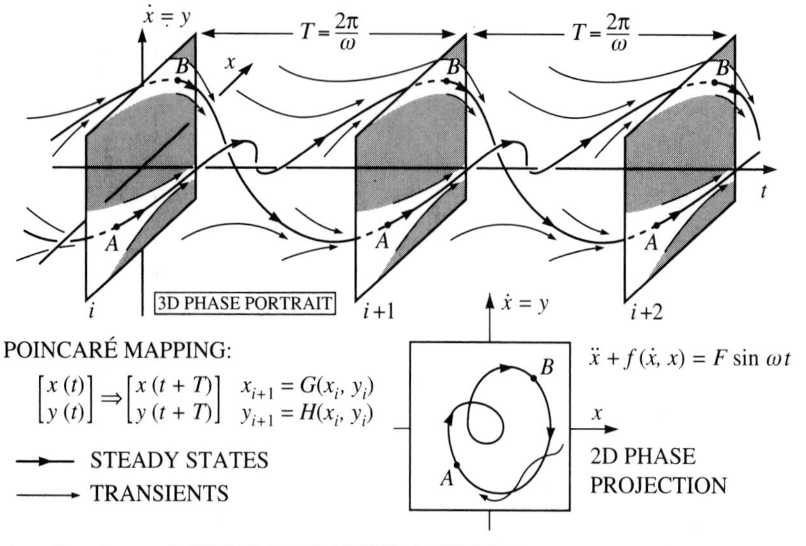

Fig. 9.1. Stroboscopic Poincaré sections for periodically driven oscillators.

will be surrounded by its own basin of attraction, as we shall see. The attractors and basins can be located very efficiently by Hsu's cell-to-cell mapping technique, which uses a cellular grid in (x, y) space to approximate the mapping action of G and H. Once the approximate cell-to-cell mapping rules are established by numerical time integrations over one forcing cycle the differential equation can be discarded, and simple sorting algorithms can quickly unravel the attractor–basin structure.

9.2 A canonical escape equation

Many of our current ideas can be illustrated very nicely by considering just the simple escape scenario illustrated in Fig. 9.2, the escape from a potential well being, of course, a recurring theme throughout physics, chemistry, and engineering. The cubic well chosen here gives in a mathematical sense the most typical escape problem, which would in fact be generically encountered by any system in the vicinity of a fold catastrophe. In structural engineering, for example, such a potential governs the lateral vibrations of axially compressed shells.

This equation can also be viewed as an archetypal capsize model for a ship or floating oil-production facility [3]. A symmetric vessel driven into resonance by lateral ocean waves can clearly capsize either to port or starboard, corresponding to the escape from a symmetric well. But wind loading and cargo imbalance mean that a worst case design scenario will

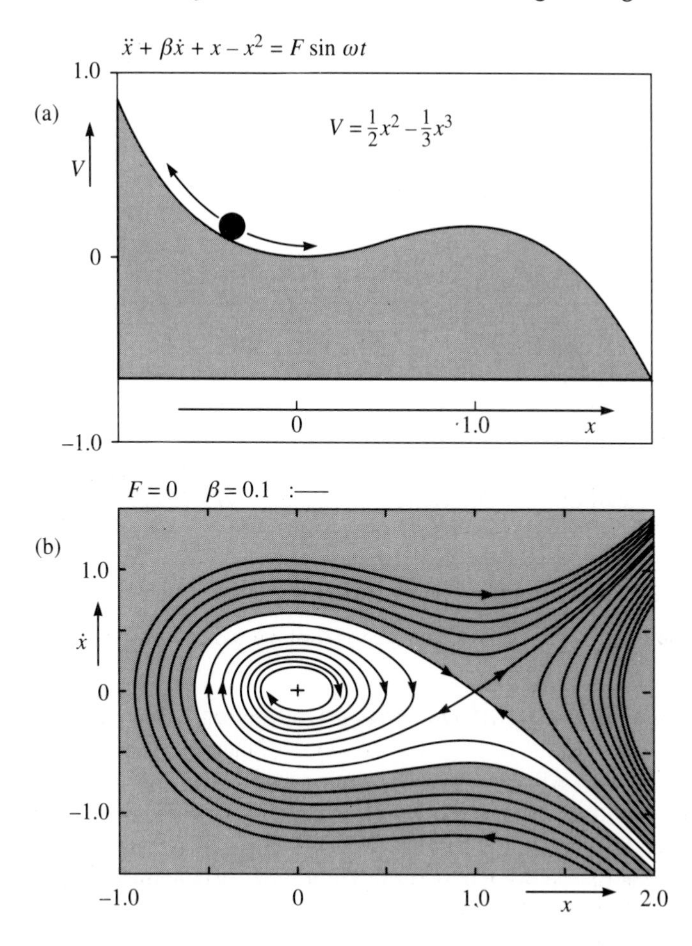

Fig. 9.2. Canonical driven oscillator governing the escape from a cubic potential well (the damping coefficient β is held at 0.1 throughout): (a) the potential well; (b) the response of the autonomous system with $F = 0$.

always involve a static bias of sufficient magnitude to ensure a one-sided escape like that of Fig. 9.2.

The lower diagram shows the behaviour of the autonomous undriven system, with F = 0, in its two-dimensional (x, y) phase space. The damping level is taken throughout as $\beta = 0.1$, which is a fairly typical value for a range of engineering applications including capsize: it corresponds to a damping ratio of $\zeta = 0.05$. We see a stable equilibrium point attractor at (0, 0), and an unstable hilltop saddle at (1, 0). Of particular interest is the basin of attraction of the stable state shown in white. Starts in this safe basin spiral into the attractor, while starts in the dotted region escape over the hilltop to the *attractor at infinity*. The basin boundary, or separatrix, is formed by the stable manifold, or inset, of the unstable saddle solution.

Points on the boundary flow under forwards time into the saddle, so computationally the boundary can be located by running time backwards from two starts close to the saddle, ideally from a two points on the incoming eigenvectors. Unstable solutions, which are often discarded as being of no physical interest, are seen to play a key role in organizing and structuring the phase space.

The basins of attraction of transient motions are in the author's opinion a much neglected feature of dynamics, and constitute the main theme of this chapter. In engineering it is common practice to establish the existence of steady-state oscillations, often by approximate analytical averaging or perturbation methods. A linearized stability analysis is then made, and it tends to be implicitly assumed that linear stability implies safety. Under real-world finite disturbances, however, the existence of a small linear neighbourhood of attraction is clearly not enough to guarantee safety. A nonlinear basin analysis is needed to ensure that there is a good sizeable robust basin around the attractor of interest. Lyapunov methods are sometimes employed, but these are not likely to be very useful in the presence of the fractal basin boundaries that we shall encounter in this chapter.

9.3 Attractor paths of the driven oscillator

We shall be interested in the response of the driven oscillator of Fig. 9.2 near to the primary resonance, with values of ω, the ratio of the forcing frequency to the linear natural frequency, close to unity. In fact, due to the softening of the system at finite amplitude, it is values of ω somewhat less than 1 that give the most interesting and practically important results.

Figure 9.3 shows the steady-state attractor paths, represented by the stroboscopically sampled x values, that would be followed by a physical system under the slow quasi-static increase of F at four different ω levels. In the lower diagram, at $\omega = 0.95$, we observe a trace of $n = 1$ harmonic oscillations that grows out of the stable equilibrium state at $F = x = y = 0$. This subsequently period-doubles at a supercritical flip bifurcation [1] leading through a complete Feigenbaum cascade to a chaotic attractor. This quickly loses its stability at a crisis, or blue sky catastrophe, at E, and the system escapes out of the potential well, over the hilltop to infinity. Notice that at this ω value chaotic motions do in fact trigger the escape: they are not some interesting side issue, but are central to the escape mechanism.

At the lower values of ω shown in the upper diagrams, this behaviour persists, but with the development of a hysteresis loop typical of nonlinear resonance. Under increasing F we would observe the jump to resonance at the saddle-node fold A, while under decreasing F we would encounter the jump from resonance at fold B. Under decreasing ω, E moves to the left, while A moves to the right, so that at $\omega = 0.8$ escape under increasing F

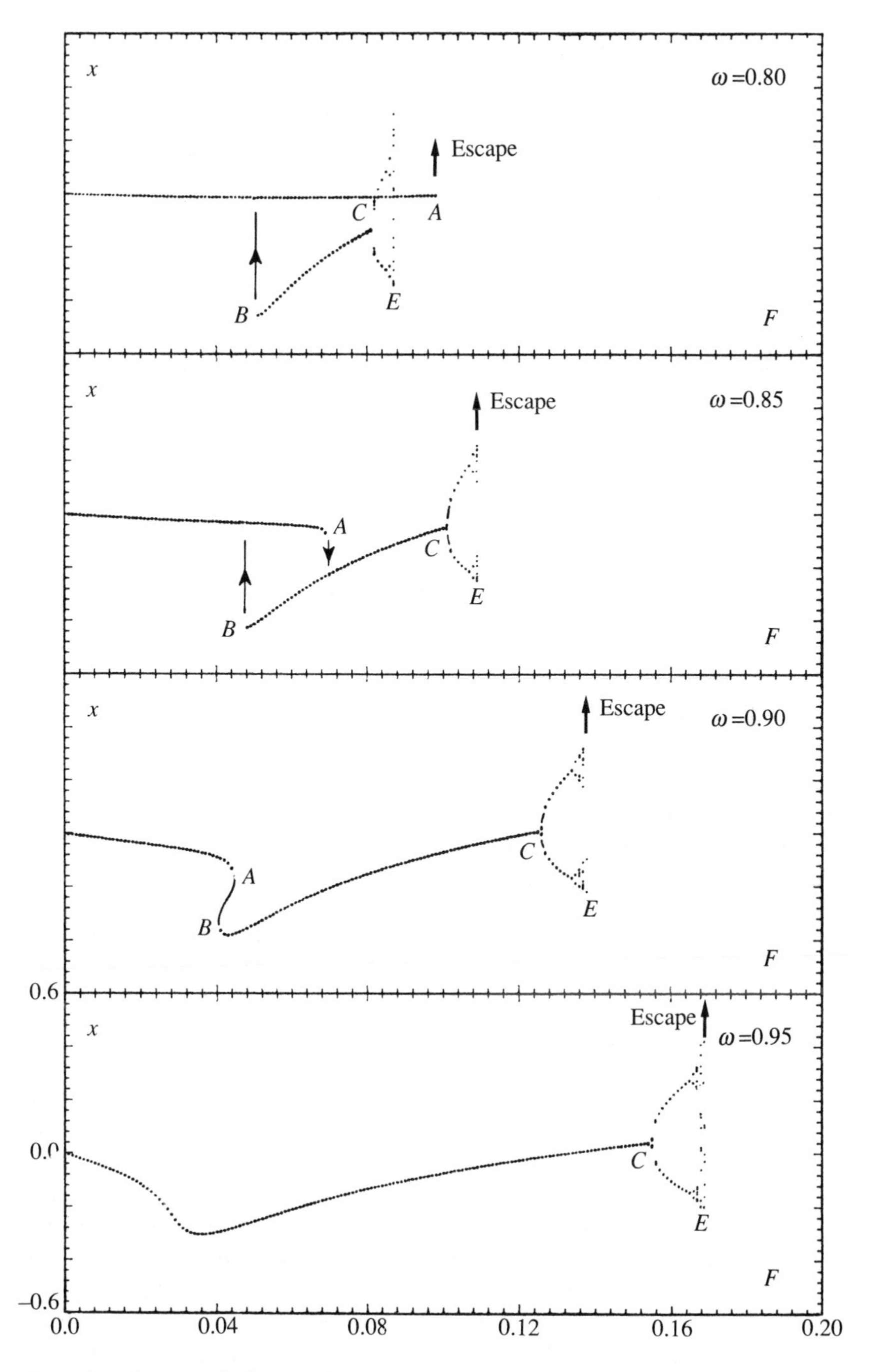

Fig. 9.3. Traces of the steady-state attractors at four ω values. The stroboscopically sampled x is plotted against the control F.

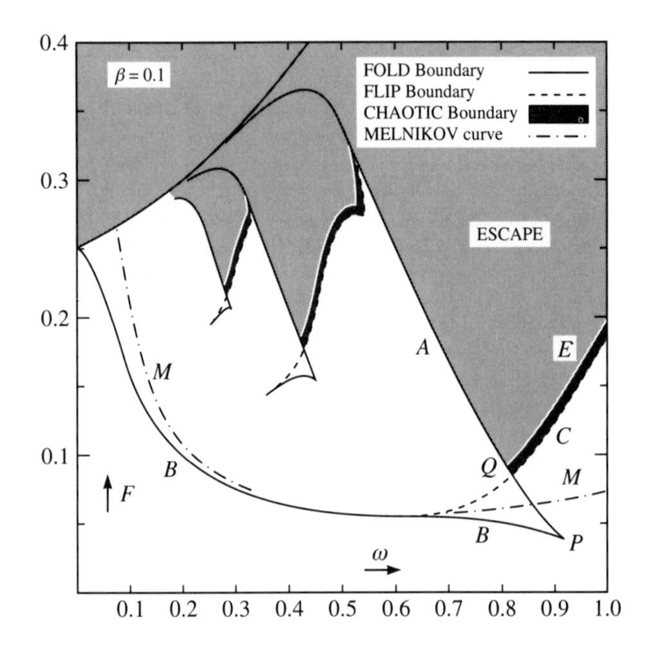

Fig. 9.4. Bifurcation diagram in the (F, ω) control space.

is directly from the fold A as shown. Clearly, between $\omega = 0.85$ and 0.8 we have passed a point of optimal escape where $F^E = F^A$ and escape can be achieved under a minimum forcing magnitude: this corresponds to the worst excitation frequency for a boat in waves. Rather than passing one another in a noninteracting neutral fashion, as would two local bifurcations, E and A cross in a complex codimension-2 bifurcation that is still a subject of our research.

The movement of the bifurcations in the (F, ω) control space is summarized in Fig. 9.4. The two folds, A and B, grow out of the cusp at P. The first period-doubling flip is on arc C, and the optimal escape is at Q where the arc E crosses the fold-line A. Interesting repetitions of the whole scenario can be seen at lower ω values, corresponding to superharmonic resonances.

Let us focus particularly on the frequency $\omega = 0.85$, which is close to the optimal escape, and a schematic view of the paths is shown in Fig. 9.5. Here the scenario of Fig. 9.3 can be seen as the path from O folds and bifurcates to give the escape at E. Also of interest, however, is the path of unstable harmonic oscillations that originates from the unstable equilibrium state $Z(F = y = 0, x = 1)$, corresponding to small oscillations about the hilltop that we shall refer to as the *hilltop saddle cycle*. This path turns back at the fold at G, to join up with the main-sequence harmonic oscillation from O that has restablized at the flip F. Constraints on the mapping eigenvalues of this dissipative system [2] do in fact prove the necessity of such a

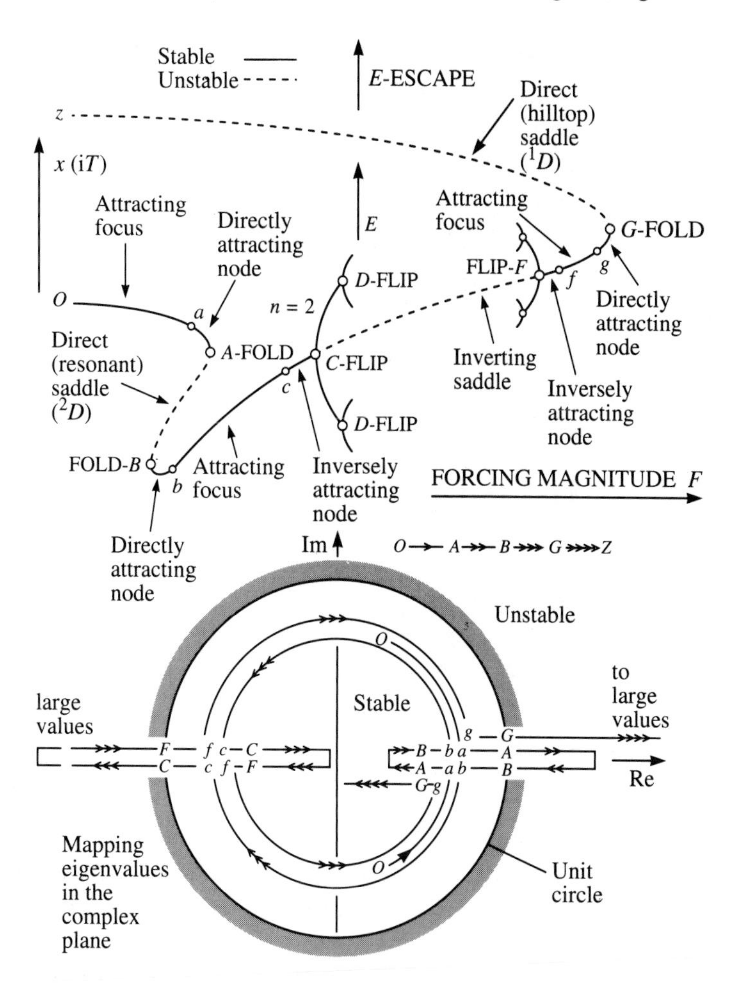

Fig. 9.5. A schematic diagram showing the variation of the steady-state stoboscopically sampled x versus F, together with the movement of the mapping eigenvalues in the complex plane.

restabilization before G. The movement of the mapping eigenvalues of the $n = 1$ oscillation in the complex plane is sketched in the lower diagram, and we recall [1] that for stability they must lie within the unit circle: an eigenvalue penetrates the unit circle at $+1$ for a fold, and at -1 for a flip. The movement reflects the constraint that at constant β and ω the product of the eigenvalues is a constant.

9.4 Escape mechanism near optimality

It is interesting to look somewhat more closely at the escape process at E. The final stages of the period-doubling cascade to chaos at $\omega = 0.85$ are

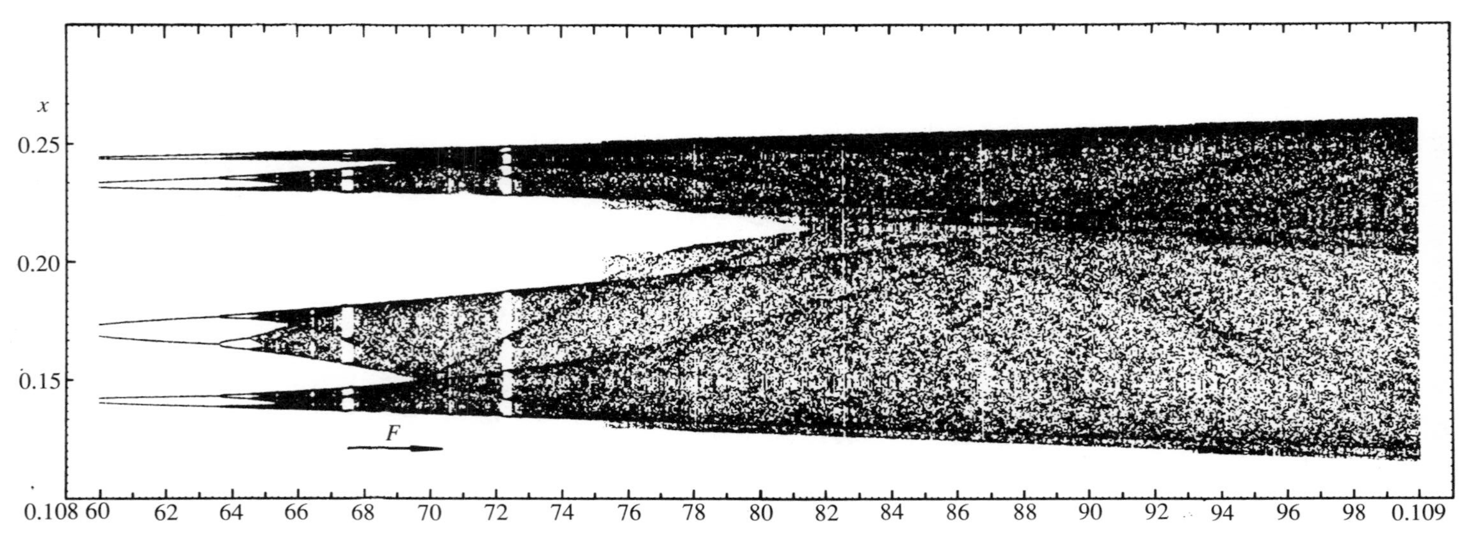

Fig. 9.6. The final period-doubling cascade to chaos and escape at $\omega = 0.85$. Here x is sampled stroboscopically at *twice* the forcing period.

shown in the blow-up of Fig. 9.6, which shows the steady state x values, stroboscopically sampled at *twice* the forcing period, under slowly incremented F. This starts with 8 lines, corresponding to a subharmonic of order 16. The periodic windows within the cascade are analogous to those of the familiar logistic map. The well-developed two-band chaotic attractor at $F = 0.109$, at the end of this cascade, is nicely displayed as a folding Möbius strip in Fig. 9.7, taken from reference [2].

As F is incremented beyond 0.109, escapes are encountered after long chaotic transients which made it extremely difficult to estimate the exact value of F^E with any precision by simple attractor-following. The first-escape is for example very sensitive to the rate of evolution of F.

The instability at E has, however, been identified as a crisis at which the chaotic attractor is in collision with a directly unstable subharmonic D^6 of

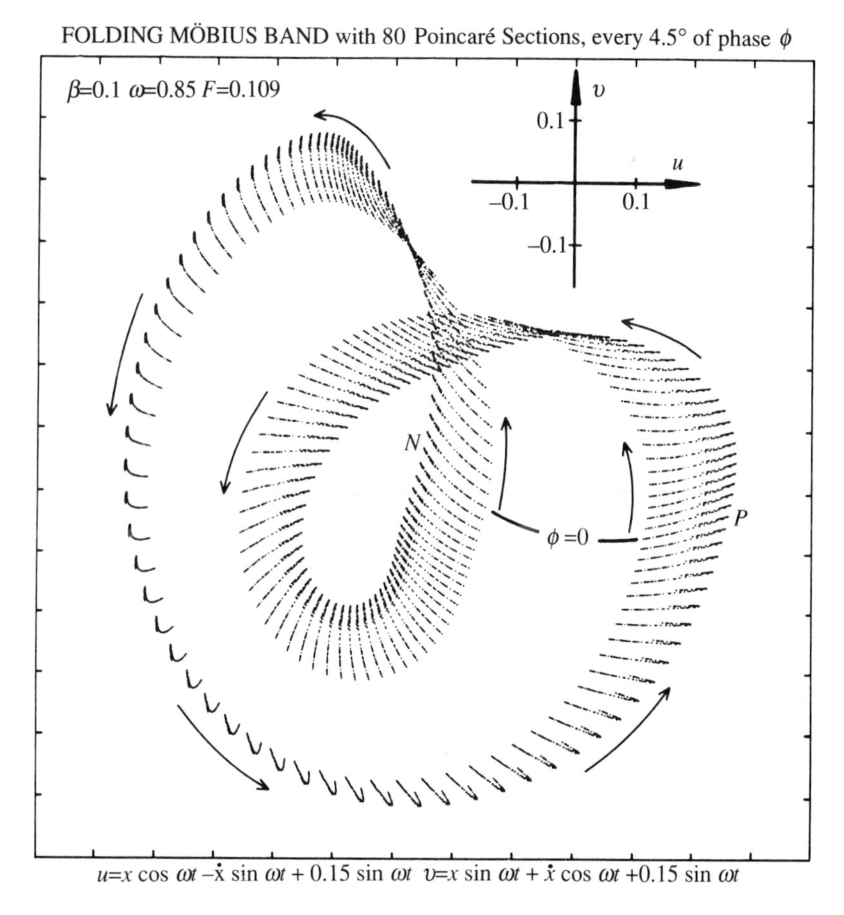

Fig. 9.7. The final chaotic attractor at $\omega = 0.85$, $F = 0.109$. In a sheared van der Pol plane, the attractor is seen as a folded Möbius strip.

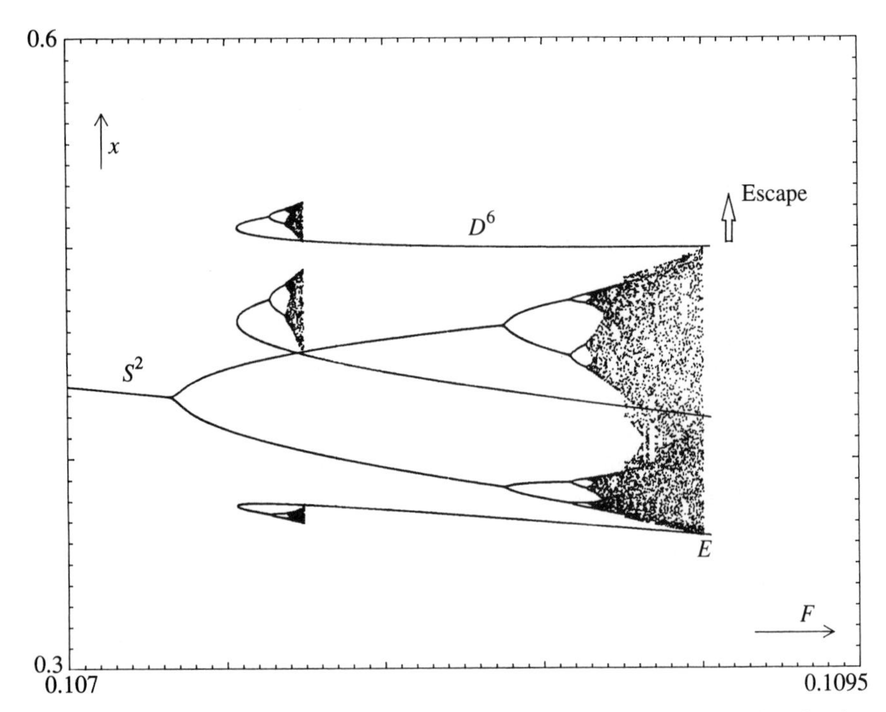

Fig. 9.8. The final chaotic-saddle instability of the chaotic attractor, showing its collision with D^6, a directly unstable subharmonic of order $n = 6$. Here x is sampled at $2T$, and S^2 is the main-sequence attractor which has already experienced one period-doubling bifurcation.

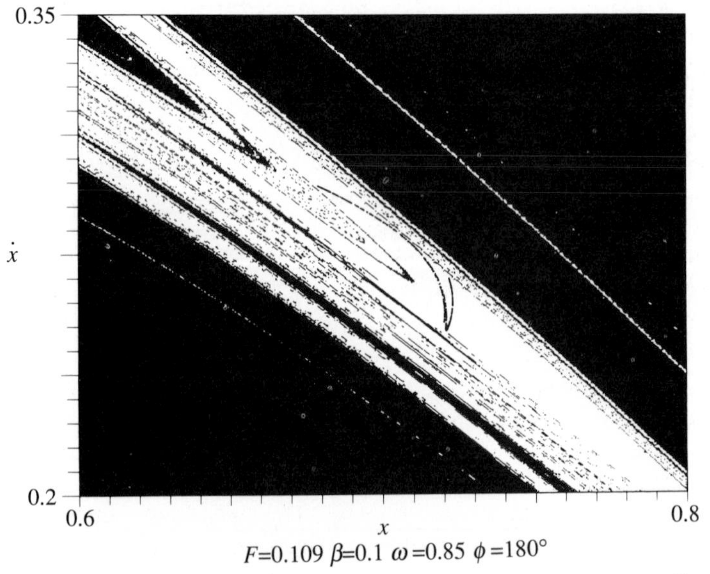

Fig. 9.9. The fractal basin boundary touching the chaotic attractor at $F = 0.109$. One band of the two-band chaotic attractor can be seen in black in this window, at phase $\phi = 180°$; white denotes its basin of attraction.

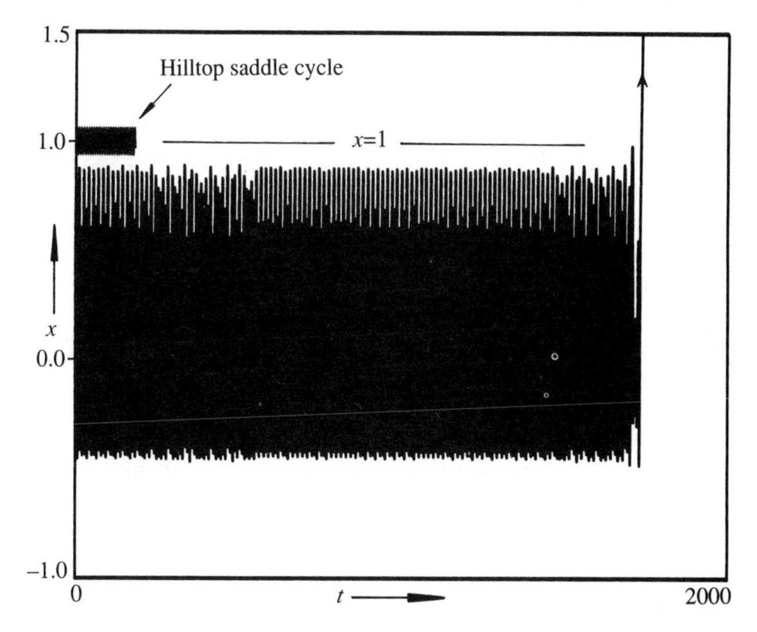

Fig. 9.10. A chaotic transient leading to escape beyond the hilltop saddle cycle at $\omega = 0.85$, $F = 0.1092$. The full $x(t)$ waveform is shown, with no stroboscopic sampling.

order 6, as shown in Fig. 9.8. Here we are again sampling at $2T$, so $n = 6$ fixed points have only 3 traces in this window. This D^6 is the remnant of a highly localized $n = 6$ fold–flip–cascade–crisis scenario which can only be located by careful and detailed numerically exploration. The fractal basin boundary around one band of the chaotic attractor at E is shown in Fig. 9.9, three mapping points of D^6 being located at the extremes of the band where it appears to touch its basin boundary.

This type of instability of a chaotic attractor, in which the attractor is in collision, not with the governing hilltop saddle cycle, but with an *accessible orbit* on the basin boundary [4] has been called a chaotic saddle catastrophe by my colleague Bruce Stewart [5]. The distance of the chaotic attractor from the hilltop saddle cycle, which forms a sort of dynamic barrier against escape, is shown in Fig. 9.10. Here a chaotic transient motion at $F = 0.1092$, which in its early form is very similar indeed to the chaotic attractor at $F = 0.109$, escapes quite suddenly after oscillating for many forcing cycles in an apparently safe manner at a rather comfortable distance from the hilltop saddle cycle. The uneasiness with which any engineer or applied scientist would view this sudden escape is compounded by the fact that the length of such an escaping transient is essentially a probabilistic quantity, depending sensitively on the starting conditions of the motion.

9.5 Homoclinic tangles and fractal basin boundaries

Having examined the behaviour of the attractors under increasing F at $\beta = 0.1$ and $\omega = 0.85$, let us now turn to the corresponding metamorphoses of the basins of attraction.

We have already observed in Fig. 9.2, how, for the autonomous system at $F = 0$, the safe basin of nonescaping starts is bounded by the stable manifold or inset of the hilltop equilibrium state. Under increasing F, this equilibrium is replaced by the unstable hilltop saddle cycle, and the stable manifold of this cycle takes on the role of the separatrix in a Poincaré section.

For a while, this separatrix has the general form of Fig. 9.2, but as F is increased the inset approaches the outset until they touch at a homoclinic tangency, as shown in Fig. 9.11 (this figure has $\omega = 1.0$, but an entirely similar diagram would be observed at $\omega = 0.85$). We should remember here that we are now dealing not with a flow but with a map: a given solution does not trickle around the manifolds but takes large strides along them. In fact point α maps to point β, illustrating the concept that if there is one tangency between the manifolds, then there must be an infinite number of tangencies. If we map forwards in time from β, there will be an infinite

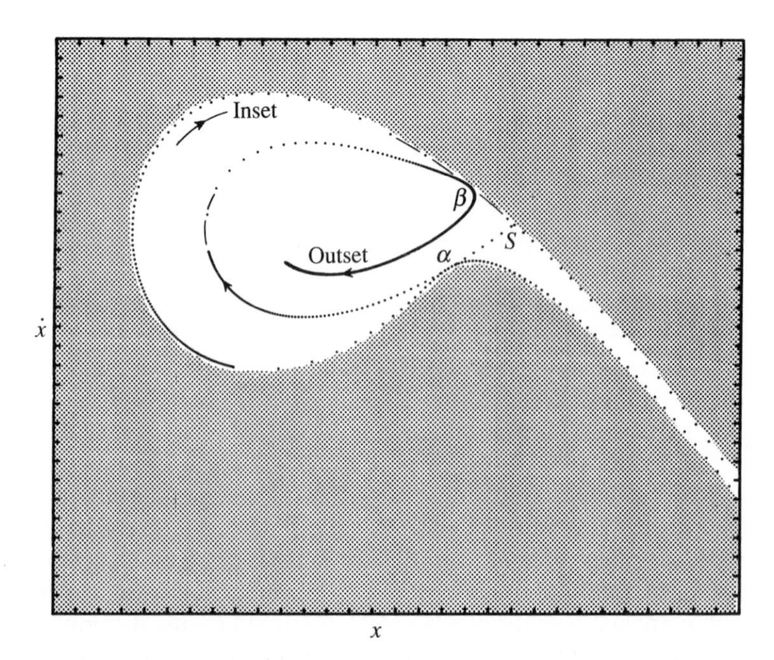

Fig. 9.11. The homoclinic tangency of the stable manifold (inset) and the unstable manifold (outset) of the hilltop saddle cycle S at $\omega = 1.0$, $F = 0.074$, $\phi = 180°$. Point α maps to point β under one forcing cycle.

number of points, lying necessarily on both the inset and the outset, as the solution tends towards the hilltop saddle S; similarly, mapping backwards in time from α there must be an infinity of tangencies between α and S. In diagrams such as this the invariant manifolds are determined, not by following just a single trajectory, but by mapping forwards and backwards in time from ladders of starts along the eigenvectors of the saddle S.

This homoclinic tangency of the stable and unstable manifolds of the hilltop saddle cycle has been predicted analytically using Melnikov's perturbation theory [2], and the resulting global bifurcation curve is shown on Fig. 9.4 as arc M. This analytical prediction is in good agreement with numerical checks in the region of current interest.

After the tangency, the manifolds intersect transversely, an infinite number of times, giving what is called a homoclinic tangle: and the boundary of the safe basin, although still defined by the continuous stable manifold, becomes fractal in nature [6] due to the complex accumulations implied by the tangling process. Some details of the highly developed tangle at the escape value of $F = 0.109$ can be seen in Fig. 9.12. The two-band chaotic attractor is indicated, and a homoclinic trajectory which lies on

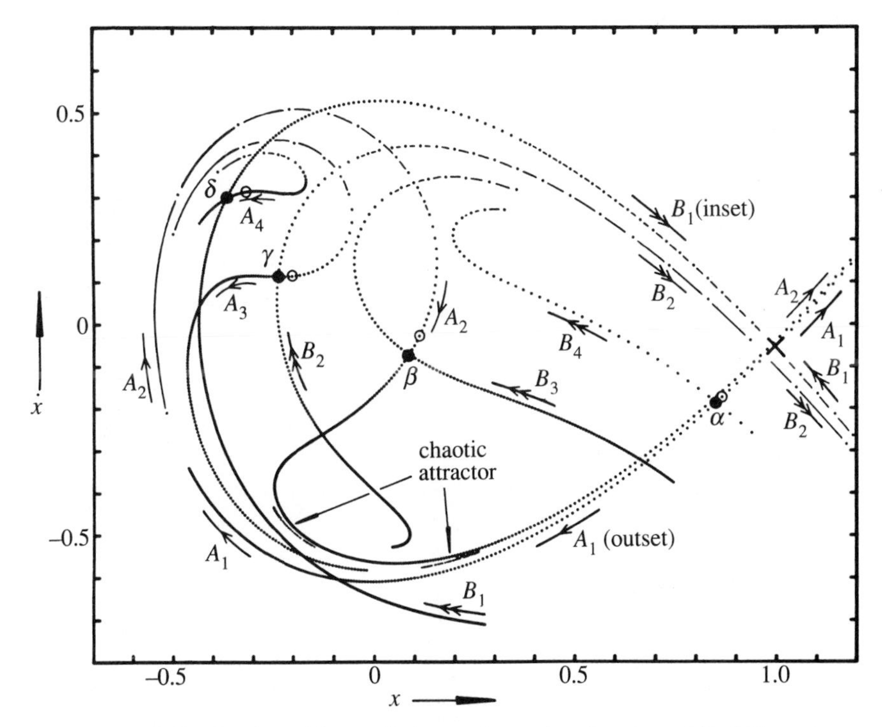

Fig. 9.12. Fully developed homoclinic tangle at the chaotic saddle bifurcation, $\omega = 0.85$, $F = 0.109$. A_n denotes the outset at n steps from the outgoing eigenvector; B_n denotes the inset at n steps backwards in time from the ingoing eigenvector. A homoclinic trajectory follows the sequence α, β, γ, δ.

both the inset and outset is illustrated by the mapping sequence *outgoing eigenvector* $\rightarrow \alpha \rightarrow \beta \rightarrow \gamma \rightarrow \delta \rightarrow$ *ingoing eigenvector*. An adjacent trajectory, which lies on the outset but not on the inset, is shown by the open circles.

9.6 Erosion of the safe basin of attraction

This homoclinic tangling, together with some subsequent heteroclinic events, generates a dramatic erosion of the safe basin of attraction as illustrated in Fig. 9.13, which charts the evolution of the safe basin in a Poincaré section at phase $\phi = 180°$ as F is increased in equal steps from 0 to 0.11, just beyond F^E. These basins were obtained by the simpler and in a sense more reliable procedure of just running computer time integrations from a grid of starts, and marking white or black according to whether or not the system escaped over the hilltop within m forcing cycles, m being here equal to 16. This finite and relatively low value of m accounts for the erroneous black region in the last figure where F is just greater than F^E: the black starts would in fact escape by chaotic transients if the simulations had been run for a longer m. Experience has confirmed, however, that

ω=0.85, β=0.1, ϕ=180°, m=16, $-0.8<x<1.2$, $-1.0<y<1.0$.

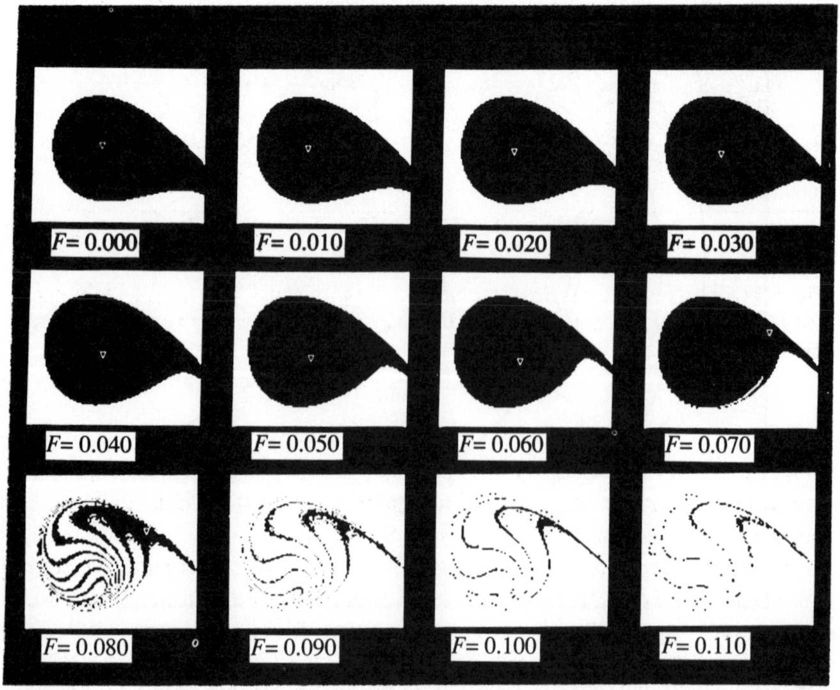

Fig. 9.13. A basin erosion sequence showing the sudden incursion of fractal fingers into the safe basin of nonescaping starts in the (x, y) Poincaré sections.

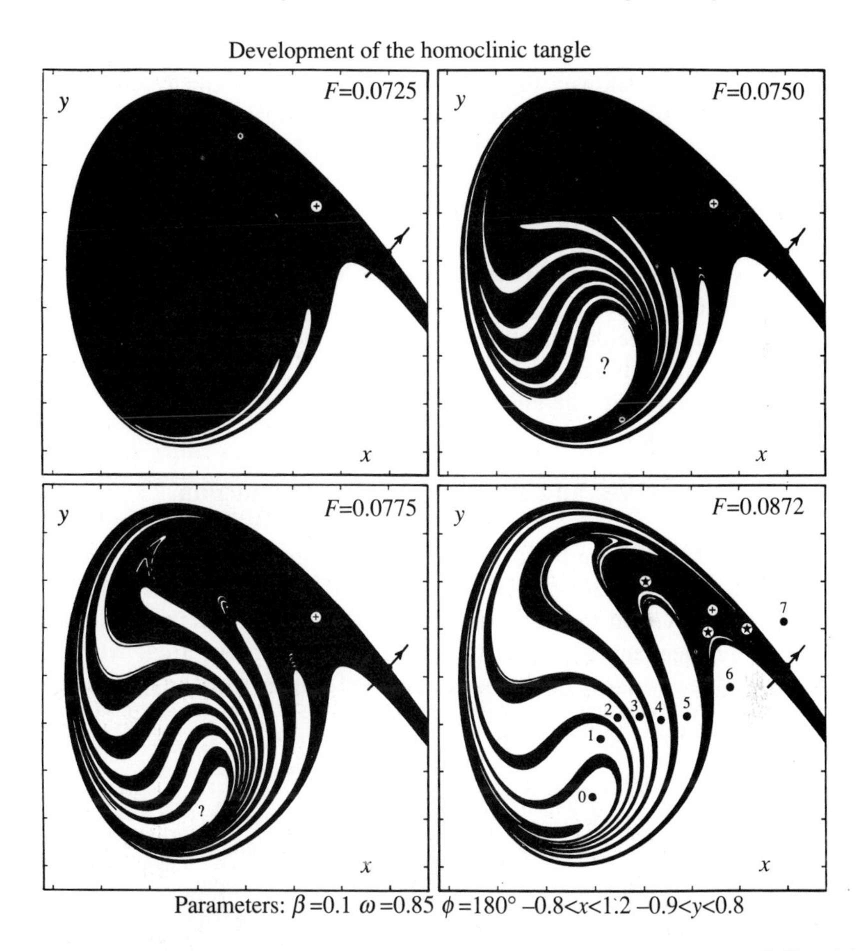

Fig. 9.14. A finer view of the incursion process. Crosses denote the main $n = 1$ attracting fixed point; stars show a coexisting $n = 3$ attractor. In the last diagram the points numbered 0 to 7 show a mapping sequence leading to escape.

$m = 16$ is sufficient to show the major features of the basin changes that are of concern to us here. The white triangle shows the Poincaré mapping point of the relevant $n = 1$ harmonic attractor

We can observe that there is a rapid incursion of the fractal fingers of the stable manifold between 0.07 and 0.08, a little after the homoclinic tangency at $F = 0.0633$. This incursion process is shown in somewhat finer detail in Fig. 9.14, obtained by mapping from the incoming hilltop eigenvectors using elaborate computer algorithms to ensure an acceptable and uniform density of points along the manifold [7]. In all four diagrams we can see the single $n = 1$ mapping point marked by a cross, and in the final diagram a coexisting $n = 3$ solution is marked by stars. The hilltop saddle solution is shown, together with its outgoing eigenvectors, the incoming eigenvectors lying along the blackwhite basin boundary. In the first diagram, we see three visible fingers, followed by a dramatic incursion in the second and third. In the final diagram the safe basin is quite substantially eroded, and the mapping sequence $0 \to 1 \to 2 \to 3 \to 4 \to 5 \to 6 \to 7 \to$ *escape* shows how a chaotic transient motion steps sequentially through the fingers before the system finally escapes over the hilltop.

It is the incursion of the boundary into the central region of the Poincaré section, and in particular its close approach to the safe attractor(s) that is of primary concern to the engineer. A nonincursive albeit fractal edge has very little influence on the engineering integrity of the system.

9.7 Integrity curves and the Dover cliff phenomenon

We have examined a number of ways of quantifying this erosion [8], using for example the distance from the boundary to the main-sequence attractor. One coarse but useful measure of basin erosion, which has the advantage of being independent of the attractors, is just the basin area, within a suitably defined window, and this can be used to generate the engineering

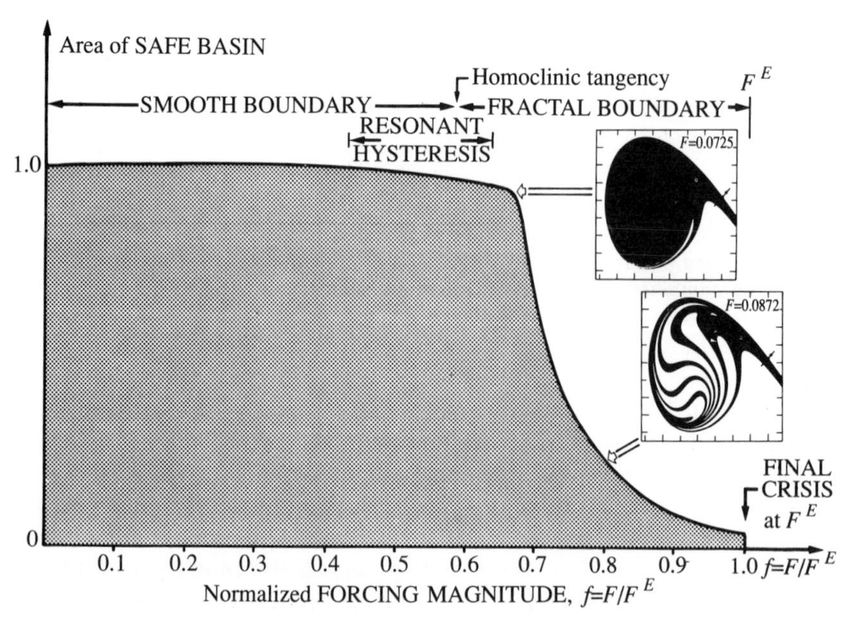

Fig. 9.15. Loss of engineering integrity, quantified by the normalized area of the safe basin of attraction. The 'Dover cliff' loss of integrity occurs just after the onset of a fractal boundary at the homoclinic tangency.

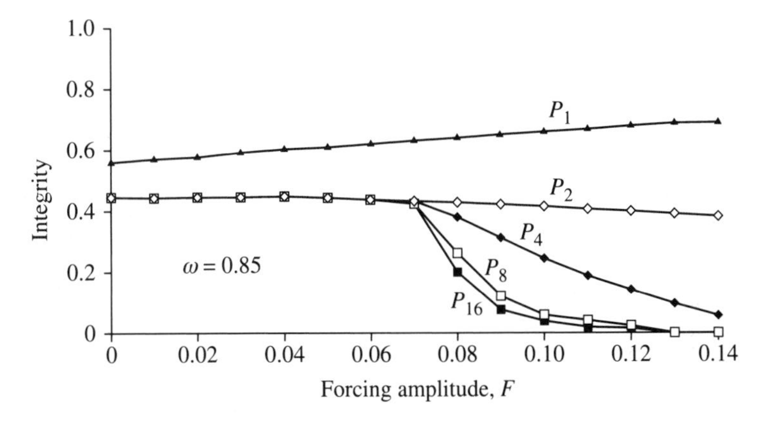

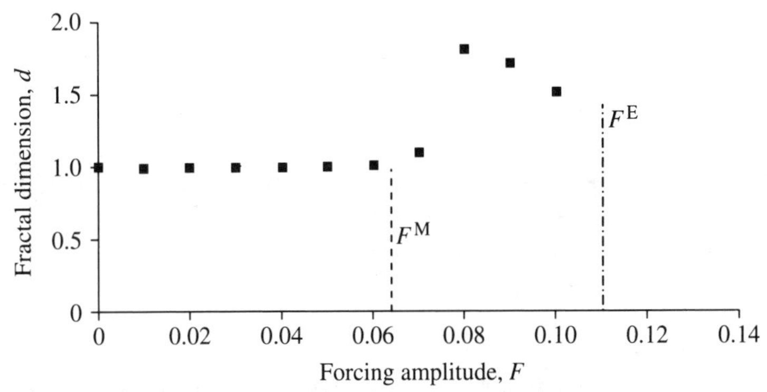

Fig. 9.16. Variation of engineering integrity, based on escape within 1, 2, 4, 8, and 16 forcing cycles, correlated with rough estimates of the fractal dimension of the basin boundary.

integrity diagram of Fig. 9.15. Here we can observe the quite sharp drop in area, associated with the incursion of the fractal fingers.

Recent research in our group at University College London has been directed towards the mechanism of this cliff-like basin erosion, which seems to proceed by a series of discrete implosions associated with orbital subharmonic motions. From a practical point of view, the sharp and well-defined cliff at about $f = F/F^E = 0.7$ could be used as a design criterion to replace the more conventional use of $f = 1$. It has the twin advantanges of being, first, more revelant to an engineering system in a noisy environment and, secondly, more easy to estimate given the complexity of the final crisis at E (Fig. 9.8).

Figure 9.16 shows the corresponding integrity diagrams that would be obtained by running simulations from a grid of starts for only $m = 1$, 2, 4, 8, or 16 forcing cycles. There is relatively little change between the P_8 and P_{16} curves, suggesting that P_{16} is a reasonably good approximation to P_∞ These integrity curves are correlated with rough estimations of the basin boundary dimension [9] in the lower diagram. Here the fractal dimension changes from unity when the boundary is smooth with $F < F^M$, to a noninteger value between 1 and 2 when the boundary is fractal with $F > F^M$.

9.8 Application to a real ship capsize

Figure 9.17 shows the basin erosion process for the *Edith Terkol*, a Danish coaster, modelled in the absence of wind loading [10]. Following its capsize, the characteristics of this vessel have been well documented in the naval architecture literature. We can see that this practical capsize problem displays the same type of fractal erosion as our canonical oscillator; and, given that a ship in waves is quite obviously never in a well-defined steady state, our *Dover cliff* criterion may well have a role to play in naval architecture.

A short pulse of regular waves directed at a ship in rather ambient condi-

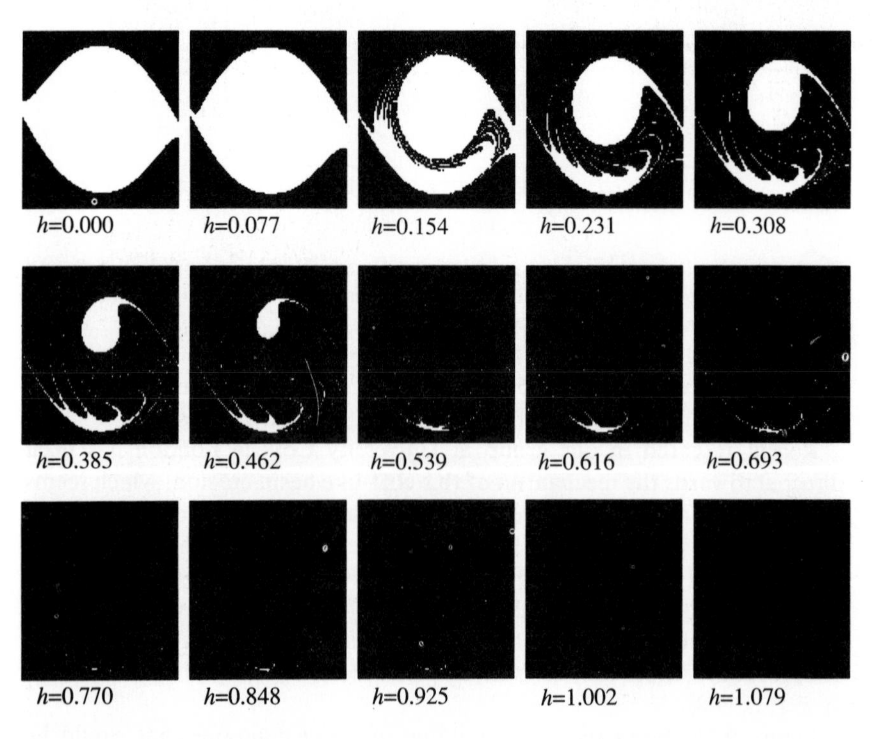

Fig. 9.17. Erosion of the safe transient basin for the *Edith Terkol*, a Danish coaster whose capsize has been extensively documented in the naval literature. The study is made in the absence of wind loading, and the wave height parameter h is normalized by the final steady-state capsize value.

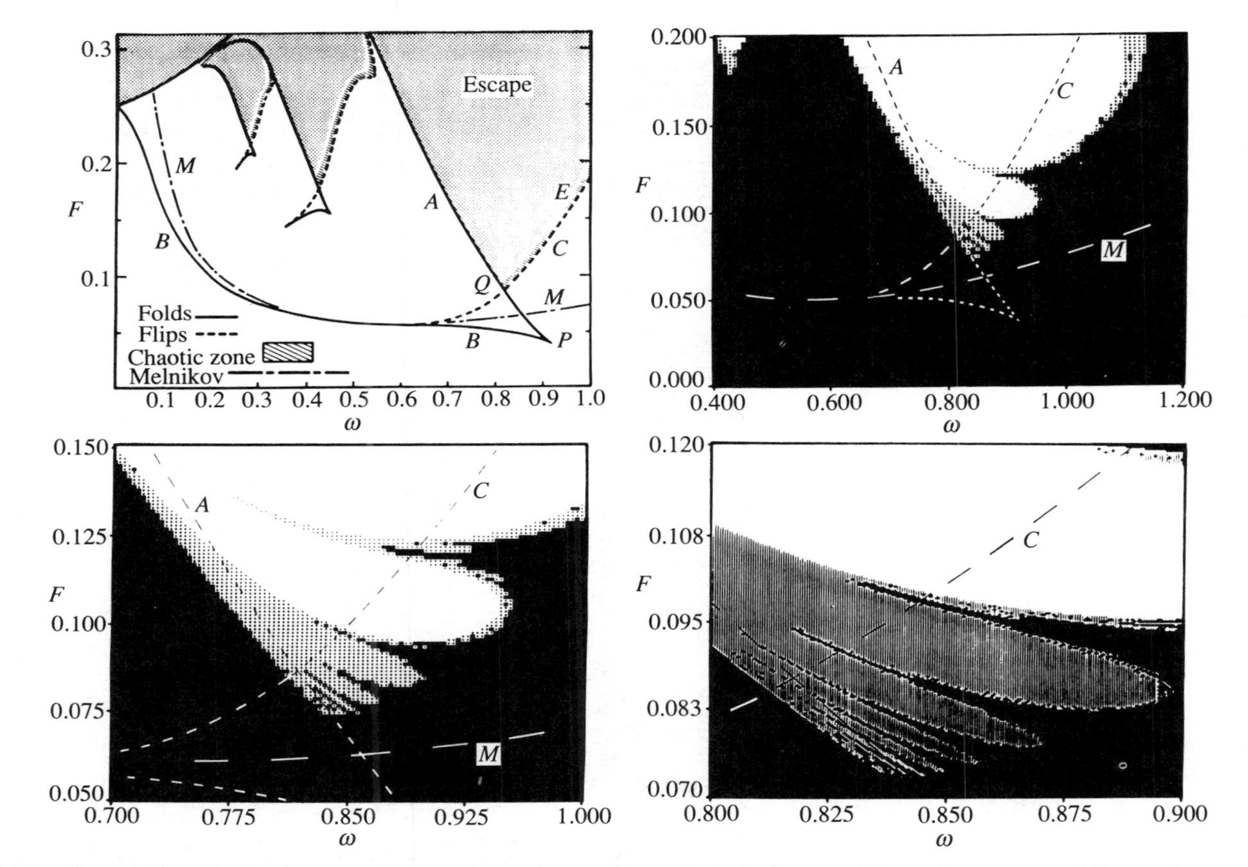

Fig. 9.18. Fractal boundaries in control space, superimposed on the steady-state bifurcation curves. All transients start at $x(0) = y(0) = 0$.

tions could well be regarded as a *worst-case* test loading, and our earlier conclusion that $P_{16} \simeq P_\infty$ tells us that if a ship does not capsize in about 16 waves it is unlikely to capsize at all. Moreover, since the incursive fingers of the (x, y) basins sweep across the whole central domain, a single run from (0, 0) is a rough and ready way of locating the cliff, and is sufficiently economical to be feasible in a laboratory test programme, rather than just in a computer analysis.

9.9 Fractals in control space

If we were to consider that a ship, or our cubic oscillator, were resting precisely in the ambient state $x(0) = y(0) = 0$ when subjected to a sudden pulse of regular sinusoidal waves of intensity F and frequency ω, we generate the fractal escape boundaries in control space shown in Fig. 9.18, taken from [9]. Here they are superimposed on the steady-state bifurcation curves.

These boundaries are easily understood to be fractal if we observe that the safe basin of nonescaping starts can be thought of as an object in the four-dimensional space spanned by $x(0)$, $y(0)$, F and ω. Our earlier basins were two-dimensional cross-sections defined by holding F and ω constant: the fractals in control space that we are now contemplating are two-dimensional cross-sections defined by hold $x(0)$ and $y(0)$ constant, at zero.

9.10 Concluding remarks

I hope this chapter conveys some flavour of the work we are doing at University College London, applying modern ideas of chaos, fractals, and more generally the geometrical concepts of phase space dynamics to practical problems arising for example in marine technology. In this area it seems that fractal basin boundaries, associated with chaotic transients, may be much more important than chaotic attractors. Regular noise-free forcing is never likely to persist long enough for the attractors to be observed. But there is evidence to suggest that the basin erosion process, based as it is on the whole ensemble of trajectories, rather than on just one steady-state motion, may be much more robust and relevant. Our more recent discoveries about the basin erosion process can be seen in references [11–14].

References

1. Thompson, J. M. T. and Stewart, H. B. (1986). *Nonlinear dynamics and chaos.* Wiley, Chichester.

2. Thompson, J. M. T. (1989). Chaotic phenomena triggering the escape from a potential well. *Proc. R. Soc. Lond.* A **421**, 195–225.
3. Thompson J. M. T., Rainey, R. C. T., and Soliman, M. S. (1990). Ship stability criteria based on chaotic transients from incursive fractals. *Phil. Trans. R. Soc. Lond.* A, **332**, 149–67.
4. Grebogi, C., Ott, E., and Yorke, J. A. (1987). Basin boundary metamorphoses: Changes in accessible boundary orbits. *Physica* D, **24**, 243–62.
5. Stewart, H. B. (1987). A chaotic saddle catastrophe in forced oscillators. In: *Dynamical systems approaches to nonlinear problems in systems and circuits* (ed. F. Salam and M. Levi). SIAM. Philadelphia.
6. McDonald, S. W., Grebogi, C., Ott, E., and Yorke, J. A. (1985) Fractal basin boundaries, *Physica* D, **17**, 125–53.
7. Alexander, N. A. (1989). Production of computational portraits of bounded invariant manifolds. *J. Sound Vib.*, **135**, 63–77.
8. Soliman, M. S. and Thompson, J. M. T. (1989). Integrity measures quantifying the erosion of smooth and fractal basins of attraction. *J. Sound Vib.* **135**, 453–75.
9. Thompson, J. M. T. and Soliman, M. S. (1990). Fractal control boundaries of driven oscillators and their relevance to safe engineering design. *Proc. R. Soc. Lond.* A, **428**, 1–13.
10. Soliman, M. S. and Thompson, J. M. T. (1991). Transient and steady state analysis of capsize phenomena. *Appl. Ocean Res.* **13**, 82–92.
11. Soliman, M. S. and Thompson, J. M. T. (1992). Global dynamics underlying sharp basin erosion in nonlinear driven oscillators. *Phys. Rev.* A, **45**, 3425–31.
12. Thompson, J. M. T. (1992). Global unpredictability in nonlinear dynamics: capture, dispersal and the indeterminate bifurcations. *Physica* D, **58**, 260–72.
13. McRobie, F. A. and Thompson, J. M. T. (1992). Invariant sets of planar diffeomorphisms in nonlinear vibrations. *Proc. R. Soc. Lond.* A, **436**, 427–48.
14. Thompson, J. M. T., Rainey, R. C. T., and Soliman, M. S. (1992). Mechanics of ship capsize under direct and parametric wave excitation. *Phil. Trans. R. Soc. Lond.* A, **338**, 471–90.

10
Applications of chaos to meteorology and climate

Peter L. Read

Chaos and meteorology have had a close relationship for a number of years. The Lorenz attractor is well known as a very influential mathematical model whose properties were first studied in detail in the 1960s by the meteorologist/mathematician Ed Lorenz, working at MIT. It became a prototype for chaotic dynamics in Lorenz's celebrated paper [1], published in 1963 in the *Journal of Atmospheric Sciences*. The most significant point in the present context is that it was originally motivated by a meteorological problem.

One of the main goals of meteorology, probably more than for any other physical science, is to predict the future. It is not too surprising, therefore, that meteorology has become associated in the minds of many with chaos. But to assert that the Lorenz model is valid and appropriate as a model of the atmosphere is a considerable overstatement. It was actually derived by Barry Saltzman [2], a collaborator of Lorenz working at Yale University, as an extreme idealization of Rayleigh–Bénard convection in a box (hardly a realistic model of atmospheric circulation!). Saltzman arrived at the model by carrying out a Fourier expansion of the convection problem in space and retaining only the gravest modes, thereby effectively turning the partial differential equations (PDEs) of fluid motion into a finite set of ordinary differential equations (ODEs). This is, of course, to do considerable violence to the original problem. As is well known, the truncated model produces results of great beauty and complexity [3] (see Fig. 10.1), though it is a matter for concern as to how valid and universally applicable this approach might be.

In considering fluid motion, it is quite clear that we are dealing with a very different kind of problem to that discussed by Professor Thompson in Chapter 9. With most mechanical structures, there is usually little doubt that the system can be described by a relatively small number of ODEs; we are clearly dealing with a system with relatively few degrees of freedom. When we consider the atmosphere (or any other fluid system), on the other hand, such a system can (at least in principle) possess a near-infinite number of degrees of freedom. It is far from being self-evident that we can apply the ideas of so-called 'simple chaos' discussed in this book to such a system.

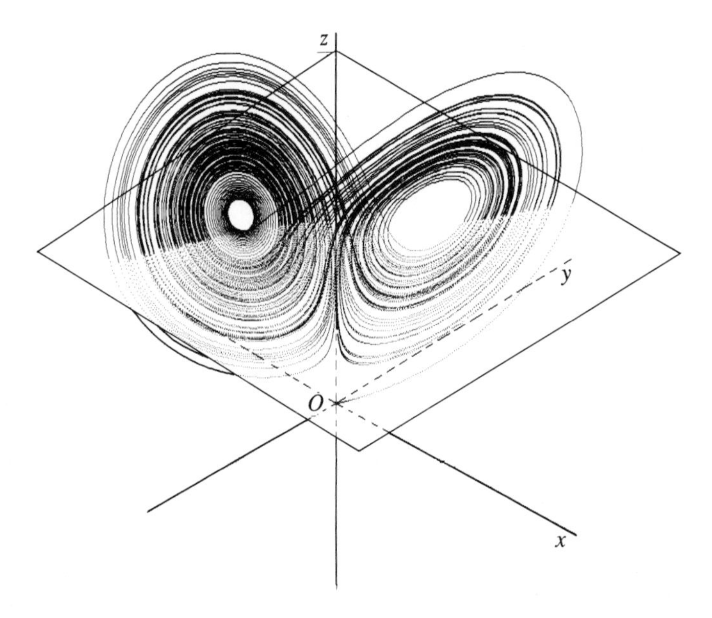

Fig. 10.1. Phase plane portrait of the Lorenz attractor [1, 3], obtained by solving the three coupled ODEs (i) $\dot{X} = \sigma(Y - X)$, (ii) $\dot{Y} = -XZ + r_a X - Y$, (iii) $\dot{Z} = XY - bZ$, with $\sigma = 10$, $b = 8/3$, and $r_a = 28$.

10.1 Chaos and 'turbulence'?

Before considering the details of atmospheric behaviour, it is important to be reminded of the essential properties (and limitations) of the modern concept of chaos. Chaos as discussed in this book is essentially concerned with the behaviour of a rather pure sort of system—a deterministic autonomous dynamical system. We may represent the state of such a system by a state vector $\boldsymbol{X}$ in its phase space, and its behaviour in time by the time derivative $\dot{\boldsymbol{X}}$. In an autonomous dynamical system, $\dot{\boldsymbol{X}}$ is just a function of the current state vector $\boldsymbol{X}$ itself, thus

$$\dot{X} = F(X).$$

In talking about *deterministic chaos*, therefore, we should not lose sight of the fact that we are dealing with a rather special type of system which happens to be capable of exhibiting complex, often aperiodic, behaviour in time. A critical property of this type of complex motion is its systematic sensitivity to initial conditions, which can be characterized by a spectrum of Lyapunov exponents, as discussed in earlier chapters, with one or more exponents greater than zero.

Systematic sensitivity to initial conditions (SIC) is a property central to the modern theory of chaos, but is this property unique to deterministic chaos? It is likely, for example, that stochastically perturbed systems (which are not chaotic in a formal sense) may also mimic systematic SIC, at least as manifest in datasets of finite duration. The other essential feature of chaos as studied in this book is its relatively low attractor dimension, with SIC arising primarily because of the attractor's infinitely folded and stretched fractal structure in phase space. With a small attractor dimension, there exists a kind of simplicity which underlies apparently complex, often aperiodic, behaviour in time. This might, for example, imply a degree of spatial and/or temporal order, even though the essential behaviour is predictable for only a limited period [4]. We are not, therefore, considering a completely random sort of motion, as might be implied by concepts typical of the 'turbulence' discussed by engineers and boundary-layer meteorologists which, even if consistent with the existence of an attractor, would not be confined to an attractor of low dimension.

The kinds of question which will motivate the following discussion of the relationship between modern concepts of chaos and meteorology must therefore include the following:

(1) Is the atmosphere *autonomous* and *deterministic*?

(2) Does the atmosphere exhibit systematic sensitivity to initial conditions?

(3) Is atmospheric behaviour *complex* or *complicated* (i.e. chaotic or 'turbulent' in the sense of 'high' or 'low' attractor dimension [4])?

(4) Are small (poorly resolved) scales and diverse physical processes *dominant* in governing the atmosphere's intrinsic variability and predictability?

10.2 Detection of atmospheric chaos?

The first question above verges on the realms of philosophy and psychology, especially given the recent interest in the extent to which man's activities (which are presumably not deterministic in the sense discussed here!) influence weather and climate. It is pertinent, however, to the meteorologists' approach to the formulation of models of atmospheric behaviour, as we shall see below in Section 10.3

The second question relates to the intrinsic predictability of the atmosphere. The fact that practical weather forecasts are not completely reliable is well known (indeed notorious) to everyone. As we shall see later, the typical reliability of forecasts decreases the further ahead we try to predict, which is at least consistent with the concept of SIC as discussed above. But the other crucial property of chaos is the existence of an attractor of low dimension governing the dynamical behaviour of the system, indicating an underlying simplicity which belies apparent complexity. When we look at the atmosphere, e.g. from space (see Fig. 10.2), we see an enormous range of scales of motion, ranging from individual clouds (on a scale of a few tens or hundreds of metres) to the scale of the planet itself, though with some suggestion of a degree of order on the larger scales. Is it reasonable [4] to expect such a system to behave in a way which is characterizable by an underlying simplicity?

These questions would be straightforward to answer directly from atmospheric observations, if we were able reliably to detect and characterize chaotic motion from measured data. Tom Mullin and Robert May discussed some techniques in earlier chapters which claim to detect and characterize chaos, at least from high-quality data generated in well-designed laboratory experiments. Is it possible to use such methods to distinguish chaos from

Fig. 10.2. Visible image of the Earth taken from the *Apollo 17* spacecraft, showing varied cloud patterns over Southern Africa and the South Atlantic and Indian Oceans (photograph courtesy of NASA). The Antarctic continent is also visible at the bottom of the picture.

other more complicated kinds of behaviour, in meteorological or climatological data? The principal quantity to be determined here is an estimate of the underlying attractor dimension, as a measure of the complexity of behaviour, whether or not the motion is regular and periodic or noisy and aperiodic.

One example of such a technique, which has been used in the analysis of laboratory measurements with a fair degree of success claimed for several years now, is the Grassberger–Procaccia correlation dimension algorithm [5], which relies on a method akin to box-counting familiar in the context of fractals (as discussed earlier by Tom Mullin). Several attempts have been made in recent years to apply this technique to climatological data, with various claims to have demonstrated the existence of an attractor with a dimension lying between around 5 and 9. This is rather larger than typically considered in currently studied models of chaotic motion, but is nonetheless very small compared to the near-infinite dimension obtainable (at least in principle) in a fluid. If correct, an attractor dimension of ~ 6 would offer very real grounds for hope that at least the larger scales of atmospheric motion responsible for the organization of weather systems may be capable of being reliably simulated by a relatively simple model.

Unfortunately, if we look more closely at these optimistically low estimates of atmospheric attractor dimension, the reliability of these claims would seem to be in some doubt. Experience from the application of dimension estimation methods to laboratory data shows that robust estimates require large quantities of very high quality data. In Chapter 3, Tom Mullin mentioned that the amount of data required goes up roughly exponentially with the implied attractor dimension, with the 'rule of thumb' that more than 10^D independent data points are necessary to measure a dimension $\sim D$. If we are interested in the weather time scales, this would imply that we need measurements sampling no more frequently than (say) once per day (to ensure samples are mutually independent), with a total dataset of at least a million days. This would require a uniform daily climatological record lasting at least 2500 years. Even the longest existing daily climatological records (such as maintained in Oxford since the seventeenth century) are woefully inadequate. In addition, laboratory experience indicates that data series must not only be of long duration but also have very low noise levels (ideally with a signal-to-noise ratio exceeding several hundred). This would imply measuring atmospheric temperatures to $\sim 0.01\,°\mathrm{C}$, around an order of magnitude better than currently recorded in meteorological data series.

While it may be possible that less demanding methods of characterizing chaos, such as the nonlinear prediction approach discussed earlier by Robert May, might eventually be developed to the extent that they can be applied reliably to more modest data series of inferior quality, this remains to be convincingly demonstrated in full generality. In the meantime, we have little alternative but to take a more qualitative and 'free-wheeling' approach to the possible application of the ideas of chaos to weather and climate. In the rest of this chapter, we will therefore discuss a range of phenomena which, although apparently complex and aperiodic in their behaviour, appear to exhibit a degree of underlying simplicity, and which may therefore be potential candidates for atmospheric chaos.

10.3 Atmospheric prediction and predictability

To begin with, we return to the first two questions of Section 10.1, concerning the extent to which the atmosphere is deterministic and exhibits SIC. In the absence of an unambiguous detection of atmospheric chaos and characterization of the underlying attractor, one approach to these questions is via the current practice of numerical weather prediction.

Since the late 1960s, when the advance of computer technology made the techniques practically feasible, weather prediction up to several days ahead has been made using state-of-the-art methods of computational fluid dynamics (numerical weather prediction, or NWP). At the present time, major meteorological centres, such as the UK Meteorological Office in

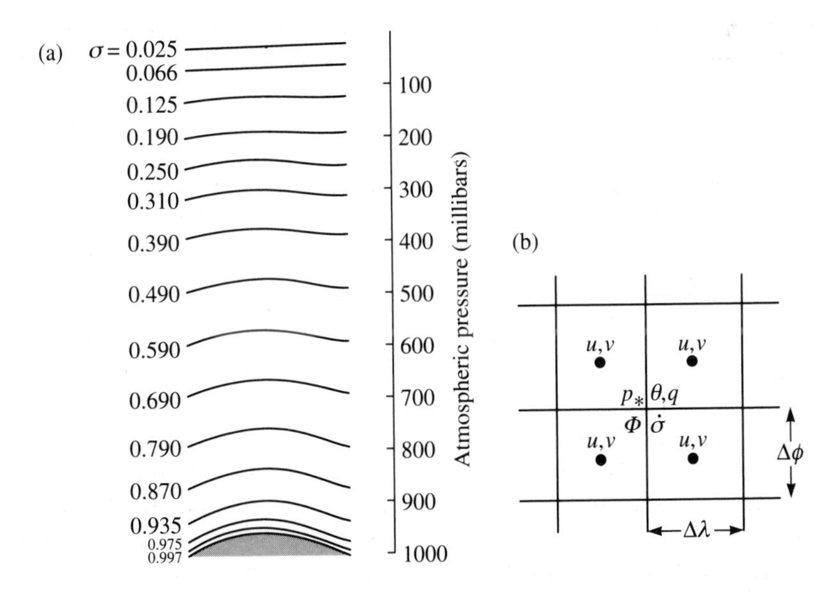

Fig. 10.3. Computational grids for a global numerical weather prediction model in use at the UK Meteorological Office after [7], reproduced with permission of the Controller of Her Majesty's Stationery Office): (a) discretization of levels in the vertical, showing terrain-following σ-coordinates; (b) discretization in the horizontal, showing the storage of horizontal velocity, pressure, temperature (Θ), vertical velocity ($\dot{\sigma}$), geopotential height of pressure surfaces (Φ) and humidity (q).

Bracknell, make use of a numerical model of the global atmosphere, in which the main physical variables (three components of wind velocity (u, v, w), pressure p, temperature T, and humidity q) are represented on a three-dimensional grid covering the whole surface of the Earth with a spacing of typically 1.5° latitude (about 150 km) in the horizontal and with 15 levels in the vertical spanning a range in altitude, from around 25 m to 25 km (see Fig. 10.3). This results in a model [6, 7] with nearly 2×10^6 variables in total, and several hundred thousand independent degrees of freedom (since not all the variables are mutually independent), allowing for the possibility of immensely complicated behaviour. Even so, the model still only resolves explicitly the larger scales of motion, i.e. those effectively influenced by the Earth's rotation.

Boundary conditions are specified to represent features such as surface orography (mountains, continents, and oceans), sea temperature, snow cover, sea ice, surface albedo, soil moisture, and thermal capacity, all specified externally or represented as a function of the explicit model variables (i.e. parametrized). Similarly, all the various processes which are not explicitly resolved or calculated in detail, such as boundary layer turbulence, solar and thermal radiation, convection, precipitation, surface exchanges, and clouds, are also parametrized with respect to the large-scale model fields. By hypothesis, therefore, the model assumes that the entire atmosphere as resolved by the model can be represented by a completely deterministic system, representable in the canonical form

$$\dot{X} = F(X)$$

(where the state vector X has $\sim 10^6$ components!). By definition, therefore, the result of a numerical model integration (provided it is long enough) is a form of chaos, though not necessarily of low dimension. Although we cannot know *a priori* that this is a rigorously correct assumption to make, we can assess by experience from the performance of the model forecasts whether the hypotheses of determinism and SIC for the atmosphere are reasonable.

In making forecasts, observations are used to define an initial state, which is fed into the model at time t. The model is then integrated forwards in time, typically up to five or six days ahead $(t + 6)$. The forecast can then be compared with observations taken on subsequent days, and the errors quantified by statistical analysis. A typical assessment of forecast errors (averaged over the Atlantic area for a year) is shown in Fig. 10.4, which shows the root-mean-square (rms) error in the pressure field as a function of forecast time [6]. Thus, rms errors in a 24 h forecast are systematically less than for 48 h, which are systematically less than for 72 h, and so on. The stars in Fig. 10.4 show how errors would grow if it were assumed that the initial pattern at time t does not change ('tomorrow's weather will be the same as today's'—the *persistence forecast*). The error of this

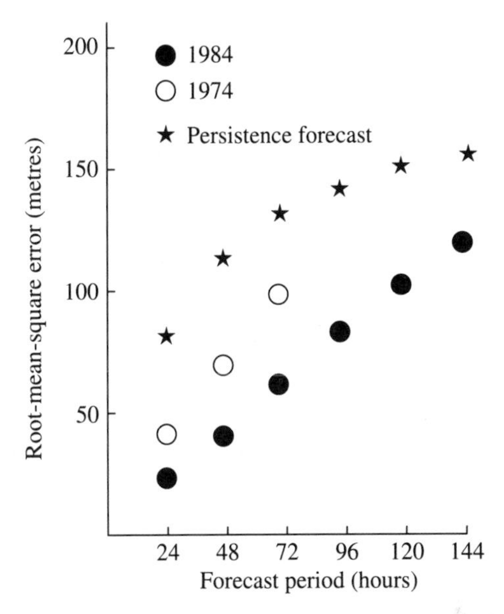

Fig. 10.4. The average growth of rms forecast error at an altitude of about 5000 m as a function of the forecast period (the time in advance of when the forecast is made) for the UK Meteorological Office numerical weather prediction model (after [6], reproduced with permission of the Controller of Her Majesty's Stationery Office).

'forecast' grows rapidly at first, then saturates at a value representative of the intrinsic rms variability of the atmosphere, i.e. the kind of error obtainable by choosing any typical atmospheric state at random. The persistence error thus indicates a measure of the relative skill of the deterministic forecast, suggesting that the model provides useful skill out to at least $t + 5$ or $t + 6$ days, which is the typical range routinely forecast by the UK Meteorological Office. The European Centre for Medium-Range Weather Forecasts (ECMWF, based in Reading) is investigating the possibility of using numerical weather forecast models to provide useful skill even further ahead, perhaps up to $t + 10$ days.

Analyses such as the one illustrated in Fig.10.4 provide some evidence that the atmosphere exhibits sensitivity to initial conditions, at least over the rotationally influenced range of scales resolved by typical NWP models. Since the model used to make the forecasts is, by definition from its formulation, dynamically deterministic, its apparent success in making short-term forecasts would also seem to suggest that the atmosphere, too, behaves as if it were also deterministic in a similar way. Furthermore, as NWP, models have been refined and extended in size and sophistication over the past $2\frac{1}{2}$ decades, the quality of forecasts has steadily improved. Figure 10.5

illustrates this by showing the rms errors for the North Atlantic region averaged over a year for 24, 48, and 72 h forecasts for each year between 1966 and 1989 from the UK Meteorological Office. For each forecast interval, the average error has decreased by a factor ~2 in the past 25 years, with perceptible jumps in the downward trend each time a new generation of model with higher spatial resolution is introduced (indicated by the arrows in Fig. 10.5). Thus, according to this (admittedly crude) measure of forecast skill, 48 h forecasts are now as accurate, on average, as 24 h forecasts were in 1970.

Such results clearly indicate that the continuing effort to improve numerical forecast models (and methods of data collection and analysis) is worthwhile at the practical level. But they also raise questions as to what particular factor has brought about this major improvement in forecasting capabilities. Is it, for example, because the models have simply increased their spatial resolution to represent a wider range of the dynamically significant spatial scales? Such an effect might be expected if atmospheric motion were truly complicated in the sense of 'classical turbulence', so that each scale behaves independently and influences every other. Alternatively, is it because new generations of models include more accurate parametrizations of unresolved physical processes, or because of some other factor,

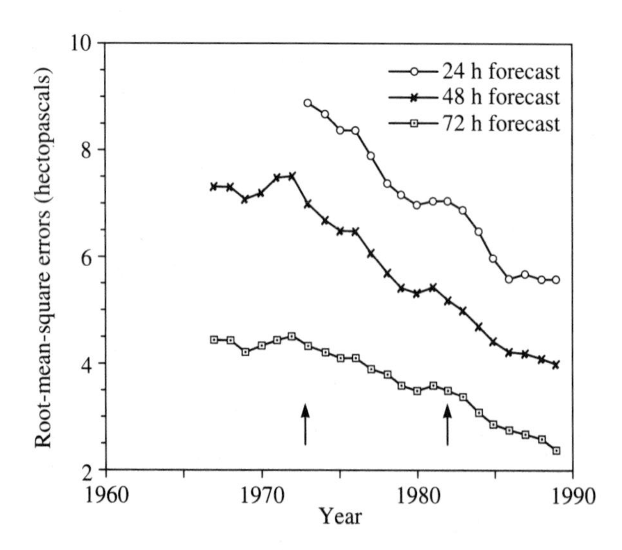

Fig. 10.5. Reduction in annual average forecast error as a function of epoch (during which NWP models were developed and improved) for (a) 24 h forecasts (squares), (b) 48 h forecasts (crosses) and (c) 72 h forecasts (circles), for successive generations of UK Meteorological Office NWP models (courtesy of C. D. Hall, Meteorological Office, with permission).

such as improved methods of atmospheric measurement and data analysis? As hinted at above, such questions are extremely difficult to resolve conclusively because of inadequacies in atmospheric measurements, and the complexity (especially the nonlinearity) and computational expense of the models themselves—it is simply too costly to run exhaustive tests of the models to investigate all the relevant factors, and in any event the available atmospheric datasets are not sufficiently accurate or complete to validate conclusively the atmospheric simulations in the first place. In seeking the ultimate source of the atmosphere's intrinsic unpredictability, therefore, we are forced to consider more circumstantial evidence.

10.4 Coherence in space: teleconnections?

It is often asserted, rather loosely, that the atmosphere is unpredictable because it is a highly complex turbulent fluid, but is this really the case?

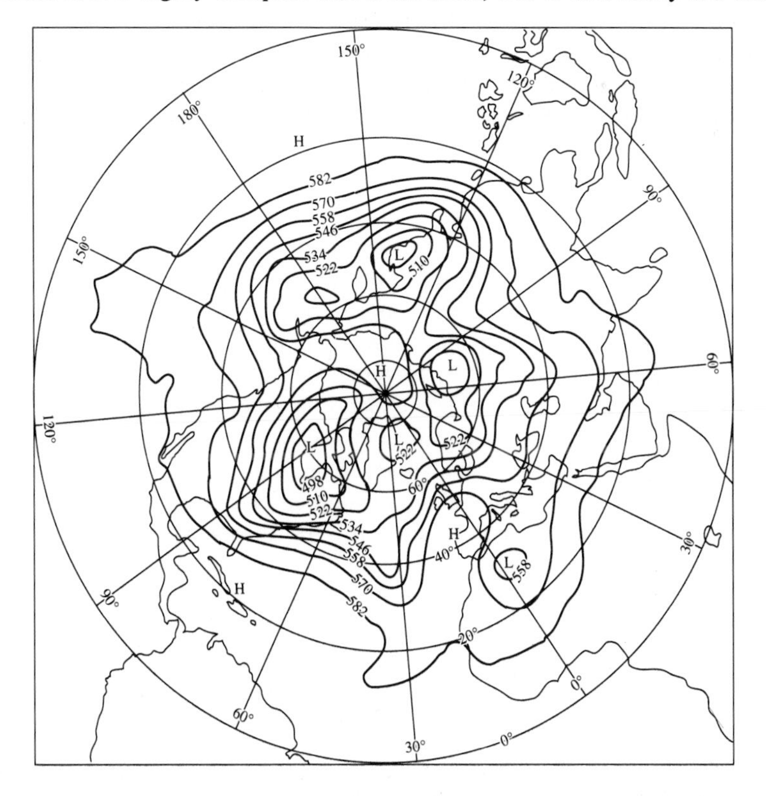

Fig. 10.6. Geopotential height contours of the 50 kPa pressure surface (effectively the pressure field at around 5000 m altitude) for a typical northern hemisphere winter day (from J. T. Houghton [8], with permission).

When physicists or engineers talk about turbulence, they normally envisage a kind of homogeneous irregularity spanning a huge range of relative length scales, with strong mixing occurring more or less continuously. It usually implies an energy cascade, in which energy is input at relatively large scales, but tends to end up in the very smallest scales where dissipation is most significant. But when we look especially at the typical, large-scale flow pattern in the atmosphere, is so-called 'classical turbulence' really an apt conceptual description?

Figure 10.6 shows a typical pressure field representative of the flow pattern near the middle of the lower atmosphere at a height of around 5 km. It is clear that the flow is dominated by a large-scale wave-like pattern. The pattern is certainly not particularly regular or symmetrical, but it is clearly not consistent with the notion of homogeneous irregularity at all scales. It has already been suggested that low-dimensional chaotic behaviour may be associated with motion involving only a restricted range of length scales. It is therefore of interest to examine the extent to which atmospheric flow acts coherently over various length and time scales.

One way of investigating the nature of coherent behaviour is to examine statistical correlations [9] between atmospheric measurements at different locations. Figure 10.7 shows an example of such an exercise, in which a reference location is chosen (in this case a point in mid-Atlantic, indicated by the small circle) and pressure fluctuations at every other location are correlated in time. Figure 10.7 then shows contours of the correlation coefficient ρ, such that $\rho = 1$ implies perfect correlation, $\rho = -1$ implies perfect anticorrelation and $\rho \sim 0$ indicates little or no correlation. Thus $\rho = 1$ at the reference point itself, and is found to decay with a characteristic length scale away from that point. It is important to note, however, that ρ does not decay uniformly to zero, but actually changes sign to give a significant anticorrelation at some locations a considerable distance from the original reference point (a so-called 'teleconnection' [9]).

If we restrict the range of time scales over which we seek correlations in space, by subjecting the original data (sampled twice daily) to a filter, some remarkable features emerge. Figure 10.7(b) shows what happens if we restrict attention to the weather-related time scales typical of middle latitude cyclones and anticyclones (familiar from daily weather maps), with periods roughly between $2\frac{1}{2}$ and 6 days. The pattern of ρ resembles a 'wave packet', aligned nearly east–west extending over more than 10 000 km from the east coast of the USA into central Europe, and with a wavelength around 4000 km. Further analysis shows this pattern to represent eastward-traveuing wave-like disturbances with a characteristic scale and structure, occurring in a preferred geographical location (sometimes referred to as a 'storm track'). This clearly indicates that mid-latitude weather systems exhibit some degree of coherent behaviour, but apparently as a sort of localized coherence within a more disordered open system—the wave-like pat-

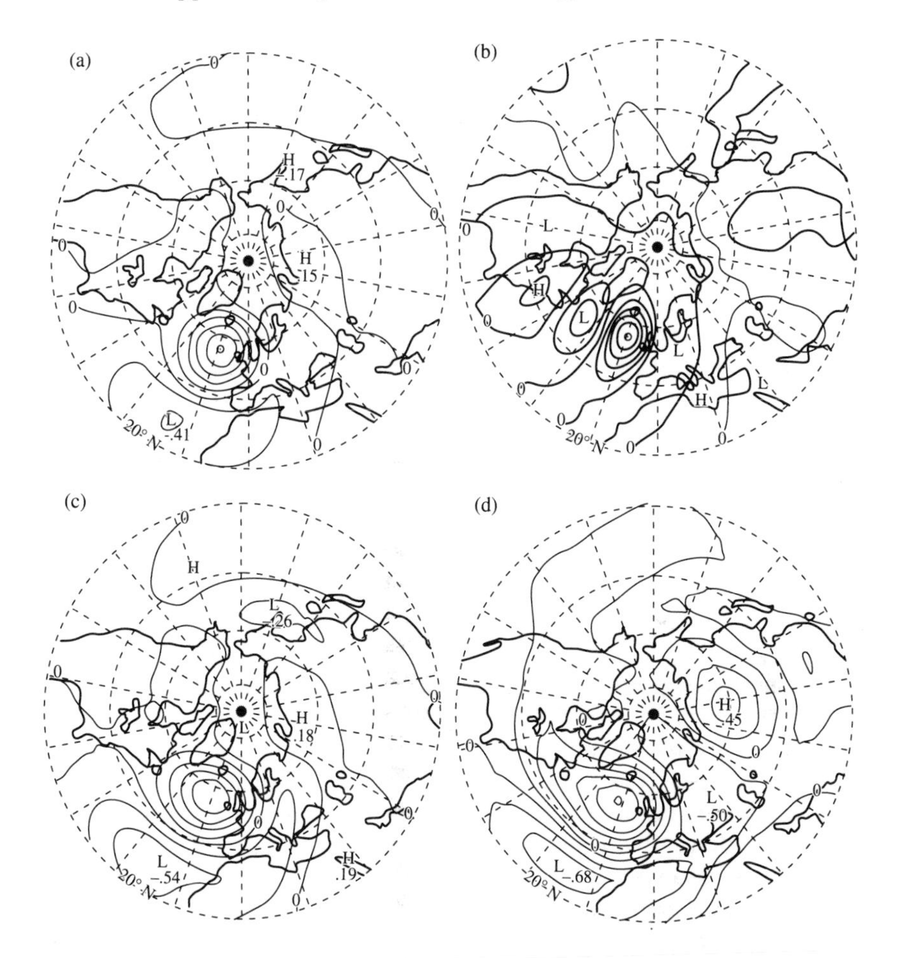

Fig. 10.7. Simultaneous temporal correlations between 50 kPa height fluctuations at 55°N, 20°W and other northern hemisphere locations during wintertime (after [9], with permission). Contours are of correlation coefficient (interval 0.2), and show (a) unfiltered twice daily data; (b) band-pass filtered data emphasizing fluctuations in the 2.5–6 day period range; (c) low-pass filtered data, emphasizing periods longer than 10 days, and (d) monthly mean data.

terns do not cover a large fraction of the globe and are not 'domain-filling'.

If we focus on rather longer, climate-related time scales, on the scale of a few weeks or longer, a rather different pattern emerges. Figure 10.7(c) and (d) show examples using low-pass filters or monthly mean data. The correlation pattern is again wave-like in character, but on a rather larger scale and extending over a large fraction of the globe. The 'waves' extend more nearly north–south along great circle paths and are found to represent quasi-stationary disturbances, somewhat akin to a standing wave with a period around a few weeks. These disturbances are believed [9] to be causally related to certain kinds of activity in the tropics. An anomalous region of unusually high temperature might occur at the surface of the tropical oceans (resulting from sporadic interactions between the circulations of the atmosphere and oceans), leading to enhanced atmospheric convection. This enhanced convection may then propagate its influence by exciting a form of inertial wave, in which motion is sufficently slow that it 'feels' strongly the influence of the Earth's rotation. Such waves tend to follow a great circle path on a sphere and nearly fill the available global domain.

Thus we find some evidence for certain kinds of phenomenon in the atmosphere which exhibit a degree of coherence in space. Simple chaos, however, is also expected to involve a limited degree of coherence in time, which we consider in the following sections.

10.5 Coherence in time: intraseasonal oscillations

Simple models of chaotic motion in fluids, such as the Lorenz attractor, typically represent the disordered behaviour in time of flow patterns comprising very simple spatial structure. Yet the disordered time dependence of such models is frequently not completely random, as revealed by a small attractor dimension, but might take the form of a nearly periodic signal perturbed by superposed 'noise' (indeed recent studies of the nature of chaos show that most chaotic attractors can be viewed as the fusion of unstable and stable periodic orbits in phase space). The previous section was concerned with atmospheric phenomena characterised by spatial coherence of relatively simple form, but whose behaviour in time (at least for relatively high frequency components) was often highly disordered. On time scales longer than a few weeks, however, phenomena are encountered in the atmosphere whose behaviour in time appears to be more ordered, often taking the form of a kind of 'noisy periodicity'.

One example of such a phenomenon is the so-called 'intraseasonal oscillation'. This manifests itself as an irregular periodic oscillatory component in series of measurements of a variety of atmospheric variables, with a period ranging from around 25–60 days. It was first discovered in the early 1970s [10] in studies of climatological time series of surface pressure and temperature from certain stations located in the tropics. This discovery excited relatively little interest until some ten years later [11, 12], when high-quality datasets of measurements covering the entire global atmosphere became available from the FGGE project (the First GARP Global Experiment, where GARP is the Global Atmospheric Research Program). With such a dataset, it became possible to obtain accurate measurements of certain properties of the global atmosphere, including its total angular momentum.

Angular momentum is a quantity of some interest and importance since, as every physicist knows, it is conserved in a closed system. The combined atmosphere–solid Earth comprises such a closed system, with the atmosphere continually exchanging angular momentum with the solid Earth via frictional interactions and pressure torques across large-scale mountain ranges, such as the Rockies and the Andes. When accurate time series of atmospheric angular momentum became available in 1979, it revealed not only the slow seasonal changes associated with the periodic north–south migration of the major eastward subtropical jet streams but also sporadic bursts of oscillatory behaviour with periods around 25–60 days. An example of such a time series is shown in Figure 10.8.

Furthermore, such oscillations were also confirmed in simultaneous measurements of the angular momentum of the solid Earth, via astrometric measurements of the instantaneous length of the sidereal day (LOD). The total angular momentum of the atmosphere, though vast in absolute terms (around $10^{26}\,\mathrm{kg\,m^2\,s^{-1}}$), is enough to change the LOD by at most around 3 ms in 24 h. Yet this tiny variation in the LOD can be measured to an accuracy of a few microseconds by the currently available techniques of astrometry and geodesy, including long-baseline interferometry and satellite and lunar laser ranging.

The observation of this oscillation in global datasets such as angular

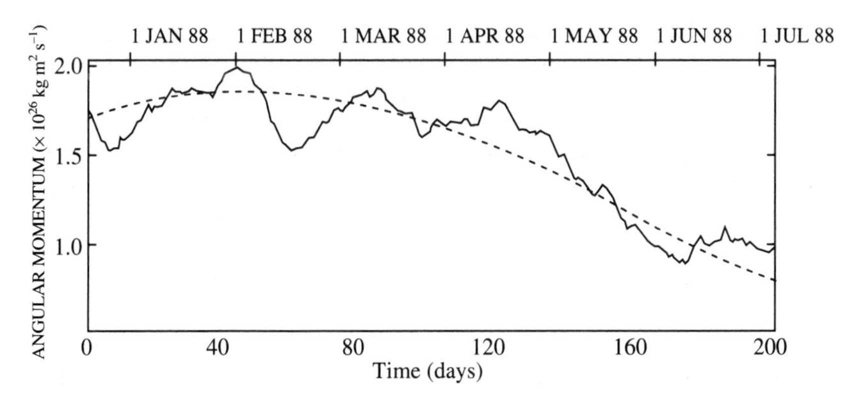

Fig. 10.8. Observed relative atmospheric angular momentum of the atmosphere (due to wind components only), plotted daily during the first half of 1988 (courtesy of M. J. Bell, Meteorological Office, with permission). The dashed curve represents the approximate seasonal variation.

momentum strongly indicates [11, 12] that the pattern of motion giving rise to the oscillation is coherent on a global scale. The original discovery of the 25–60 day oscillation came only from data taken at single locations on the Earth's surface. Subsequent studies using conventional meteorological measurements have confirmed the globally coherent nature of the intraseasonal oscillation, in which tropical and mid-latitude circulation systems apparently act in concert. Yet the causal mechanism for the oscillation remains poorly understood.

The reason for this continuing uncertainty is that a variety of quite different types of theoretical model are each capable of exhibiting irregular oscillations on this time scale. It is important to bear in mind that, because controlled experiments cannot be carried out on the atmosphere, observations alone, however accurate and global in extent, are incapable of distinguishing unambiguously between cause and effect. It is only through the use of models of atmospheric circulation, verified against observations and carefully designed to facilitate the testing of hypotheses, that progress in understanding can be made. In the case of the intraseasonal oscillation, models focusing on certain types of tropical wave motion, driven essentially by patterns of enhanced convection in the tropics themselves, seem equally capable of exhibiting 25–60 day oscillations as other types of model which focus attention instead on, for example, certain types of instability in the interaction of middle-latitude circulation systems with mountain ranges. The question as to which mechanism (or any other?) is actually responsible for the observed oscillation remains to be resolved.

10.6 Coherence in time: the quasi-biennial oscillation

On yet longer time scales, another noisy periodic phenomenon is encountered in the upper atmosphere between altitudes of 10 and 50 km (the stratosphere), strongest in the vicinity of the tropics, whose nature is somewhat better understood. When regular observations from the stratosphere started to become available in the 1950s, it soon became apparent that the east–west component of the wind near the equator varied in time in a way which did not reflect the seasonal cycle. Instead, roughly every year the wind direction would reverse, so that the general flow pattern would repeat itself around every two years (hence the near-periodic reversal became known as the 'quasi-biennial oscillation' or QBO [13]).

Further investigation [13, 14] revealed that the QBO has an intricate structure in space and time (see Fig. 10.9). The oscillation is clearly strongest close to the equator, with increasingly weaker amplitude at higher latitudes (though remaining detectable well into middle latitudes). The reversal of the zonal wind does not occur simultaneously at all heights, but starts at high altitudes and migrates downwards with time. The period of the oscillation

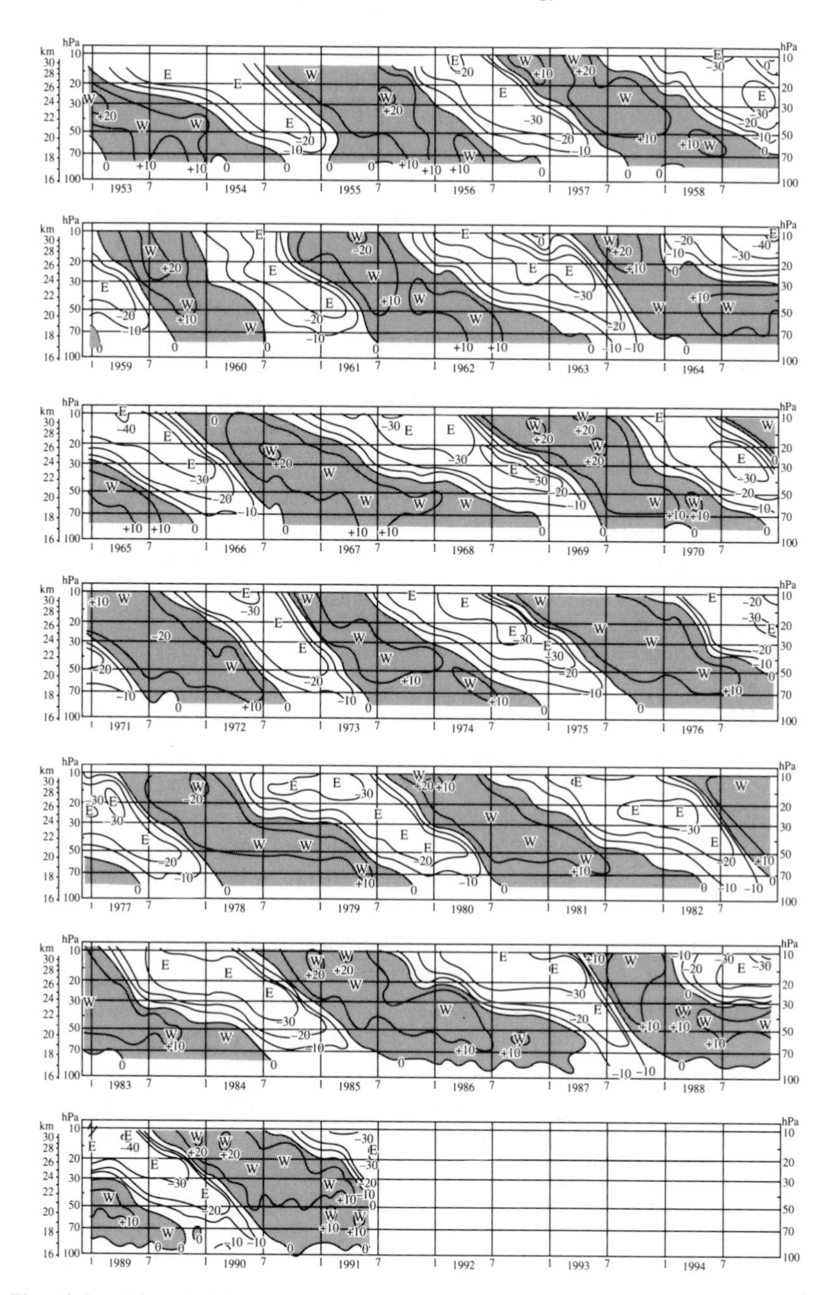

Fig. 10.9. Time-height sections of observed monthly mean east-west winds ($m\,s^{-1}$) in the stratosphere at equatorial stations between 1959 and 1991 (courtesy of B. J. Naujokat, Freie Universität Berlin, updated from [14]). Winds from the west are shown hatched, and contour interval is $10\,m\,s^{-1}$.

is closer to 28 months than precisely two years, but varies erratically between around 20 and 35 months. Such a mean period has no obvious connection with any process which might directly force such an oscillation (such as the annual seasonal cycle), though one notable curiosity is a tendency for the onset of the wind reversal to occur more commonly at certain preferred times of year. This would seem to indicate that the oscillation is at least partially phase-locked to the annual cycle, though not perfectly so.

The nature of this oscillation was a complete mystery until the early 1970s, when Jim Holton and Richard Lindzen (respectively of the Universities of Washington, Seattle, and Harvard) brought together some earlier ideas and observations and formulated them into a coherent model, which appeared to account for many of the features of the QBO. In essence, their model considered the vertical transport of angular momentum by a certain class of large-scale atmospheric wave—the so-called 'inertia-gravity wave', which owes its existence to restoring forces arising both from buoyancy forces and Corioli's accelerations (associated with the Earth's rotation). A particular subclass of these waves is confined to within a certain distance from the equator on a spherical planet, effectively propagating in an eastward or westward direction along a kind of 'waveguide' centred on the equator itself. It is significant that such waves are notoriously difficult to simulate in the types of numerical model commonly used to study atmospheric circulation and climate, because of their small scale in the vertical (requiring impractically high vertical resolution). This fact is also consistent with the continuing failure of such models to simulate a realistic QBO.

Holton and Lindzen considered the combination of a particular pair of waves, identified in the atmosphere from observations, one of which propagated in the eastward direction carrying eastward momentum (the 'Kelvin wave', after Baron Lord Kelvin, the nineteenth-century physicist who discovered a similar form of wave in the context of tides in the oceans), and the other propagated towards the west carrying westward momentum (the 'Yanai wave', after the Japanese meteorologist who discovered this form of wave from stratospheric observations in the tropics). These two modes are both excited from below by disturbances arising from cloud-scale convection in the tropical lower atmosphere, and compete for dominance.

Their ability to propagate upwards, however, is strongly influenced by the structure, and especially the direction, of the mean east–west (zonal) wind, in a mutually exclusive way. Thus, the Kelvin wave can propagate upwards, carrying eastward momentum to high altitudes, while the zonal wind is towards the west (in a frame of reference in which the wave is stationary), but not otherwise, and vice versa for the Yanai wave. As each wave achieves dominance in turn, it deposits momentum of the corresponding sign at upper levels where the wave is dissipated, causing the zonal flow to accelerate in the direction of propagation of the dominant wave. In so doing, the zonal flow is modified to reduce the ability of the dominant wave

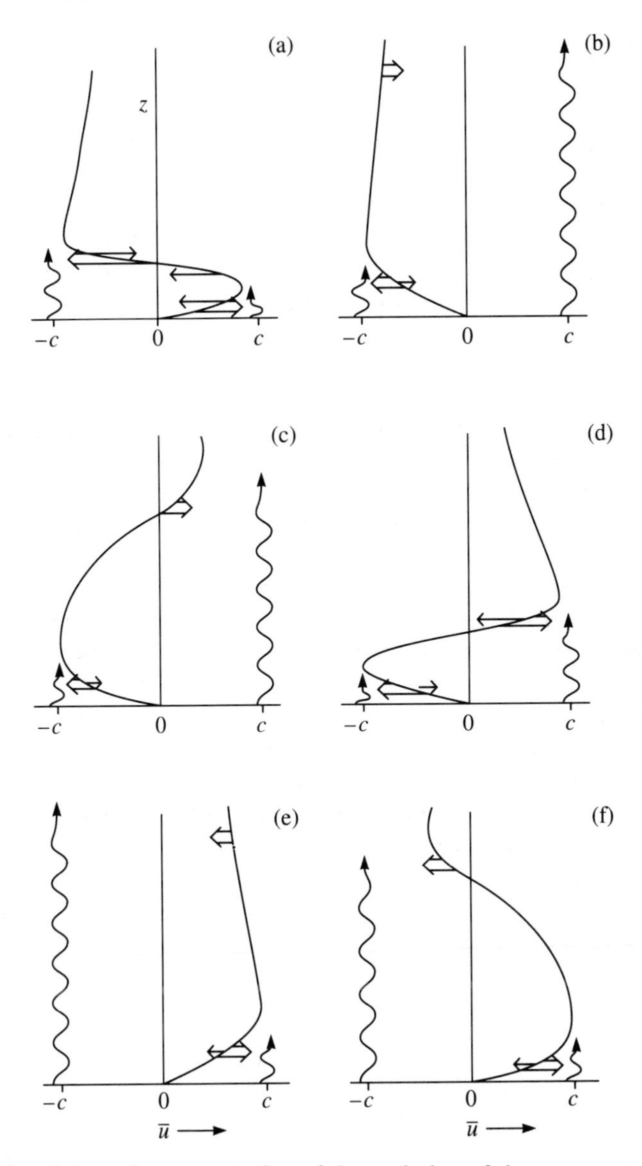

Fig. 10.10. Schematic representation of the evolution of the mean east–west flow in the Holton-Lindzen/Plumb-Bell model of the QBO. Six stages of a complete cycle are shown. Double arrows show wave-driven accelerations, and wavy arrows indicate the penetration of westward ($-c$) and eastward ($+c$) waves (after [13] with permission).

to propagate upwards effectively, eventually causing this wave mode to be suppressed. As this happens, it becomes possible for the other wave mode to propagate to upper levels, and hence to accelerate the zonal flow in the opposite direction, achieving a similar effect in reverse, and so on (see Fig. 10). The result is a sustained oscillation of the zonal wind with a period which depends largely on the forcing and dissipation rates for the waves themselves—a kind of autonomous nonlinear oscillation.

The essential features of this mechanism were captured very effectively in the 1970s in an elegant laboratory experiment carried out by Alan Plumb and Alastair McEwan [15] of CSIRO in Australia (see Fig. 10.11). Their experiment differed from the atmosphere in not involving a rotating frame of reference. Instead, a standing gravity wave was excited in a salt-stratified fluid contained in a cylindrical annular container by oscillating motions of a segmented lower boundary. A standing wave can always be regarded as the superposition of two waves with the same wavelength and frequency but equal and opposite phase velocities, thus forming an analogue of the Kelvin and Yanai waves in the stratosphere. The fluid was started from rest, and when the oscillatory forcing at the boundaries was switched on, a periodic oscillation in the azimuthal flow developed in an entirely analogous way to the QBO, as each component of the standing wave achieved dominance in turn.

The success of Plumb and McEwan's experiment greatly enhanced theoreticians' confidence in the Holton and Lindzen model, and encouraged much further theoretical work. This culminated in the time-dependent

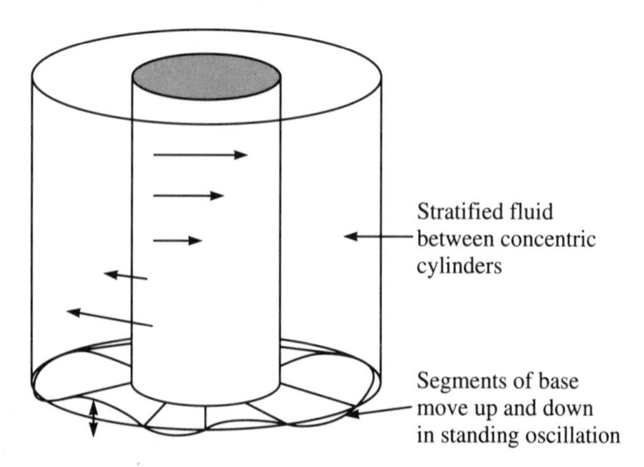

Fig. 10.11. Schematic diagram of the laboratory experiment used by R. A. Plumb and A. McEwan to study an analogue of the QBO (after [13] with permission). The fluid is density stratified in the vertical using a variable concentration of a salt solution, and standing gravity waves are excited by a flexible segmented bottom boundary.

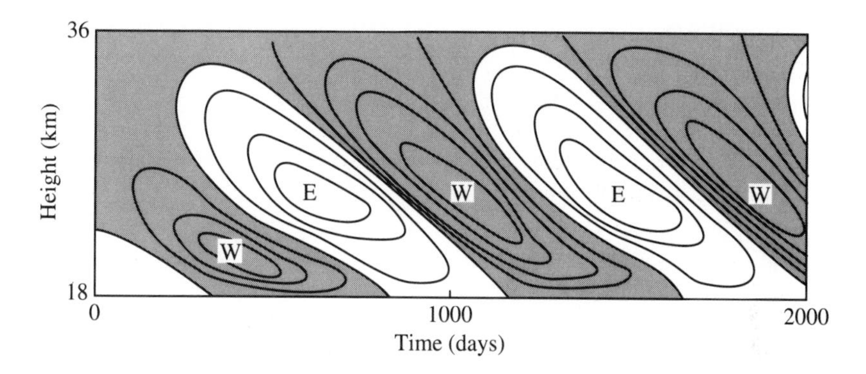

Fig. 10.12. Time-height section of mean east-west wind at the equator from a two-dimensional (2 wave/zonal flow) model of the QBO similar to [16] (courtesy of *X.* Li, University of Oxford). Contours are at $2\,\mathrm{m\,s^{-1}}$ intervals, and flow from the east is shaded.

numerical model of Alan Plumb and Robert Bell [16] (also of CSIRO). Their model again considered the interaction of only two equatorial wave modes and the zonal flow, but included a more realistic structure for the flow in latitude and height. Such a conceptually simple model was able to reproduce many detailed features of the observed QBO, including the descending reversal of the zonal wind with time and other aspects of the structure in latitude as well as in height (see Fig. 10.12 for an example).

One of the most remarkable features of this model, however, is that, in common with virtually every other model of this phenomenon studied so far, the oscillation it produces is almost perfectly periodic—the oscillation is much less chaotic than the stratosphere appears to be! Yet the model contains many of the prerequisites for chaos, viz. more than three degrees of freedom, with nonlinear coupling. It remains to be established conclusively whether the Holton–Lindzen/Plumb–Bell type of model as so far investigated is simply not sufficiently nonlinear to exhibit realistically chaotic behaviour, or whether some other factor (such as the seasonal cycle or effects akin to stocliastic perturbations) is necessary to account for the observed irregularities.

10.7 Coherence in time and space: the El Niño–Southern Oscillation

Shifting attention to even longer time scales, this time into the lower atmosphere which we inhabit, we encounter another phenomenon which exhibits a form of coherence in time (in the form of a noisy periodicity), and which

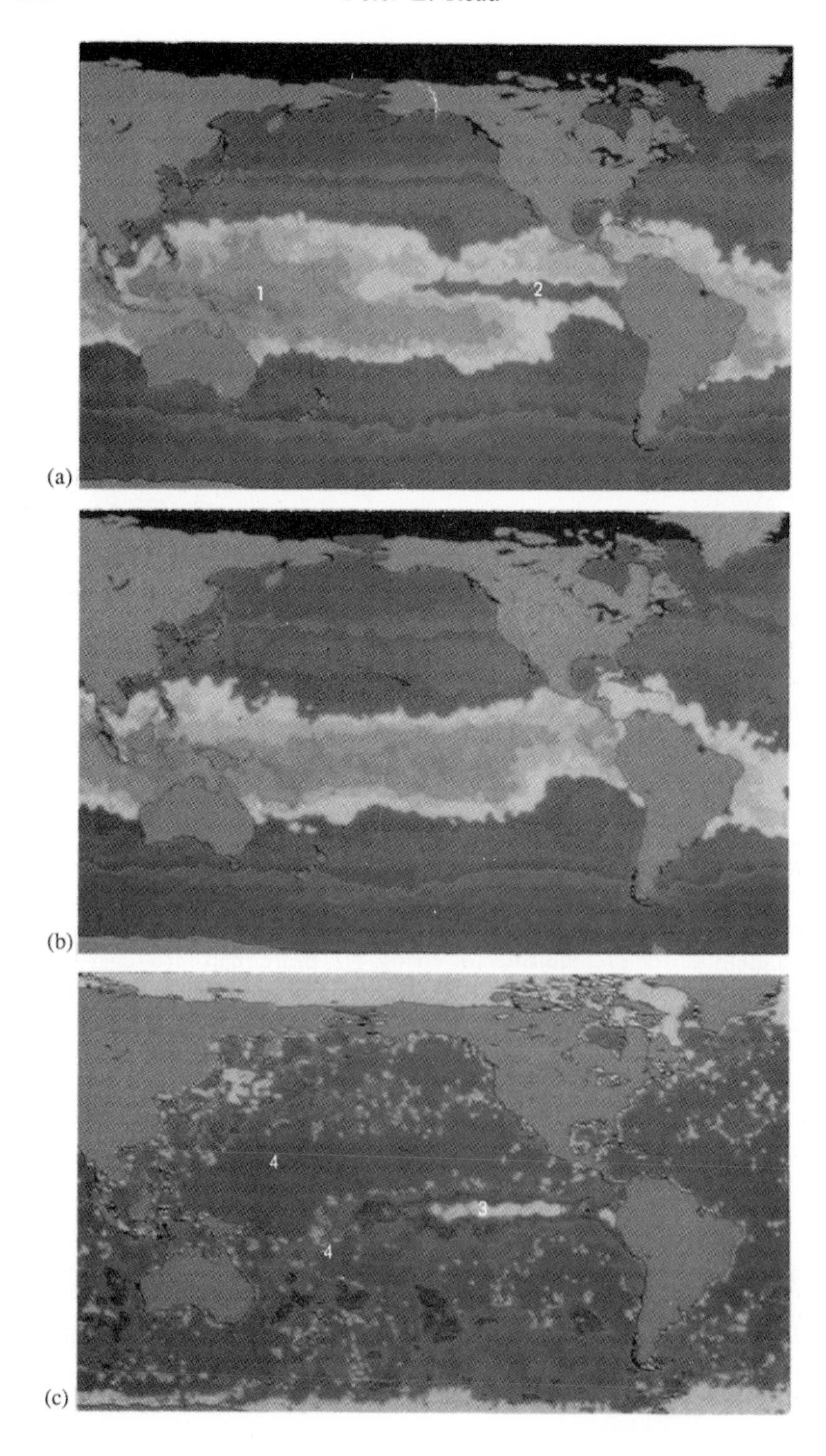
1
2
(a)
(b)
4
3
4
(c)

is also coherent over immense distances across the entire tropical Pacific Ocean and beyond. The phenomenon is known as the El Niño–Southern Oscillation (ENSO). Although there are many different aspects of its manifestation [17, 18, 19] the most essential feature is the sporadic disappearance of a tongue of cold, nutrient-rich surface water off the west coast of South America (especially Peru and Colombia; see Fig. 10.13(a)), resulting in the catastrophic loss of anchovy fisheries for several months at a time. This phenomenon often seems to occur around Christmas time, hence the name 'El Niño' (the 'Christ child').

Under normal conditions, the ocean surface temperature adjacent to South America is relatively cold (see Fig. 10.13(a)). This is associated with the upwelling of deep water in this region, maintained by a circulation within the ocean driven by a persistent pattern of winds in the tropical lower atmosphere (the 'trade winds'; see Fig. 10.14(a)). For the same reason, the surface of the western Pacific, near Indonesia, is usually relatively warm, hence maintaining a pattern of strong convection in the atmosphere and very high rainfall. The distribution of sea surface temperature in turn influences the main pattern of circulation in the equatorial atmosphere, which therefore takes the form of a huge overturning cell oriented east–west, with rising motion over the western Pacific and sinking over the east–the so-called 'Walker circulation' (see Fig. 10.14(a)) after its discoverer, Sir Gilbert Walker, the eminent nineteenth-century explorer and meteorologist. The prevailing wind over the warm water near Indonesia (and the corresponding ocean surface current) is therefore mainly from the east, hence causing the cold water along the eastern equatorial Pacific extending to the coast of Peru.

During an El Niño event, for some reason the prevailing wind near Indonesia weakens, eventually reversing its direction to blow towards South America. This causes the warm pool of surface water in the west to begin to migrate eastwards, carrying with it the associated pattern of enhanced atmospheric convection into the mid-Pacific. This alters the whole circulation in both the atmosphere and ocean (see Fig. 10.14(b)), shutting off

Fig. 10.13. Maps of sea surface temperature in the Pacific Ocean, obtained using infrared data from the Advanced Very High Resolution Radiometer on board the NOAA-7 polar orbiting satellite (by Richard Legeckis of NOAA-NESDIS). Images have a spatial resolution of 100 km and shading indicates temperature contrasts: (a) the normal temperature distribution, with warm water around Indonesia and the west Pacific (1) and a cold tongue of water next to the coast of Peru (2) extending into the equatorial Pacific. (b) shows the situation on 20 January 1983, during the peak of the 1982–83 El Niño event. Note that warm water now extends right across the Pacific to South America. The difference between the two (the 'anomaly') is shown in (c), clearly highlighting the anomalously warm tongue of water in the western Pacific (3) which causes major changes in the atmospheric circulation and climate all around the Pacific basin and beyond.

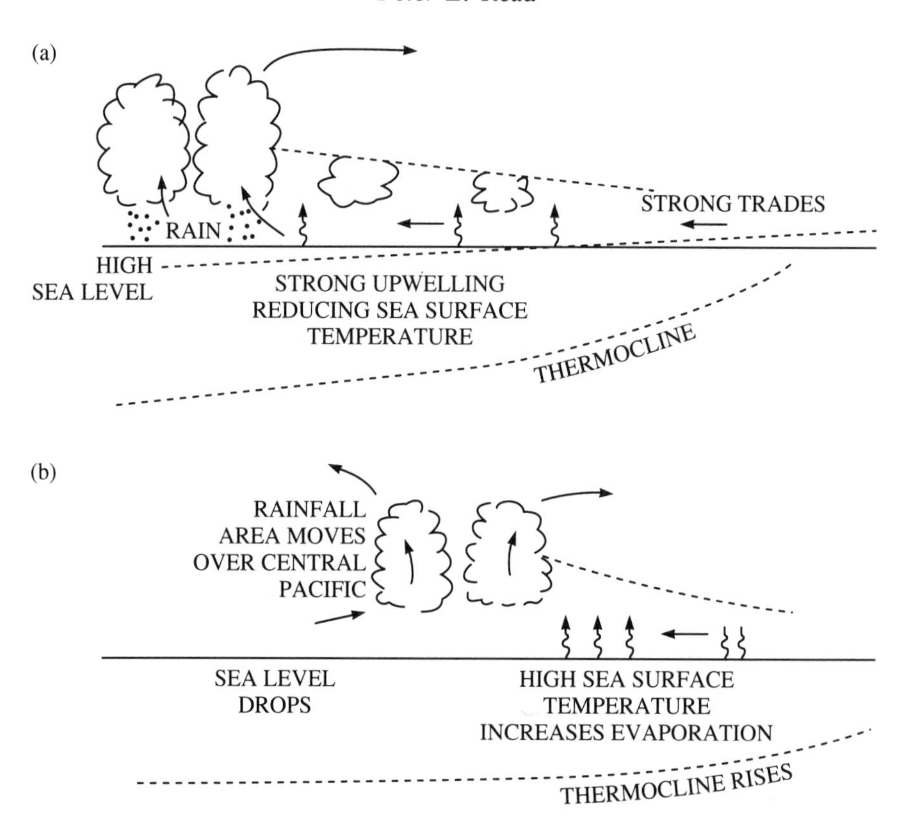

Fig. 10.14. Schematic cross-sections along the equator of atmospheric and oceanic features in the Pacific basin: (a) the normal phase of the El Niño–Southern Oscillation (the Walker circulation); (b) the anomalous (warm) phase of the El Niño circulation. (after [18] with permission).

the upwelling of cold water near the coast of Peru. At the peak of El Niño, warm water stretches almost all the way across the equatorial Pacific (see Fig. 10.13(b)). In addition to its effects on Peruvian coastal fisheries, the changes in the ocean-atmosphere circulation during El Niño can have significant impacts over a large portion of the globe, enhancing rainfall in certain areas (e.g. in North and South America) and causing drought conditions in others (such as in Indonesia and Australia), with some influence being felt as far afield as the Atlantic and Indian Ocean basins. The phenomenon therefore has a significant social and economic impact on a very large number of people, and its prediction is widely regarded as of considerable practical importance.

The reason why the wind direction over the western Pacific begins to reverse at the onset of El Niño is far from obvious, but appears to arise at least partly [20] from a natural instability intrinsic to the coupled atmospheric and oceanic circulations in this region. El Niño events occur roughly every 3–6 years and last several months at a time, suggesting that they are associated with a kind of autonomous oscillation. The state of the oscillation can be characterized by various kinds of index, defined from climatological data, such as the mean sea-level pressure difference between Tahiti (in the mid-Pacific) and Darwin (on the north-east coast of Australia). The resulting time series (Fig. 10.15) shows the irregular noisy periodicity of the oscillation, with the index remaining positive or near zero most of the time, but with occasional short-lived large negative excursions indicative of El Niño events. Thus, the event in 1982–3 is seen to have been the most extreme on record, when the phenomenon and its climatic impacts achieved considerable publicity.

Because the ENSO phenomenon appears to entail an oscillation intrinsic to the coupled atmosphere–ocean circulation in the Pacific basin, it would seem that in order to understand and predict its behaviour it would be necessary to construct an immensely complicated and realistic model [20] of both atmospheric and oceanic circulations. This approach is very much

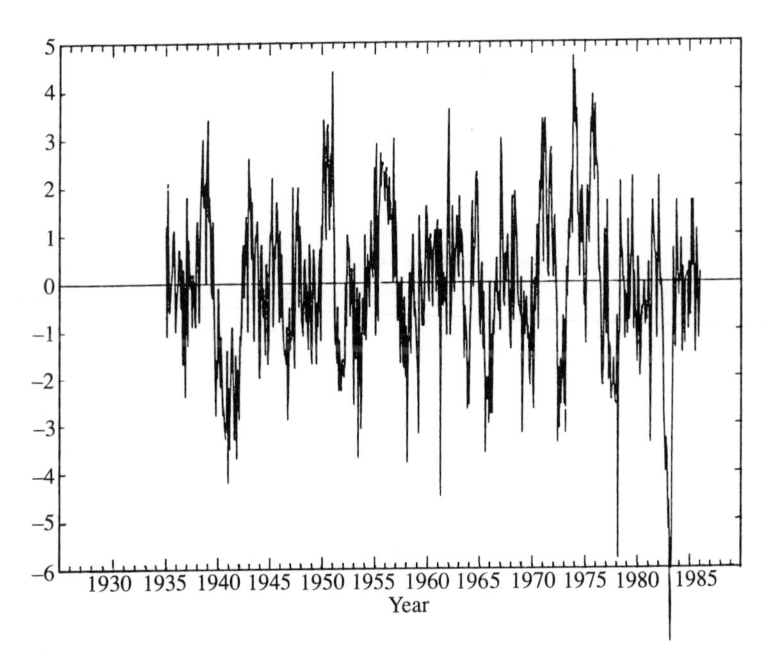

Fig. 10.15. Time series of the Southern Oscillation Index, an indicator of the El Niño–Southern Oscillation (measured in standard deviations from the climatological monthly mean sea level pressure difference between Tahiti in mid-Pacific and Darwin on the northern Australian coast), between 1933 and 1985 (after [18] with permission).

at the current forefront of climatological research, such as carried out at the Robert Hooke Institute in Oxford and the Hadley Centre for Climate Prediction and Research in Bracknell. In the context of simple chaos, however, alternative approaches have been suggested by workers such as Geoff Vallis [21, 22] of the Scripps Oceanographic Institution in California.

Vallis [21] attempted to deduce the simplest possible kind of model which could represent a kind of dynamical behaviour reminiscent of El Niño. Using some *ad hoc* arguments, he reduced the problem to one of determining the behaviour of just a few variables, representing the interaction of ocean and atmosphere circulations. These comprised notional mean values of the ocean surface temperatures in the east Pacific (T_E) and west Pacific (T_W), and the surface wind velocity (u—partially driven by the temperature difference $T_E - T_W$). The resulting three coupled nonlinear ODEs obtained were

$$\frac{\mathrm{d}u}{\mathrm{d}t} = B\,\frac{T_E - T_W}{2\Delta x} - C(u - u^*(t)),$$

$$\frac{\mathrm{d}T_W}{\mathrm{d}t} = u\,\frac{\bar{T} - T_E}{2\Delta x} - A(T_W - T^*),$$

$$\frac{\mathrm{d}T_E}{\mathrm{d}t} = u\,\frac{T_W - \bar{T}}{2\Delta x} - A(T_E - T^*),$$

where A, B, C, and T^*, together with a measure of the deep ocean temperature $\bar{T}$, are externally imposed constants. Here $u^*(t)$ represents an externally defined forcing of u. The model is clearly seen (deliberately) to have many aspects in common with the well-known Lorenz attractor, and differs in essential mathematical structure only by virtue of the 'trade wind' forcing term u^*. This introduces a bias towards one of the fixed points (depending upon the sign of u^*) about which the system tends to orbit.

With plausible values of the main parameters and with u^* = constant (representing steady westward 'trade winds'), u is seen to exhibit an oscillatory behaviour for a number of cycles covering several model years, but in which u remains continuously negative (westwards), followed by occasional large positive (eastward) excursions towards the other fixed point which are relatively short-lived (see Fig. 10.16(a)). These sporadic positive excursions in u are accompanied by a less prominent positive excursion in $T_E - T_W$, and are clearly identified as the model analogue of El Niño events. This pattern of near-regular oscillations, followed by occasional 'El Niños', has a number of qualitative similarities to the observed sequence of ENSO indices, but is clearly much more ordered and regular than the observed El Niño. It is of interest, however, that if the model is made slightly more complicated by the addition of a periodic component to u^*, representing the seasonal variation of the 'trade wind' forcing, then the

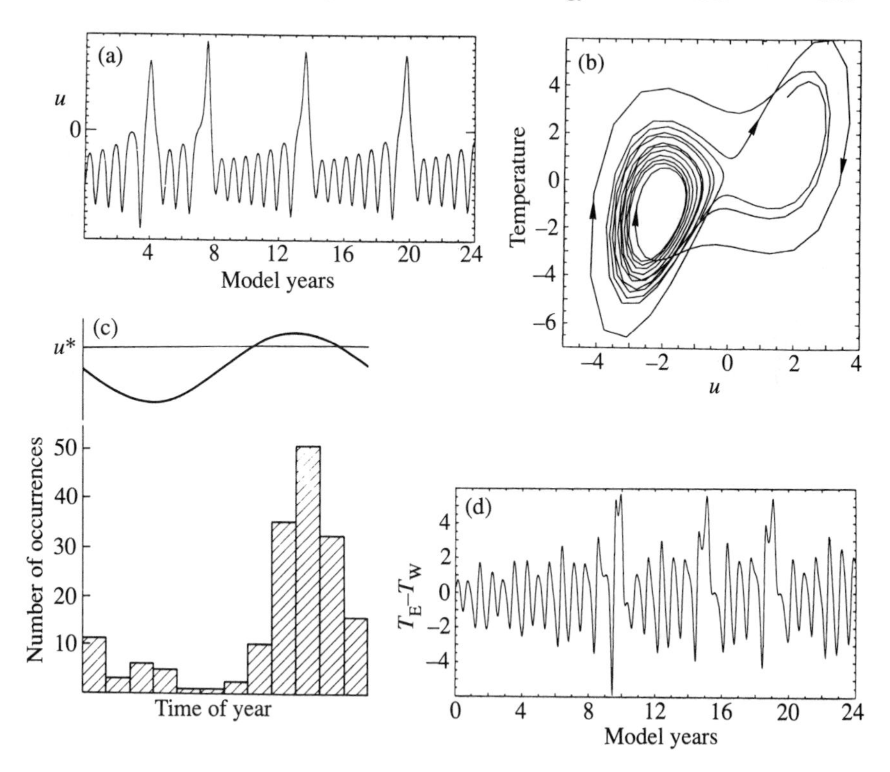

Fig. 10.16. Typical time series from the simple model of the El Niño–Southern Oscillation of Vallis ([21], copyright 1986 by the AAAS): (a) the zonal wind variable u, and (b) a $[u,(T_E - T_W)]$ phase portrait with steady u^* forcing; (c) typical time series of $(T_E - T_W)$ with seasonal u^* forcing; (d) periodic (seasonal) forcing of u^* and the frequency of onset of El Niño-like events as a function of time of year.

model behaviour becomes significantly more erratically variable, but still retains the sporadic occurrence of El Niño-like events (see Fig. 10.16(b)). The onset of El Niño-like events is also partially phase-locked to the period of $u^*(t)$, apparently like the real ENSO (see Fig. 10.16(c)).

Vallis's models [21, 22] are illuminating in showing at least one way in which a conceptually very simple model can reproduce some of the qualitative features of what intuitively would seem to be a highly complicated phenomenon. Intrinsic instabilities cause the system to oscillate irregularly between two unstable equilibria, one representing the 'normal' circulation state and the other the anomalous 'El Niño'. Such a model is, however, still a long way from accounting for all significant aspects of El Niño, let alone the prospect of producing useful practical forecasts of the ENSO cycle. It is important to stress, for example, that this model is not derived systematically from any realistic model of the atmosphere–ocean circulation, and omits several processes which observations suggest may be important, such as the slow transfer of information between the east and west Pacific via the propagation of waves (notably the 'Kelvin' and 'Rossby' modes) in the ocean. These waves may take about a year to cross from one side of the Pacific basin to the other. This effect builds a significant time delay into the coupling of the circulations across the ocean basin, influencing the time scale on which events may occur.

Such effects can be incorporated into another class of simple conceptual models, which then take on various forms of nonlinear delayed oscillator equation, somewhat akin to the Mackey–Glass equation well known in the context of the regeneration of blood cells in the body. One simple example of an oscillator model which has been applied to ENSO has the form

$$\frac{\mathrm{d}T}{\mathrm{d}t} = cT - bT(t - \delta) - eT^3.$$

where c, b, e and δ are constants and the second term on the right involving δ respects a feedback with a fixed time delay (see e.g. [20]). Such (nominally infinite-dimensional) models may also exhibit low-dimensional oscillations of simple periodic or (less commonly under plausible parameter values) irregularly chaotic form, depending (rather sensitively) upon the parametric conditions, though no model has yet produced oscillations apparently as irregular as the observed ENSO index. It remains unclear as to whether this enhanced irregularity is due simply to insufficient nonlinearity in any of the simple models (including the Vallis attractors [22]), or to extra perturbations arising from other circulation systems centred elsewhere in the atmosphere and oceans and which may also act on a similar time scale to ENSO (perhaps including the QBO discussed in the previous section).

While such simple models should not be expected to offer much useful skill in making practical forecasts of El Niño, their apparent qualitative success would seem to suggest that at least some aspects of ENSO may involve a chaotic 'strange attractor'. One might expect, therefore, that certain aspects of the real ENSO phenomenon might be capable of qualitatively realistic simulation in a deterministic model of intermediate complexity, without the necessity (at least in principle) of highly detailed and realistic circulation models of the atmosphere and ocean in the Pacific basin. In fact several groups around the world are actively involved in attempts to use such intermediate models to simulate the observed El Niño. Mark Cane and his co-workers at Lamont–Doherty Geophysical Observatory in New York, for example, have even claimed some degree of success in actually forecasting the onset of El Niño up to two years in advance [23]. This is a very much longer time scale than is commonly regarded as the typical predictability horizon for weather prediction in the atmosphere, offering some hope that certain kinds of behaviour on climate-related time scales may exhibit a kind of much extended predictability. It may be that a form of 'simple chaos', as discussed in this book, may be acting as an 'organizing centre' for some aspects of ENSO and partially underlies the apparently extended predictability of this phenomenon.

10.8 Chaotic models of long-term climate

When we consider climate variations on the astronomical time scales—the time scale of relevance to palaeoclimatology, comparable with the advance and retreat of the Ice Ages on the scale of tens or hundreds of thousands of years—it becomes quite impractical to simulate the entire climatic variation using existing generations of numerical simulation models. In addition, the available observational information on these very long time scales is extremely sparse and often highly indirect, making use of various proxy measurements of climatic variables, such as oxygen isotope ratios to infer ocean temperatures, which are also effectively averaged in area and time.

In attempting to come to grips with the question of long-term climate variability, researchers during the past 25 years have made extensive use of conceptually very simple mathematical models, similar in formulation to Vallis' ENSO model, which handle only the grossest properties of the atmosphere and oceans to describe the climate. Such models are often referred to [24] as 'energy balance climate models' (EBCMs), since the primary model equation often represents time variations in the energy budget of the climate system: the balance between input of solar energy, transport within the climate system and the energetics of phase transitions (especially involving the ice phase of water), outward radiation of energy in the thermal infrared, and consequent variability of mean atmospheric temperature. Such models vary in complexity from a single nonlinear ODE to systems of several coupled ODEs.

One example of a relatively sophisticated form of EBCM is provided by a recent study of Barry Saltzman [25] (the same Saltzman as contributed to the formulation of the Lorenz attractor model). This model was intended to be sufficiently comprehensive to offer direct comparisons with palaeoclimatological measurements, and is formulated to predict the variations of global ice mass I at the Earth's surface, the atmospheric CO_2 concentration μ, and the mean deep ocean temperature θ, on timescales up to a million years. The model reduces to three coupled ODEs of the form

$$\frac{dX}{dt} = -\alpha_1 Y - \alpha_2 Z - \alpha_3 Y^2,$$

$$\frac{dY}{dt} = -\beta_0 X + \beta_1 Y + \beta_2 Z - (X^2 + \beta_3 Y^2) Y + F_{Y1} + F_{Y2},$$

$$\frac{dZ}{dt} = X - \gamma_2 Z,$$

where the variable X is related to I, Y is related to μ, and Z to θ. The functions F_{Yn} represent externally specified forcing, including that due to variations in the eccentricity of the Earth's orbit and rotation parameters such as the obliquity and precession index (the variations highlighted by Milankovitch in his pioneering work on climate cycles and the Ice Ages), and the sources and sinks of CO_2. The behaviour of the orbital and rotation parameters during the past million years, derived from astronomical measurements and dynamical computations which take into account perturbations of the motion of the Earth due to the presence of the Moon and planets, is illustrated in Fig. 10.17.

The resulting model again bears more than a passing resemblance to the Lorenz attractor model, though is of somewhat greater complexity. When integrated forwards in time, it is not surprising to find that chaotic variations occur, and these variations appear to bear some resemblance to climatic variations deduced from actual measurements. Saltzman even goes

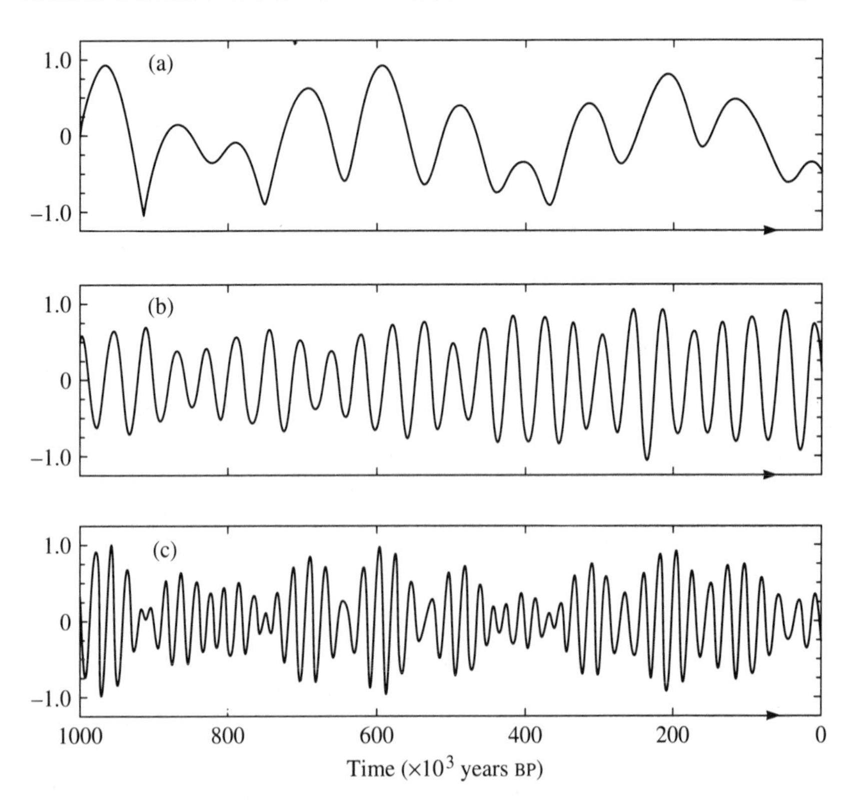

Fig. 10.17. Variation of the Earth's orbital parameters during the past million years: (a) eccentricity, (b) obliquity, and (c) precession index (after [26] with permission of the American Meterological Society). Positive values represent conditions believed to result in net warming in the Northern Hemisphere.

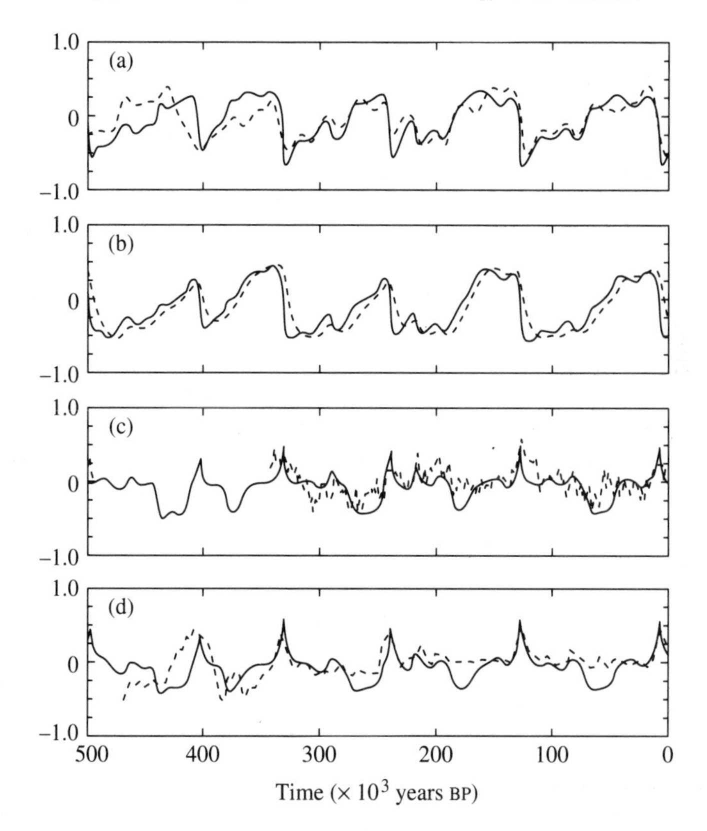

Fig. 10.18. Solutions to the simple energy balance climate model of Saltzman (after [25], with permission), compared with palaeoclimate measurements: (a) solution for planktonic ^{18}O fraction δ^* (solid line) compared with inferred δ^{18}O from palaeoclimate data (δ_s^*, dashed line); (b) solution for global ice mass I* (solid line) and mean deep-ocean temperature θ^* (dashed line); (c) solution for atmosphereic CO_2 concentration μ^* compared with the CO_2 variations inferred from δ ^{13}C measurements; (d) solution for mean ocean surface temperature τ^* (solid line) compared with observed surface water temperature for the mid-Indian Ocean T_s^* (dashed line).

so far as to attempt to fit his model (by varying the 8 free parameters) to palaeoclimatological data series, to obtain the time sequences shown in Fig. 10.18. Solid lines indicate the model predictions, and dashed lines show the climatic observations. The apparent agreement is remarkably good, given the relative simplicity of the model, but it is arguable whether fitting such an idealized model with so many free parameters to measurements of a natural system is a valid procedure, and whether the apparent agreement with observations is merely fortuitous. It is nonetheless intriguing to find a class of problems in meteorology and climatology for which the construction of models based on simple, deterministic dynamical systems is the main theoretical tool in the researcher's armoury.

10.9 Assessing the validity of chaos: a role for laboratory experiments?

Much of the above discussion has presented examples of phenomena which appear to have some features which can be represented in terms of simple chaos. In several cases, however, the formulation of the models used can be criticized as being somewhat arbitrary, e.g. entailing the use of an arbitrarily truncated Fourier representation (cf. the Lorenz attractor as a model of chaotic convection). Doubts concerning the validity of simple chaotic models as representing fluid motion are further reinforced by the common result [27] that the occurrence and nature of chaos in such models often varies markedly with the assumed model truncation. In some cases, the chaotic behaviour found in a very low resolution model containing only a few degrees of freedom can disappear altogether [27] as more components are included. The extent to which the model behaviour converges upon the flow of a real fluid as resolution increases is an important question which is often hard to answer using numerical studies alone. The role of unresolved scales and processes in models of the atmosphere or oceans in determining the ultimate predictability of the atmosphere is a further question of considerable practical importance, but which is also extremely difficult to resolve given numerical model and/or atmospheric data alone.

The task of attacking these questions is severely aggravated by our inability (thankfully!) to carry out controlled experiments on the atmosphere. It is certainly possible to construct numerical models of varying complexity, which can be used as a 'proxy atmosphere' on which to carry out controlled experiments. However, as discussed in Section 10.2 above, this approach has several potential pitfalls, not least of which is the implicit assumption that the problem can effectively be reduced to a finite (albeit large), deterministic set of ODEs, with the effect of unresolved scales 'enslaved' to the larger scales of motion.

An alternative approach which overcomes some of the drawbacks of numerical models is to study the behaviour of a real fluid in the laboratory under rigorously controlled conditions. Through careful design, some dynamical similarity with certain scales of motion in the atmosphere or oceans can be achieved, resulting in a flow system to which theoretical models, originally intended to describe atmospheric motion, can be applied with equal validity [28]. With care, such laboratory experiments can be controlled to high precision over a wide range of parameters.

10.10 A laboratory analogue of large-scale motion in atmospheres and oceans

The simplest kind of system which includes the essential physical attributes of the large-scale atmospheric circulation consists [28, 29] of a tank of fluid which can be heated and/or cooled in different (horizontally separated)

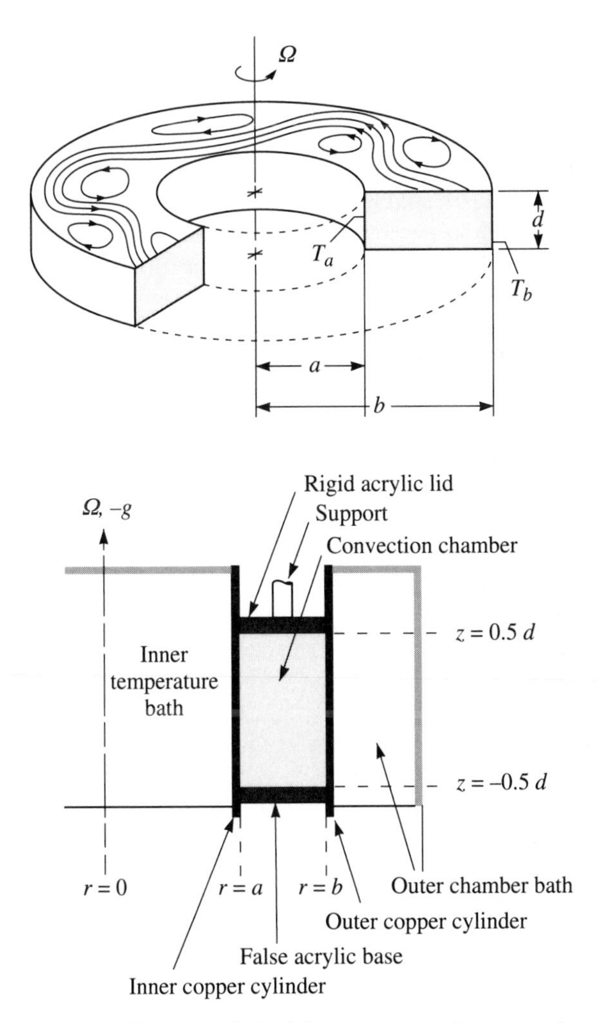

Fig. 10.19. Schematic diagram of the laboratory experiment used to study flow regimes analogous to the large-scale circulation of planetary atmospheres (the 'rotating annulus' [28, 29]): (a) perspective view of annular convection chamber, with isothermal side boundaries at $r = a$ and $r = b$, rotated about the symmetry axis (indicated Ω); (b) cross-section of annular gap.

locations, and operated in a rotating frame of reference. One very straightforward way of realizing this is to use an annulus. The fluid is contained between two upright metal cylinders with a common axis of symmetry. Heating and cooling of the fluid is achieved by maintaining the two cylinders at different temperatures; for similarity with the Earth's atmosphere, we make the inner cylinder cool (analogous to the polar regions) and the outer cylinder warm (the tropics). The entire system is then placed on a turntable and rotated about the axis of symmetry (see Fig. 10.19).

Depending upon the precise values of thermal contrast and rotation speed, a whole variety of different flow regimes can be obtained [29] (see Fig. 10.20). At low rotation rates, for example, we typically obtain a steady axisymmetric flow with an overturning circulation, akin to the Hadley circulation in the tropical atmosphere, in which fluid rises near the heated wall and sinks near the cold wall, and an azimuthal swirl brought about by the action of Coriolis accelerations and the tendency for the fluid to conserve its angular momentum. At higher rotation rates, this flow becomes

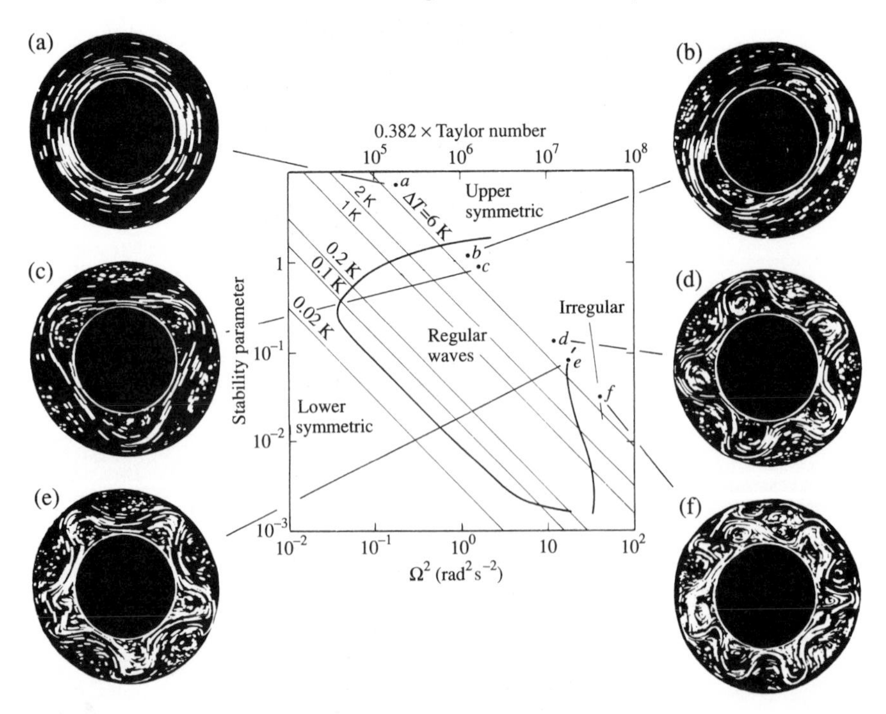

Fig. 10.20. Principal flow regimes in the rotating annulus, (after [28]), classifying flows according to the two main dimensionless numbers dependent upon imposed temperature contrast ΔT and rotation rate Ω. Typical flows are illustrated by horizontal 'streak photographs', showing the onset of baroclinic waves, the increase in wavenumber with increasing Ω, and the ultimate breakdown to disordered flow ('geostrophic turbulence').

unstable to wave-like perturbations which carry heat between the hot and cold boundaries, releasing potential energy in a form of thermal convection (often called 'sloping convection' [29], because typical fluid trajectories follow sloping paths relative to the geopotentials and isotherms). The wave-like states are widely understood to be dynamically similar to the dominant cyclone-scale waves in the atmosphere at middle latitudes, as responsible for organizing our day-to-day weather patterns.

At high rotation rates in the experiments, the wave pattern is highly disordered and resembles typical meteorological flows. At lower rotation rates, however, an intriguing regime occurs in which the wave-like pattern becomes highly regular and symmetrical, often dominated by a single wave-number that can be either steady (apart from a slow drift) or periodically

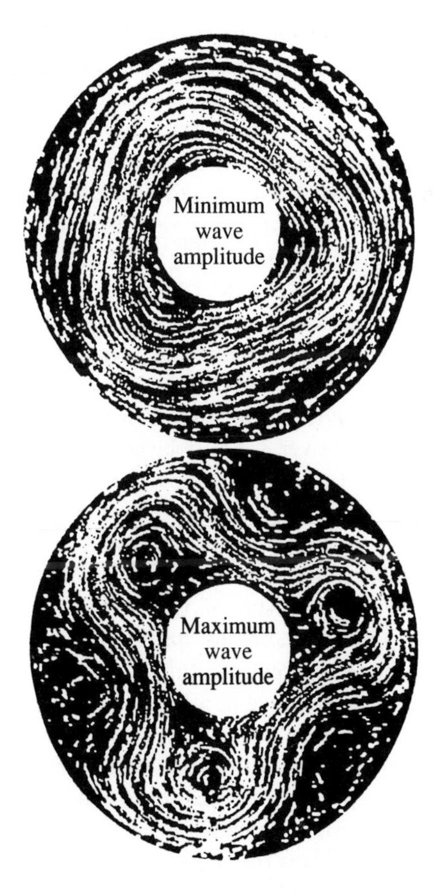

Fig. 10.21. 'Streak photographs' of the two extreme states of a typical periodic amplitude-modulated flow in the rotating annulus. The flow is dominated by an azimuthal wavenumber 3 pattern, which preserves its form while moving steadily around the apparatus and periodically growing and decaying in amplitude.

modulated in amplitude (see Fig. 10.21) or shape. Closer examination of the flows reveals a very rich variety of possible behaviour, approaching the complexity and propensity for multiple equilibria found for the Taylor–Couette system and discussed by Tom Mullin in Chapter 4. In the present context of atmospheric predictability, we might expect that studies of the ways in which the laboratory waves make transitions from regular to more disordered flows might provide insights which are helpful in understanding the nature of the atmosphere's unpredictability, especially if any of these order–disorder transitions entail simple chaos.

One example of a possible candidate for simple chaotic behaviour in the annulus experiment arises through a bifurcation from a periodic amplitude-modulated state (see Fig. 10.21). These flows are illustrated in Fig. 10.22, in which we show phase portraits, derived using some of the methods discussed earlier by Tom Mullin, to measurements of temperature in the flow. We start from a quasi-periodic state, in which a single wavenumber is present and oscillates periodically in amplitude while moving at a steady rate around the apparatus (see Fig.10.21). This flow can be represented in Fig. 10.22(a) as motion on a torus, with the poloidal oscillation representing the phase propagation of the wave past the (fixed) measurement point and the transverse (toroidal) oscillation representing the periodic amplitude modulation. The Poincaré section in Fig. 10.22(b) clearly shows the thin 'skin' of the torus, demonstrating the doubly periodic nature of the flow.

If we change the conditions slightly, we obtain an apparently 'fuzzy' torus, in which the walls acquire a finite thickness, even though the toroidal structure is still strongly apparent. On applying some of the quantitative measures of chaotic behaviour discussed above, we obtain an apparent attractor dimension of just over 3, and a largest Lyapunov exponent which is significantly positive (in the previous, quasi-periodic, case, λ_1 was approximately 0). This leads us to suspect that at least some of the fuzziness we find on the surface of our torus is actually a manifestation of a kind of fractal structure implicit in the dynamics. In practice, this new disordered flow appears to arise as the result of the presence of one or more new wavenumber components which are not harmonically related to the original wave—a kind of symmetry-breaking bifurcation. The new components evidently try to achieve dominance in a kind of nonlinear competition, but never quite make it.

This kind of nonlinear competition between a small number of spatial modes could, at least in principle, occur in an atmosphere (not necessarily that of the Earth) under some circumstances, though it is not intended to suggest that this phenomenon is the only (or even the most important) way in which the atmosphere becomes intrinsically unpredictable. By comparison with theoretical models put forward to account for atmospheric behaviour, however, the experiments are at least able to assist in determining whether the models are valid when applied to a real fluid, and it is an

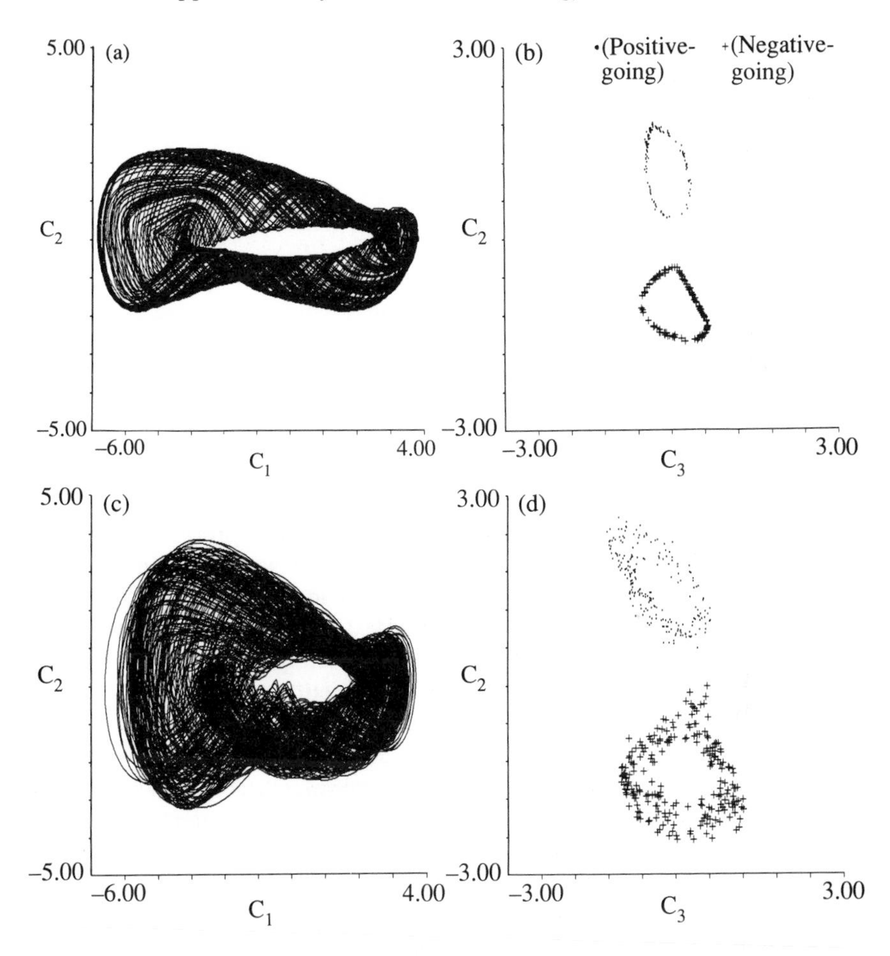

Fig. 10.22. Typical phase portraits and Poincaré sections derived from temperature measurements at the centre of the annular convection chamber for (a) and (b) periodic amplitude-modulated wave ($m = 3$; cf. Fig. 10.21), and (c) and (d) chaotic doubly modulated wave ($m = 3$) (after [30]).

approach which needs to be developed much further in the future. The application of the ideas of simple chaos to our laboratory 'proxy atmosphere' is still only at an early stage, and much more remains to be done concerning both model verification and in the development of methods of data analysis to distinguish the properties of chaos from physical measurements. In the context of atmospheric measurements, we should further remark that, in order to assess whether the atmosphere itself is chaotic in the restricted sense discussed here, it is highly desirable that the techniques used should have been thoroughly tried and tested in the laboratory, preferably (as here) on flows with some dynamical connection to the atmospheric phenomena under investigation.

10.11 Concluding remarks

In this chapter, we have shown that, although it is evidently not yet feasible to characterize atmospheric behaviour unambiguously as true chaos, the concept of 'simple chaos' in autonomous dynamical systems does appear to be helpful in beginning to interpret and understand a wide range of (especially aperiodic) atmospheric phenomena, ranging from everyday weather to long-term climate. It is important, nevertheless, constantly to bear in mind the limitations of the current theory of chaos, especially as applied to fluid systems.

Recent modelling and data studies of weather-related processes in meteorology, for example, have identified phenomena which are intrinsically unpredictable, but for which ideas of 'simple chaos' do not appear to apply. Certain types of cyclonic storm in middle latitudes which develop very rapidly over the oceans (such as the storms which hit the shores of the UK in October 1987 and in January 1989) are sometimes found (at least for a while) to increase their amplitudes more rapidly than exponentially, in the sense that the time scale for them to double their amplitude actually decreases with time [31] (their explosive development leads them sometimes to be referred to as 'cyclonic bombs'!). In this situation, concepts such as the attractor dimension or the spectrum of Lyapunov exponents are not helpful, since the coefficients are not invariants but are evidently variable in time.

'Cyclonic bombs' would appear to offer a graphic example of a phenomenon which is always in a transient state. It is important to remember that chaos can only describe the post-transient statistically stationary behaviour of a system. Yet it is becoming increasingly recognized [32] that many real physical systems may only be observable in transient states, and must therefore require a different kind of model to describe their behaviour from that capable of being provided by the theory of chaos in its current form. In studying chaos, and the theoretical developments which are sure to succeed it in the future, therefore, it is clear that atmospheric phenomena are assured of an important role in continuing to motivate studies of nonlinear dynamical phenomena and in providing a challenging 'test bed' for the resulting hypotheses and predictions.

Acknowledgements

It is a pleasure to thank friends and colleagues in Oxford, at the Meteorological Office, and elsewhere for numerous discussions which have contributed to my appreciation of the subject of chaos and the material discussed above, and for their comments on the manuscript. Particular mention should be made of Drs Mike Bell, Raymond Hide, and David Anderson. Material for some of the illustrations was generously provided courtesy of Dr Mike Bell and Mr Chris Hall of the Met Office and Drs Mike Davey, Peter Killworth, and X. Li of the Robert Hooke Institute in Oxford.

References

1. Lorenz, E. (1963). Deterministic non-periodic flow. *J. Atmos. Sci.*, **20**, 130–41.
2. Saltzman, B. (1962). Finite amplitude free convection as an initial value problem. *J. Atmos. Sci.*, **19**, 320–41.
3. Sparrow, C., (1982). *The Lorenz Equations.* Springer, New York.
4. Procaccia, I., (1988). Complex or just complicated? *Nature*, **333**, 498–9.
5. Grassberger, P. and Procaccia, I. (1983). Characterization of strange attractors. *Phys. Rev. Lett.*, **50**, 346–9.
6. Flood, C. R., (1985). Forecast evaluation. *Meteorol. Mag.*, **114**, 254–260.
7. Gadd, A. J., (1985). The 15-level weather prediction model. *Meteorol. Mag.*, **114**, 222–6.
8. Houghton, J. T. (1987). *The physics of atmospheres*, Cambridge University Press.
9. Wallace, J. M. and Blackmon, M. L. (1983). Observations of low-frequency atmospheric variability, In *Large-scale dynamical processes in the atmosphere* (ed. B. J. Hoskins and R. P. Pearce), pp. 55–94. Academic Press, London.
10. Madden, R. A. and Julian, P. R. (1971). Detection of a 40–50 day oscillation in the zonal wind in the tropical Pacific. *J. Atmos. Sci.*, **28**, 702–8.
11. Hide, R., Birch, N. T., Morrison, L. V., Shea, D. J., and White, A. A. (1980). Atmospheric angular momentum fluctuations and changes in the length of the day. *Nature*, **286**, 114–17.
12. Rosen, R. D. and Salstein, D. A. (1983). Variations in atmospheric angular momentum on global and regional scales and the length of day. *J. Geophys. Res.*, **88**, 5451–70.
13. Andrews, D. G., Holton, J. R., and Leovy, C. B. (1987). *Middle atmosphere dynamics*. Academic Press, Orlando.
14. Naujokat, B. (1986). An update of the observed quasi-biennial oscillation of the stratospheric winds over the tropics. *J. Atmos. Sci.*, **43**, 1873–7.
15. Plumb, R. A. and McEwan, A. D. (1978). The instability of a forced standing wave in a viscous stratified fluid: a laboratory analogue of the quasi-biennial oscillation. *J. Atmos. Sci.*, **35**, 1827–39.
16. Plumb, R. A. and Bell, R. C. (1982). A model of the quasi-biennial oscillation on an equatorial beta-plane. *Quart. J.R. Met. Soc.*, **108**, 335–52.

17. Enfield, D. B. (1989). El Niño, past and present. *Rev. Geophys.*, **27**, 159–87.
18. Bigg, G. R. (1990). El Niño and the Southern Oscillation. *Weather Mag.*, **45**, 2–8.
19. Philander, S. G. H. (1990). *El Niño, La Niña and the southern oscillation.* Academic Press, New York.
20. McCreary, J. P. and Anderson, D. L. T. (1991). An overview of coupled ocean-atmosphere models of El Niño and the Southern Oscillation. *J. Geophys. Res.*, **96**, 3125–50.
21. Vallis, G. K. (1986). El Niño: a chaotic dynamical system? *Science*, **232**, 243–5.
22. Vallis, G. K. (1988). Conceptual models of El Niño and the Southern Oscillation. *J. Geophys. Res.*, **93**, 13979–91.
23. Cane, M. A., Zebiak, S. E., and Dolan, S. C. (1986). Experimental forecasts of El Niño. *Nature*, **321**, 827–32.
24. Ghil, M. and Childress, S. (1987). *Topics in geophysical fluid dynamics: atmospheric dynamics, dynamo theory and climate dynamics.* Springer, New York.
25. Saltzman, B. (1988). Modelling the slow climatic attractor. In *Physically-based modelling and simulation of climate and climatic change*, Part II (ed. M. E. Schlesinger), pp. 737–54. Kluwer, Dordrecht.
26. Saltzman, B., Hansen, A. R., and Maasch, K. A. (1984). The late Quaternary glaciations as the response of the three-component feedback system to Earth-orbital forcing. *J. Atmos. Sci.*, **41**, 3380–9.
27. Curry, J. H., Herring, J. H., Loncaric, J. and Orszag, S. A. (1984). Order and disorder in two- and three-dimensional Bénard convection. *J. Fluid Mech.*, **147**, 1–38.
28. Read, P. L. (1988). The dynamics of rotating fluids: the 'philosophy' of laboratory experiments and studies of the atmospheric general circulation. *Meteorol. Mag.*, **117**, 35–45.
29. Hide, R. and Mason, P. J. (1975). Sloping convection. *Adv. Phys.*, **24**, 47–100.
30. Read, P. L., Bell, M. J., Johnson, D. W. and Small, R. M. (1992). Quasi-periodic and chaotic flow regimes in a thermally-driven, rotating fluid annulus. *J. Fluid Mech.*, **238**, 599–632.
31. Farrell, B. F. (1990). Small error dynamics and the predictability of atmospheric flows. *J. Atmos. Sci.*, **47**, 2409–16.
32. Tavakol, R. K. and Tworkowski, A. S. (1988). Fluid intermittency in low dimensional deterministic systems. *Phys. Lett.*, **126**, 318–24.

11
Nonlinearity and chaos in atoms and molecules

M. S. Child

The concepts of nonlinearity and chaos have been applied to two areas of atomic and molecular science. One, which is the subject of a recent text by Gray and Scott [1], relates to the sensitivity of certain complex reaction schemes to small changes in pressure, temperature, and concentration. The other, to which this chapter is devoted, concerns the internal dynamics of atoms and molecules at energies which approach their ionization or dissociation thresholds.

To set the historical perspective, it has been known for many years that molecules undergo orderly normal mode vibrations in their lowest energy states [2]. On the other hand, modern theories of chemical dissociation [3] assume a statistical or chaotic distribution over all degrees of freedom. Hence a transition from order to chaos must have occurred at some intermediate energy. Experimental investigation of this interesting transitional region was until recently notoriously difficult but the advent of laser spectroscopy (see e.g. [4]) has completely changed the picture. This chapter attempts to bring together the main theoretical insights stimulated by these new experiments.

The exposition is largely within the context of classical mechanics, because the quantum mechanical implications of classical chaos are treated in Chapter 12. To simplify the discussion, attention is restricted to systems involving just two degrees of freedom, and the theory is expounded at a largely descriptive level. Readers seeking a more detailed theoretical background are referred to the texts by Lichtenberg and Lieberman [5] and Tabor [6], to reviews by Percival [7] and Noid, *et al.* [8], and to conference proceedings edited by Berry, *et al.* [9].

The discussion is illustrated by use of Poincaré sections, which are briefly introduced in Section 11.1, in a separable context. The account thereafter proceeds to outline different stages in the nonlinear transition to chaos as the energy in a molecule or an atom increases. The first topic to be treated (in Section 11.2) is the transition from normal mode to local mode behaviour in a system of two coupled equivalent anharmonic oscillators, as symptomatic of an isolated nonlinear resonance. Consequences for the vibrational spectrum of H_2O and other small hydride molecules are addressed.

Aspects of the onset of chaos, in terms of breaking the separatrix between different parts of the phase space, are then described in Section 11.3, and applications to phase space bottlenecks relevant to the molecular dissociation are outlined in Section 11.4. Finally, Section 11.5 shows that the quantum mechanical behaviour of classically chaotic systems may still carry orderly imprints of particular relatively unstable periodic classical orbits. The Rydberg spectrum of atomic hydrogen in a strong magnetic field is shown to provide a striking example.

11.1 Poincaré sections

Aspects of the classical dynamics are conveniently revealed by use of Poincaré sections, which are most simply introduced in the context of the separable Hamiltonian

$$H = \frac{1}{2m}(p_1^2 + p_2^2) + V_1(q_1) + V_2(q_2), \tag{11.1}$$

where the p_i and q_i are momenta and coordinates respectively. One picture of the motion is displayed in Fig. 11.1 in the form of a Lissajous figure corresponding to a particular division of the total energy between the q_1 and q_2 modes, for a model in which

$$V_1(q_1) = D(1 - e^{-aq_1})^2, \qquad V_2(q_2) = \tfrac{1}{2}q_2^2. \tag{11.2}$$

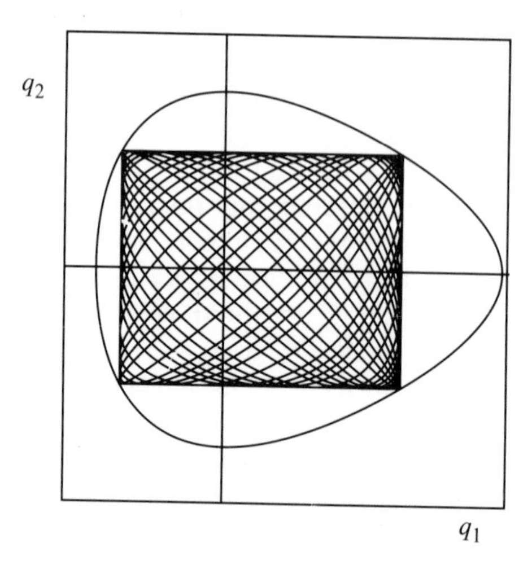

Fig. 11.1. A Lissajous figure for the Hamiltonian in equation (11.1) with $D = 12.5$ and $a = (2D)^{-1/2}$. The bounding curve is the potential energy contour.

Notice that separability automatically ensures that the caustics which bound the figure are straight lines parallel to the q_1 and q_2 axes. Secondly these caustics touch the energy contour at four corners, whose positions vary with the energy division between the oscillators. Two extremes could occur such that the motion is restricted to one or other of the coordinate axes; such trajectories are termed periodic, in the sense that they repeat themselves after set intervals. Other periodic orbits could also occur if the two frequencies were rationally related ($\omega_1 : \omega_2 = k_1 : k_2$, where k_i are integers), but the typical motion is 'quasi-periodic' in the sense that it is governed by two frequencies (which vary from trajectory to trajectory), but never precisely repeats itself.

Poincaré sections of Fig. 11.1 are constructed by plotting the coordinate and momentum in one variable, while holding the other coordinate fixed and restricting its conjugate momentum to either positive or negative values. Thus the (q_1, p_1) section in Fig. 11.2(a) was obtained by marking the points corresponding to 20 upward ($p_2 > 0$) crossings of the trajectory through the line $q_2 = 0$ in Fig. 11.1. Similarly Fig. 11.2(b) shows the (q_2, p_2) values arising from 20 intersections of the line $q_1 = 0$ with $p_1 > 0$. Thus an infinitely long quasi-periodic trajectory would yield closed curves precisely equivalent to orbits of the two isolated oscillators. The family of such orbits, in say Fig. 11.2(a), corresponding to different energy fractions in the two oscillators, would be concentric on a fixed point at the centre of the phase plane, which is the Poincaré image of the periodic motion along q_2. This connection between periodic orbits and fixed points in the Poincaré section is central to what follows in Sections 11.2–11.4.

Of course, the sections shown in Fig. 11.2 could readily be constructed for this separable model without recourse to the trajectory, because the functional dependences of p_1 on q_1 and p_2 on q_2 are analytically determined

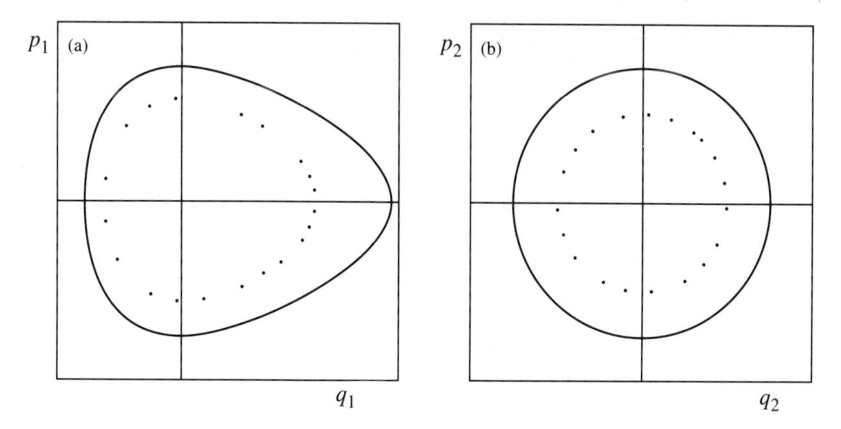

Fig. 11.2. Poincaré sections of Fig. 11.1: (a) at $q_2 = 0$, $p_2 > 0$ and (b) at $q_1 = 0$, $p_1 > 0$. The outer curves are limits fixed by the total energy.

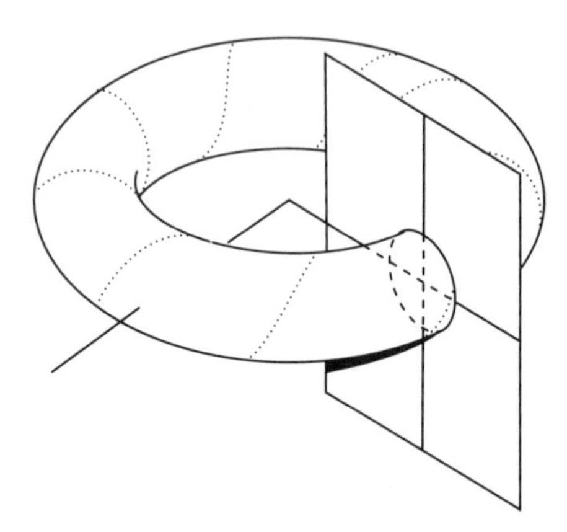

Fig. 11.3. An invariant torus, showing construction of the Poincaré section.

once the energies in the two modes are fixed. The value of the surface of section is, however, more obvious in the nonseparable case, when analytical separation becomes impossible. The caustic boundaries of the Lissajous figure become curved, but the topology of the Poincaré section and even its enclosed area prove to be independent of the sectioning line provided that it cuts opposite caustic boundaries and that the defining trajectory is quasi-periodic [7]. These properties have been extensively exploited in the semiclassical quantization of nonseparable systems [8]. Their relevance to the transition from regularity to chaos is illustrated in the sections that follow.

An alternative view of this quasi-periodic motion is by means of an invariant torus, as illustrated in Fig. 11.3, which combines the cyclic motions in the two parts of Fig. 11.2 into a single figure. Any torus may be circumscribed by two topologically distinct paths, one of which is supposedly followed by motion in the (q_1, p_1)-plane and the other by motion in the (q_2, p_2)-plane. Notice that restriction of the motion to a two-dimensional surface in a four-dimensional space (q_1, q_2, p_1, p_2), implies two constraints, which may be taken for convenience as the two enclosed areas in Fig. 11.2, or as the energy and one of these areas. The term 'invariant' refers to invariance of the area cut out by a typical interrupting surface, with respect to changes in the shape of the surface. This property is, however, restricted to regular or quasi-periodic motions. As we shall see in Section 11.3, the signature of chaos is a blurring in the definition of the torus.

11.2 Normal to local mode bifurcations

One of the best-characterized nonlinear processes to affect molecular vibrations is the transition from normal mode to local mode behaviour in coupled equivalent anharmonic oscillator systems [10]. The stretching vibrations of the H_2O molecule provide the simplest example. Suppose the coordinates q_i represent changes in the bond lengths from their equilibrium values. One knows that sufficient extension will cause them to break, a property that is readily incorporated by use of the Morse potential energy function

$$V(q_i) = D(1 - e^{-aq_i})^2, \tag{11.3}$$

where D is the bond dissociation energy. The two bonds do not, however, vibrate independent by because the motions of one will transfer energy to the other either by shaking the central oxygen atom, or because the strength of one bond varies with the length of the other. These two effects are roughly equally important in the H_2O molecule, but it is sufficient for illustrative purposes to include only the former in the model Hamiltonian

$$H = \frac{1}{2m}(p_1^2 + p_2^2) + V(q_1) + V(q_2) + gp_1p_2. \tag{11.4}$$

The second type of coupling would involve a leading term of the form fq_1q_2.

Small-amplitude vibrations governed by this Hamiltonian are well described by approximating $V(q_i)$ as quadratic in q_i, in which case there is a ready transformation to separable normal coordinates $q_\pm$, such that

$$q_\pm = \frac{1}{\sqrt{2}}(q_1 \pm q_2), \qquad p_\pm = \frac{1}{\sqrt{2}}(p_1 \pm p_2). \tag{11.5}$$

The resulting normal mode frequencies may be shown to be

$$\omega_\pm = (k/\mu_\pm)^{1/2}, \tag{11.6}$$

where $k = 2a^2D$ and

$$\frac{1}{\mu_\pm} = \frac{1}{m} \pm g. \tag{11.7}$$

The diagrams below are presented in a scaled system, with energies measured in units of $\hbar\bar{\omega}$, where $\bar{\omega} = (k/m)^{1/2}$, coordinates in units of $(\hbar/m)^{1/2}\bar{\omega}^{-1}$ and momenta in units of $(\hbar m)^{1/2}\bar{\omega}$, in which case the potential parameters in equation (11.3) are related by $a = (2D)^{-1/2}$.

Trajectories of the same system at somewhat higher energies are shown with their Poincaré sections in Fig. 11.4. The motions in and q_+ and q_- are no longer strictly separable, as evidenced by the curved caustic boundaries, but the general behaviour is 'normal' in the sense that the Lissajous figures have the full symmetry of the potential function, whose contour bounds the figure. Similarly the Poincaré section shows the normal pattern of concentric orbits around a central fixed point.

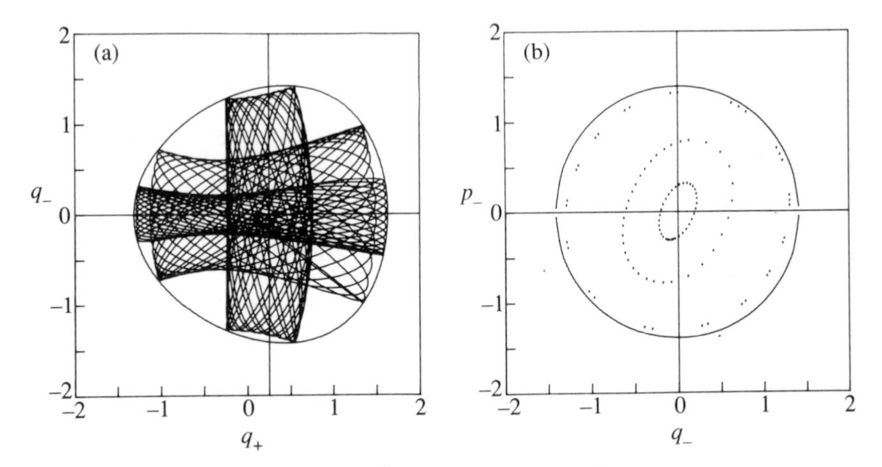

Fig. 11.4. (a) Lissajous figures and (b) $q_- = 0$, $p_- > 0$ Poincaré sections for three trajectories of the Hamiltonian given by equations (11.3) and (11.4) at $E = 2$.

Figure 11.5, on the other hand, demonstrates a complete change in behaviour, on increasing the energy further. There are now two distinct coexistent trajectory types: one (in Fig. 11.5(a)) is termed 'normal' because it carries the full symmetry of the potential, while the other (in Fig. 11.5(b)), which has a mirror image counterpart, is called 'local' because the amplitude of motion is predominantly localized in one of the two equivalent bonds. The corresponding Poincaré section, shown in Fig. 11.5(c), is seen to divide into two parts separated by the dashed figure-of-eight separatrix, such that the images of the normal and local trajectories lie respectively outside and inside the separatrix. The vacant lobe of the figure-of-eight is occupied by the Poincaré section of the symmetry-related counterpart to the local mode depicted in Fig. 11.5(b).

The symmetry-breaking bifurcation responsible for the transition from Fig. 11.4 to Fig. 11.5(c) is associated with a change in character of the periodic orbit along the q_+ axis in Figs. 11.3 and 11.5 whose image is the central fixed point in the Poincaré sections. The point is an elliptic (or stable) one in Fig. 11.4, as evidenced by the roughly elliptical orbits around it, but it is hyperbolic (or unstable) in Fig. 11.5(c) because neighbouring points lie along the arms of hyperbolae which diverge from it [11]. Nevertheless, the motion remains regular because the change from stability to instability at the centre of the section is accompanied by the appearance of two new elliptic points which act as foci for the local mode sections. The generic term for such transitions is a pitchfork bifurcation.

Another more physically intuitive way to visualize this normal mode to local mode bifurcation is in terms of the resonant response of one

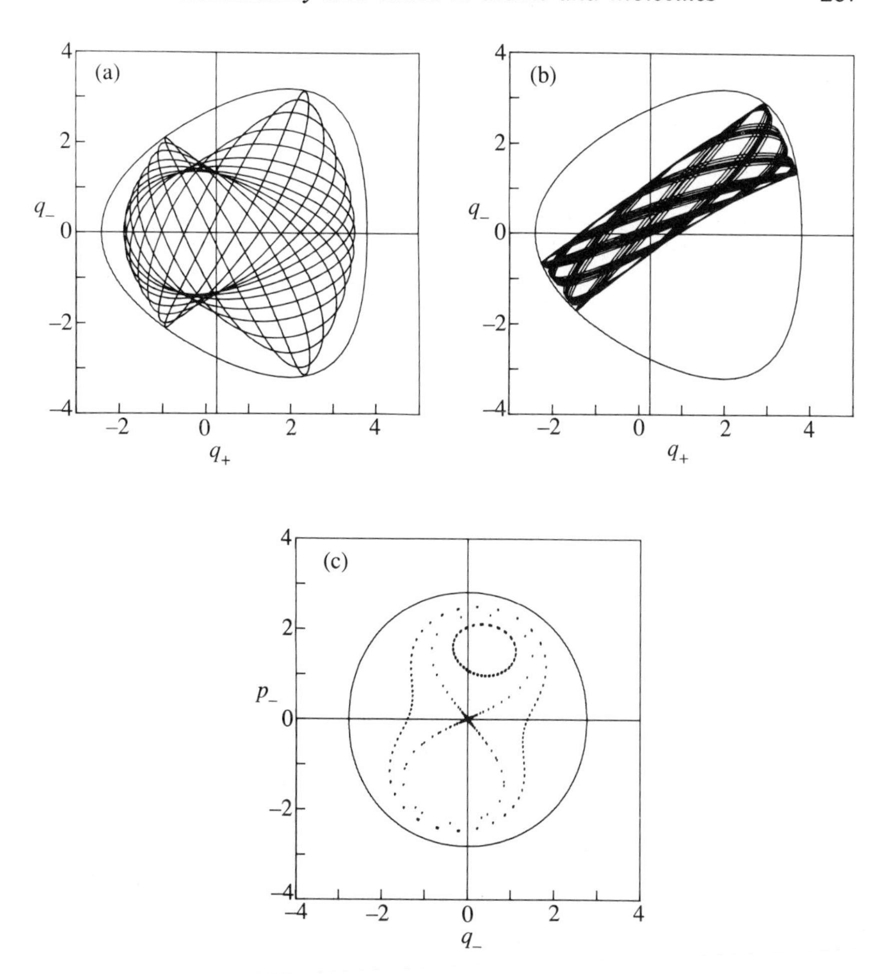

Fig. 11.5. Trajectories and the $q_- = 0$, $p_- > 0$ Poincaré section at $E = 4$ showing (a) a normal mode (b) a local mode and (c) the Poincaré section. The figure-of-eight section defines the separatrix between the inner local mode and outer normal mode parts of the phase space.

bond vibration to the other. At low enough energies, where the harmonic approximation of equation (11.5) applies, each bond oscillates with the same frequency regardless of its amplitude. Hence the oscillations of one are tuned to the natural frequency of the other and energy flows backwards and forwards at the frequency difference $|\omega_+ - \omega_-|$ implied by equation (11.6); it cannot be predominantly locked in one bond or the other. At higher energies, the anharmonicity of the bond potential comes into play, causing the frequency to decrease with increasing amplitude, and thereby inhibiting the possibility of a resonant response of one bond to the other.

Table 11.1 Local mode $(v, 0)$ doublets in the vibrational spectrum of H_2O spectrum (taken from [12]).

v	E_v (cm^{-1})	ΔE_v (cm^{-1})
1	3657.0	98.9
	3755.9	
2	7201.5	48.3
	7249.8	
3	10599.7	13.7
	10613.4	
4	13828.3	2.6
	13830.9	
5	16898.4	0.4
	16898.8	

In other words, the local motion in Fig. 11.5(b) arises from the fact that the amplitudes of the q_1 and q_2 motions are so disparate that any initial in phase excitation of q_2 by q_1 is immediately quenched as they slip into antiphase. The net effect is that neither oscillator is seriously disturbed by the other. On the other hand, if, at the same energy, the two bond amplitudes are more nearly equal, their frequencies are more closely matched and energy flows between them to produce the normal mode trajectory shown in Fig. 11.5(a).

The most immediate practical consequence of this symmetry-breaking bifurcation is to produce pairs of symmetry-related and hence classically degenerate local modes. This classical degeneracy is lifted by classically forbidden quantum effects (see [10] for a recent review), but the trend towards local mode doubling with increasing excitation is clearly apparent from the results in Table 11.1. The $(v, 0)$ bands, which are the strongest features in the overtone absorption spectra, arise from direct excitation of one bond or the other, modified by quantum mixing of the resulting states. Similar trends are also apparent in the weaker nearby $(v - 1, 1)$, $(v - 2, 2)$, etc., bands [12].

11.3 The onset of chaos: breaking the separatrix

The motions described above are strictly regular, as indicated by the smooth one-dimensional nature of the Poincaré sections of different trajectories. The bifurcation described in Section 11.2 is, however, an important step in the transition towards chaos, because the next observation, as the energy is increased, is that the Poincaré map of a trajectory initiated near the

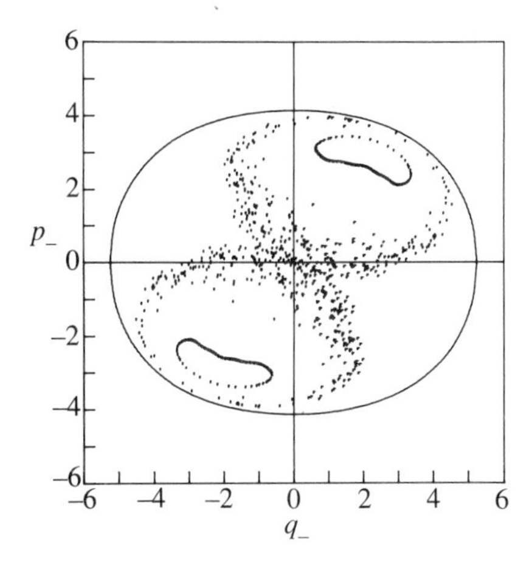

Fig. 11.6. Onset of chaos around the separatrix at $E = 8$ with coexistent regular motion.

separatrix in Fig. 11.5 consists of the scattered pattern of dots shown in Fig. 11.6, although other trajectories still remain on invariant tori, as exemplified by the regular island in the upper half of the figure. The origin of this chaotic scatter is traced below to a failure of the separatrix to join up with itself [13, 6].

Figure 11.7, which illustrates this failure, was generated by following trajectories from a ladder of points arbitrarily close to the unstable central point P. Forward propagation in time yields the so-called unstable manifolds H^-, while backward propagation yields the stable manifolds H^+, and their failure to meet poses awkward questions. Consider the consequences of some initial so-called homoclinic intersection X between H^+ and H^- in Fig. 11.8(a), remembering that any given trajectory appears as a set of discrete points in the Poincaré map. Forward time propagation of the point X^- on H^- would yield TX^- also on H^-, while a point at X^+ on H^+ would notionally appear at TX^+, although (since H^+ can be generated by only by backward time integration) X^+ is better regarded as the operation of T^{-1} on TX^+. The question arises: what happens to X? Its forward time image TX must lie on H^- at such a point that the backward time operation T^{-1} must take it back along H^+ to X. In other words TX must also be an intersection of H^- and H^+ (Fig. 11.8(b)), with similar consequences for T^2X, T^3X, etc. (Fig. 11.8(c)). A further restriction imposed by the area-preserving property of Hamiltonian dynamics is that the areas of the resulting loops between H^+ and H^- must be equal. A second consideration is that H^- cannot cross itself without destroying the determinacy of classical mechanics, with the result that its crossings with H^+ must become arbitrarily close on approaching P. Hence the homoclinic

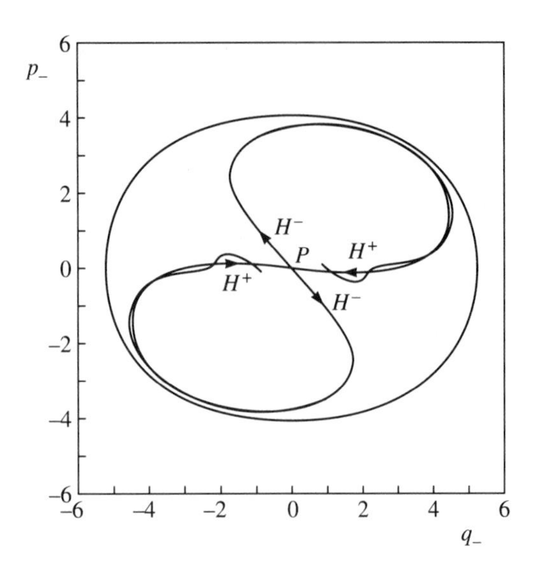

Fig. 11.7. The broken separatrix responsible for the incipient chaos in Fig. 11.6.

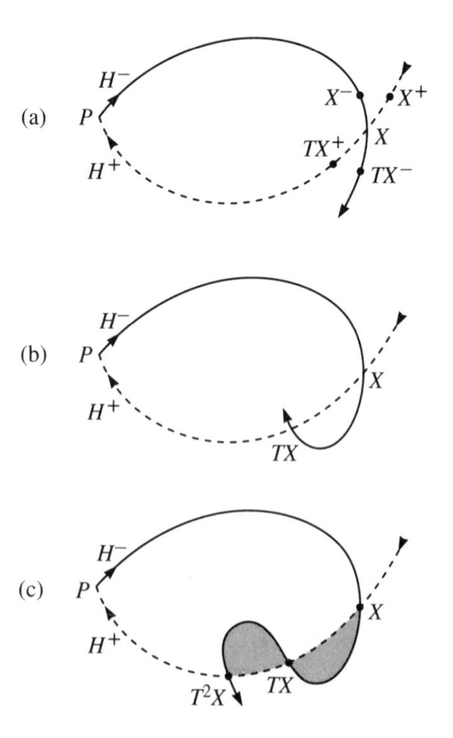

Fig. 11.8. Stages in the homoclinic tangle: (a) the first homoclinic intersection X; (b) propagation of X to TX; (c) further propagation to T^2X. H^+ and H^- are termed the stable and unstable manifolds respectively. Forward propagation in time takes X^- to TX^- while backward propagation takes TX^+ to X^+. The hatched areas in (c) are equal.

tangles become longer and longer and thinner and thinner. Similar behaviour also applies to H^-.

The consequences of this behaviour are illustrated by following some steps of the chaotic trajectory in Fig. 11.6 in more detail. Fig. 11.9(a) shows a sequence of about 50 steps, from the initial point marked with a square, that all lie outside the broken separatrix. Another crucial sequence of ten steps is shown in Fig. 11.9(b) against a more fully developed picture of the joint tangles of H^+ and H^-: those of H^- run SE to NW and those of H^+ almost E to W. The ultimate thinness of the tangles makes it impossible to show every detail, but it is clear that points 2 and 3 lie in successively narrower tangles of H^+ and broader tangles of H^- while point 4 lies in

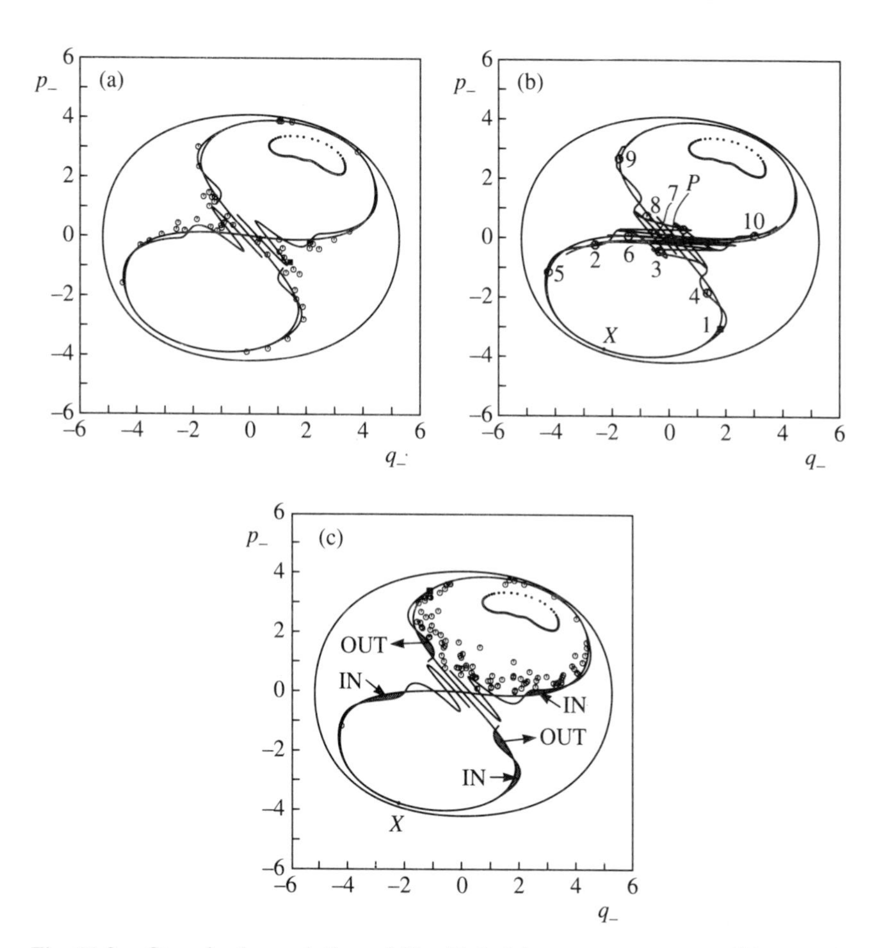

Fig. 11.9. Steps in the evolution of Fig. 11.6: (a) an outer sequence; (b) a transitional sequence; and (c) an inner sequence. Successive steps are numbered in (b). The arrows IN and OUT in (c) indicate the turnstiles.

the last broad tangle of H^+ and what would (if it were shown) be an even narrower tangle of H^-. Point 5 then appears outside the smooth union of H^- from P to the primary homoclinic intersection X, and H^+ from X to P (a so-called KAM curve). The next points 6, 7, 8, 9 lie in progressively narrower tangles of H^- and broader tangles of H^+ and points 10 and 11 are similarly controlled inside the upper inner lobe of the KAM curve. Finally, Figure 11.9(c) shows a long sequence over which the trajectory remains inside the upper inner lobe.

The crucial stages in this process may be seen to be steps 1–2, 4–5, and 9–10 in Fig. 11.9(b) that take the trajectory across the broken separatrix, all of which start from a 'last broad loop of H^+' and end in a 'first broad loop of H^-'. Thus the shaded regions in Fig. 11.9(c) and their images through the centre in the upper lobe constitute 'turnstiles' [13] between the inner and outer regions. Any point, such as 1, that appears in the IN side of the H^+ turnstile will next appear in the first IN loop of H^-, while points such as 4 in the OUT side of H^+ turnstile will be carried to the marked OUT loop of H^-. Consequently the sequence in Fig. 11.9(c), which started at the point marked IN, will remain inside until it finds its way to the doorway marked OUT. The average escape probability is therefore given by the ratio of the area of the turnstile loop to that of the chaotic fraction of the inner lobe. Notice, however, that the inward passage from 1 to 2 in Fig. 11.9(b) is followed after only two more steps by escape from the lower inner lobe by steps 4 to 5, whereas a similar entry in Fig. 11.9(c) is followed by a very much longer internal sequence. The difference lies in the fact that, once through the inward door of H^- the subsequent fate of the trajectory depends on its simultaneous choice of a normally unseen vanishingly thin loop of H^+. In Fig. 11.9(b) point 2 lies in only the third loop of H^+ so it is readily ejected, whereas the initial point in Fig. 11.9(c) could perhaps be in the hundredth loop of H^+.

11.4 Phase space bottlenecks to chemical reaction

Two types of application of these ideas to molecular systems have been given. The first, due to Davis [14], relates to theories of intramolecular energy transfer and unimolecular dissociation, which are cast in terms of probabilities of crossing phase space bottlenecks equivalent to the broken separatrix described above. To appreciate their full force, it must be recognized that the unstable point and associated manifolds $H^\pm$ in Figs. 11.8 and 11.9 may be one of many. Any unstable periodic orbit, with frequency ratio $n:m$, gives rise to a related set of n unstable points in one Poincaré section and m in the other, and the members of each such set are linked by separatrices, or KAM curves, some of which are much more resistant to break up than others. The most resistant are found to be those whose frequency ratio most nearly approaches a value [13]

$$\alpha_k = k + \gamma, \tag{11.8}$$

where k is an integer and γ is the golden mean

$$\gamma = \tfrac{1}{2}(\sqrt{5} - 1) \tag{11.9}$$

The span of possible frequency ratios, at a given energy, normally allows only one such golden-mean-related KAM curve, which acts as a phase space bottleneck to 'intramolecular energy flow' in the sense that it restricts the motion for long periods to a limited volume of phase space. The inner curve in Fig. 11.10 shows a bottleneck of this type, for a model of the weakly bound HeI_2 complex [15], and the dots in Fig. 11.10(a) show that trajectories can be trapped inside it for long periods. Its turnstile is more readily apparent in Fig. 11.10(b).

The outer boundary in Fig. 11.10 is a second, so-called 'intermolecular', phase space bottleneck which is closely approximated by the 'last bound' phase curve that terminates on an unstable fixed point at $Q \to \infty$ and this clearly has a larger turnstile. Hence, once the trajectory reaches the intermediate region in Fig. 11.10(b), it has a higher probability of escape into the dissociative region than returning through the inner bottleneck; the line of dots above the outer curve shows that the trajectory does indeed escape. These ideas have been extended to other dissociative systems by Gray and co-workers [16–18].

Similar concepts apply to exchange reactions of the type

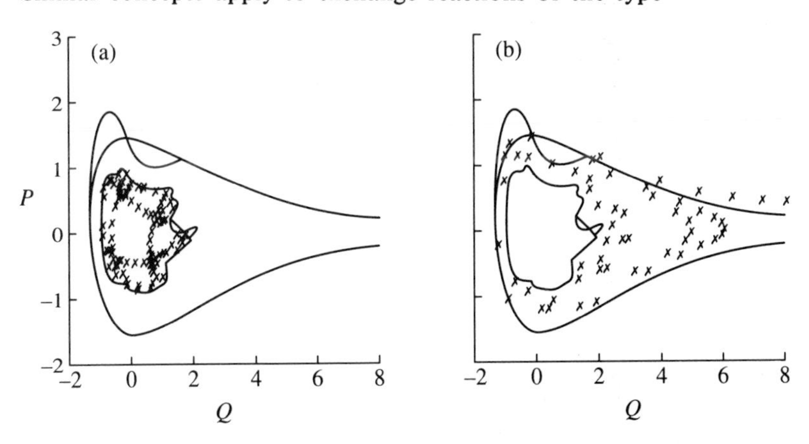

Fig. 11.10. Phase space bottlenecks to the dissociation of HeI_2: (a) trapping by the intramolecular bottlenecks; (b) trapping and ultimate passage through the intermolecular passage through the intermolecular bottleneck. Note the turnstiles, which are most clearly evident in (b). (Reproduced from 15], with permission.)

$$A + BC \rightarrow AB + C.$$

An extensive account of the role of periodic orbits in dividing reactants A + BC from products AB + C has been given by Pollak [19]. In many circumstances they provide the true 'transition states' or 'activated complexes', so familiar in the chemical literature [20]. Nonlinear effects directly related to the turnstiles have been described by Davis [21] and Skodje and Davis [22].

11.5 Imprints of order in the chaotic regime

The scenario in Section 11.3 concerns the break up of separatrices of KAM curves associated with unstable periodic orbits of the system. Figure 11.6 shows, however, that the resultant chaos initially leaves islands of order around other periodic orbits that are stable, and it is found that the most stable of these are typically the ones with the simplest frequency ratios, i.e. those which complete their periods in the shortest time. Recall, by contrast, that the most stable KAM curves are those with golden-mean-related ratios

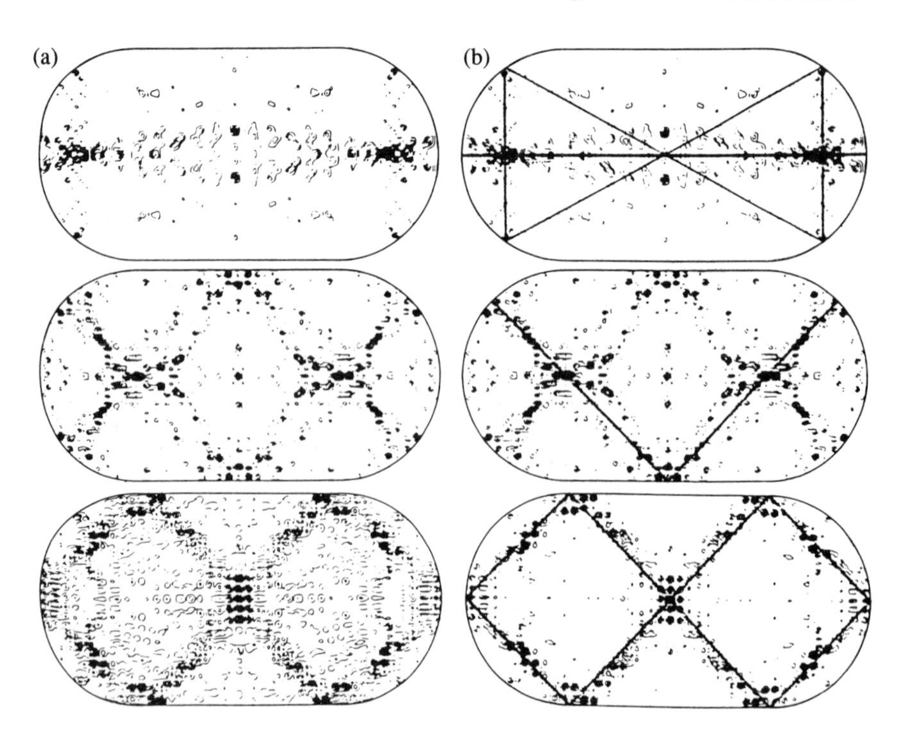

Fig. 11.11. (a) Three scarred quantum mechanical wavefunctions of the classically chaotic stadium; (b) the corresponding periodic classical orbits. (Reproduced from [23], with permission.)

given by equation (11.8)

Ultimately the chaos becomes global as even these 'most stable' orbits become unstable, but despite this instability they have been found to impose a 'scarring' imprint on the quantum mechanical behaviour of the system. The first compelling demonstration was given by Heller [23] in the context of the globally chaotic stadium problem, for which certain of the wave-functions are seen in Fig. 11.11 to carry clear scars of relatively short periodic orbits.

A different aspect of this behaviour is predicted in the spectrum of magnetically perturbed hydrogen atoms, subject to the Hamiltonian

$$H = \tfrac{1}{2}\boldsymbol{p}\cdot\boldsymbol{p} - \frac{1}{r} + \tfrac{1}{2}\gamma L_z + \tfrac{1}{8}\gamma^2(x^2 + y^2), \qquad (11.10)$$

where $\gamma = B/B_c$ is the magnitude of the field (along the z-axis) measured in atomic units ($B_c = 2.34 \times 10^5$ T). This system has been subject to close theoretical and experimental investigation; full details are collected in the conference proceedings edited by Nicolaides *et al.* [24]. Notice that the angular momentum L_z around the field direction is conserved, and hence that the interesting motion involves only two degrees of freedom.

The transition to chaos arises from a competition between the Coulomb and magnetic fields. At low magnetic fields the Coulomb forces dominate and the energy levels, shown on the left of Fig. 11.12, are those of a

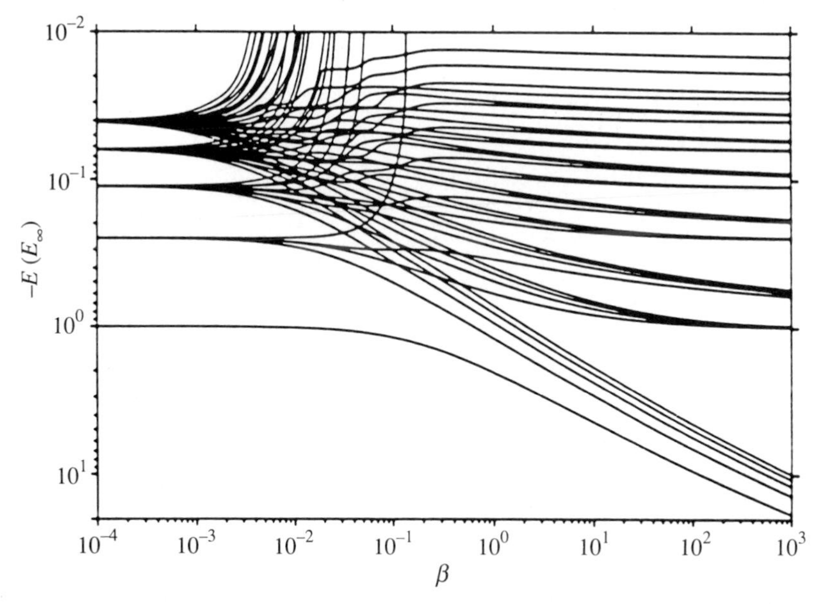

Fig. 11.12. Energy levels of the magnetically perturbed hydrogen atom as a function of the magnetic field strength (Reproduced from [32], with permission.)

hydrogen atom subject to a weak Zeeman effect. At very high fields, on the other hand, the magnetic field predominates and the levels are best seen as multiplets associated with a one-dimensional hydrogen atom (moving along the z-axis) with superimposed vibrations (for low-energy members of the multiplet) or quasi-Landau rotations (at higher energies) in the (x, y)-plane. Calculations by Wunner *et al.* [25] showing this transition for the lowest few energy states are shown in Fig. 11.12, but a complicated pattern of curve intersections and avoided intersections should also be

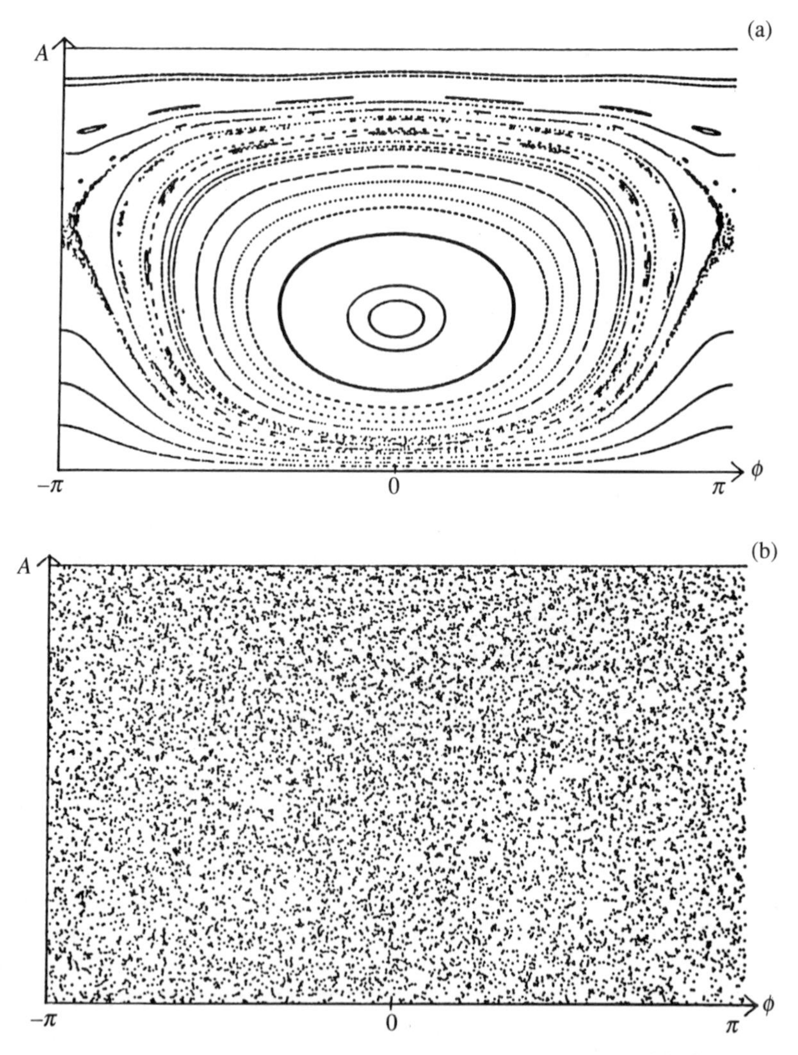

Fig. 11.13. Poincaré sections for the scaled magnetically perturbed hydrogen atom at (a) $\beta = 0.80$ and (b) $\beta = 70$. The central fixed point belongs to the periodic quasi-Landau orbit. (Reproduced from [29], with permission.)

noted. Notice also that the magnitude of the magnetic field (4.7×10^8 T) required to dominate the lowest states is of the order of those that exist at the centre of neutron stars.

However, the observation of similar effects in Earth bound spectra become practicable when it is realized that the classical mechanics under the Hamiltonian in equation (11.10) scales with the dimensionless parameter [29]

$$\beta = \gamma^2/(-2E)^3. \tag{11.11}$$

Hence a field of 5×10^8 T at $E = -100\,000\,\text{cm}^{-1}$ is classically equivalent to one of 500 T at $E = -10\,\text{cm}^{-1}$. Details of the calculations are beyond the scope of this chapter, but Fig. 11.13 shows two Poincaré sections reported by Delande and Gay [29] in a particular representation that displays coexistent rotational (quasi-Landau) and vibrational modes of the high-energy multiplet at the dimensionless parameter value at $\beta = 0.08$. Evidence of incipient chaos is also apparent in the scattered dots around the separatrix in Fig. 11.13(a). The effect of increasing β is to increase the chaotic fraction of the phase space. It is found that the periodic orbits at the centre of the vibrational parts of the manifold (i.e. the upper and lower bounding curves in Fig. 11.13(a)) become unstable at $\beta \simeq 3$, while the central fixed point associated with the quasi-Landau states remains stable until $\beta \simeq 60$, after which the classical motion is globally chaotic (Fig. 11.13(b)).

The interesting feature for this chapter is that the last stable (quasi-Landau) periodic orbit still imposes an imprint on the quantum mechanical spectrum at energies beyond the chaotic limit. As an example, Fig. 11.14 shows a simulated $L_z = 0$ even parity spectrum at a field of 8 T ($g = 3.4 \times 10^{-5}$) at energies between -132 and $44\,\text{cm}^{-1}$. The arrow indicates the transition to global chaos. Panel (a) shows squared projections of the quantum mechanical eigenfunctions onto a quasi-Landau subspace. Isolated wavefunctions in the partially regular regime to the left of the arrow are seen to have almost pure quasi-Landau character and their energy separations are determined by the frequency of the periodic orbit. There is then a sharp change in passing to the globally chaotic regime, where the central orbit becomes unstable, but the picture still has some order in the spacing between clusters of energy levels. Figure 11.14(b) shows a similar picture in terms of the oscillator strengths of transitions from the $(2p, m = 1)$ state, except that the clustering in the chaotic regime is less well marked. Nevertheless, the regular quasi-Landau spectrum may be recovered (Fig. 11.14(c)) by convoluting the lines in Fig. 11.14(b) with a Gaussian lineshape function with a width of $\Delta\bar{\nu} = 4\,\text{cm}^{-1}$ (0.2×10^{-1} a.u.). An intuitive interpretation of this final spectrum is that the order is preserved over a time $\Delta\tau = 2\pi c\Delta\bar{\nu}^{-1} \simeq 1.3 \times 10^{-12}$ s.

The transition from order to chaos arose in this case at an energy at which the energy level separations are comparable with the magnitude of the

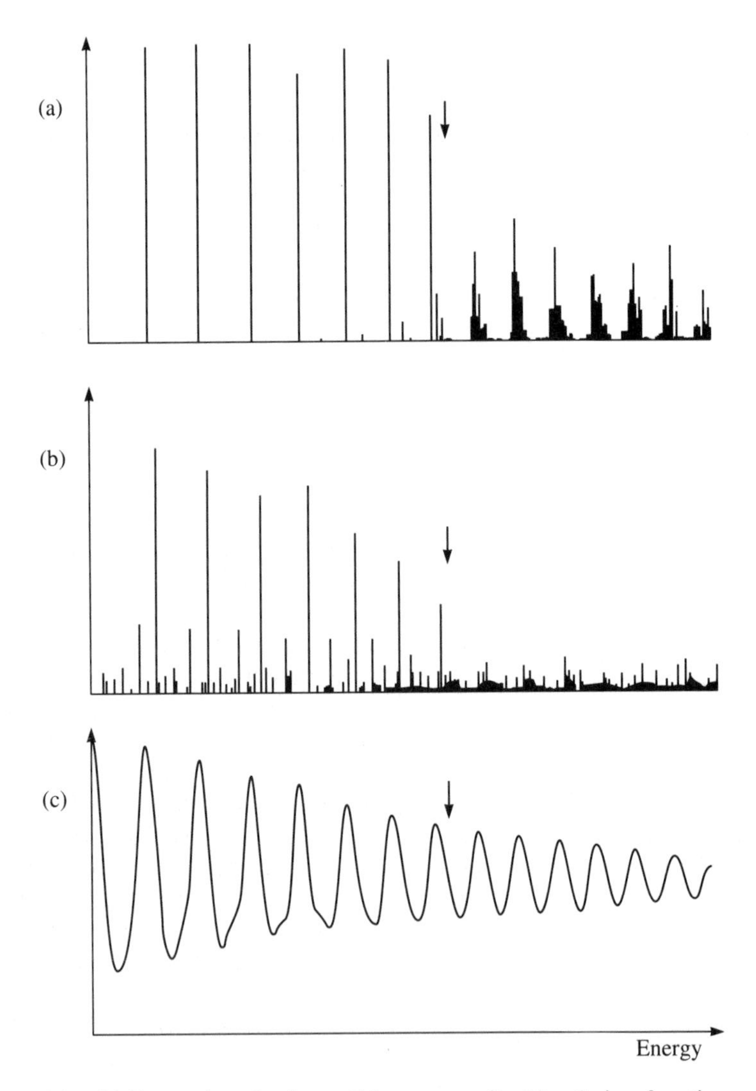

Fig. 11.14. (a) Squared projections of the even parity $M = 0$ eigenfunctions onto the quasi-Landau subspace; (b) oscillator strengths from the $(2p, M = -1$ state; (c) a convolution of (b) with a lineshape function with width 0.00002 a.u. ($4\,\mathrm{cm}^{-1}$). (Reproduced from [33], with permission.)

magnetic perturbation. The Rydberg states of molecules have somewhat similar properties, even in the absence of external fields, because resonances can arise between the (fixed) frequency of molecular rotational motion and that of the Rydberg electronic motion which varies as n^{-3}, where n is the principal quantum number. Lombardi *et al.* [30] give a revealing analysis of the Rydberg spectrum of Na_2 from this point of view.

11.6 Summary and conclusion

Aspects of the transition from order to chaos in Hamiltonian dynamics with two degrees of freedom have been discussed in relation to chosen atomic and molecular systems. Emphasis was placed on the characters of particular periodic classical orbits, which appear as important fixed points in the Poincaré section.

At the simplest level the two modes of motion are separable (Figs. 11.1 and 11.2) with each giving rise to a stable periodic orbit. The familiar low-energy normal modes of molecular vibration [2] typify this situation. Interesting situations arise when one of the periodic orbits becomes unstable. The example discussed in Section 11.3 involved an unstable resonance between two equivalent anharmonic oscillators, modelled on the stretching vibrations of H_2O, which was seen (Fig. 11.5) to lead to a bifurcation of the phase space into local and normal mode regions, and to experimentally observed local mode near degeneracies in the vibrational spectrum. Fermi resonances [2] between vibrational modes with rationally related frequencies lead to similar effects [31].

Early bifurcations of these types may still be associated with regular motion, but the presence of an unstable fixed point in the Poincaré section was found to herald the eventual onset of chaos, as evidenced by a complex homoclinic tangling of the separatrix between different parts of the phase space (Figs. 11.6 and 11.7). The existence of these tangles was found to result in a 'turnstile' from one phase region to the other (Fig. 11.9). Evidence not presented here shows that the most resistant separatrices (or KAM curves) to this homoclinic break up are those associated with 'least rational, unstable periodic orbits, i.e those with frequency ratios related to the golden mean by equations ((11.8) and (11.9). Applications of the rate of turnstile flow across these most rigid KAM curves to theories of chemical reaction were outlined.

Finally the 'scarring' imprints left behind after the disappearance of the last stable periodic orbit were described. Their most direct manifestation was as scars on the quantum mechanical wavefunction (Fig. 11.11), but it was also seen by reference to the magnetically perturbed hydrogen atom that their frequencies could be reflected in a 'clumping' of the quantum mechanical spectrum, under situations where the classical motion was chaotic, particularly if the linewidths are artificially broadened to simulate a limited lifetime in the resonance region.

References

1. Gray, P. and Scott, S. K. (1990). *Chemical oscillations and instabilities*. Oxford University Press.

2. Herzberg, G. (1945). *Infra-red and Raman spectra*. Van Nostrand, New York.
3. Robinson, P. J. and Holbrook, K. A. (1972). *Unimolecular reactions*. Wiley, New York.
4. Crim, F. F. (1984). Selective excitation studies of unimolecular reaction dynamics. *Ann. Rev. Phys. Chem.*, **35**, 657.
5. Lichtenberg, A. J. and Lieberman, M. A. (1983). *Regular and stochastic motion*. Springer, New York.
6. Tabor, M. (1989). *Chaos and integrability in non-linear dynamics*. Wiley, New York.
7. Percival, I. C. (1977). Semiclassical theory of bound states. *Adv. Chem. Phys.*, **36**, 1.
8. Noid, D. W., Kosykowski, M. L., and Marcus, R. A. (1981). Quasiperiodic and stochastic behaviour in molecules. *Ann. Rev. Phys. Chem.*, **32**, 267.
9. Berry, M. V., Percival, I. C., and Weiss, N. (1987). *Dynamical chaos*. Princeton University Press.
10. Child, M. S. and Halonen, L. (1984). Overtone frequencies and intensities in the local mode picture. *Adv. Chem. Phys.*, **57**, 1.
11. Percival, I. C. and Richards, D. (1982). *Introduction to dynamics*. Cambridge University Press.
12. Child, M. S. and Lawton, R. T. (1982) Local mode degeneracies in the vibrational spectrum of water. *Chem. Phys. Lett.*, **87**, 217.
13. Mackay, R. S., Meiss, J. D., and Percival, I. C. (1987). Resonances in Hamiltonian maps. *Physica D*, **27**, 1.
14. Davis, M. J. (1985). Bottlenecks to intramolecular energy transfer and the calculation of relaxation rates. *J. Chem. Phys.*, **83**, 1016.
15. Davis, M. J. and Gray, S. K. (1986). Unimolecular reactions and phase space bottlenecks. *J. Chem. Phys.*, **84**, 5389.
16. Gray, S. K. and Rice, S. A. (1986). The photofragmentation of simple van der Waals complexes. *Faraday Discuss. Chem. Soc.*, **82**, 307.
17. Gray, S. K., Rice, S. A. and Noid, D. W. (1986). The classical mechanics of vibrational predissociation. *J. Chem. Phys.*, **84**, 3475.
18. Gray, S. K., Rice, S. A., and Davis, M. J. (1987). Bottlenecks to unimolecular reactions and an alternative form for classical RRKM theory. *J. Phys. Chem.*, **90**, 3470.
19. Pollak, E. (1985). Periodic orbits and the theory of reactive scattering. In *Theory of chemical reactions* (ed. M. Baer). CRC Press, Boca Raton, Florida.
20. Laidler, K. J. (1987). *Chemical kinetics*, 3rd edn. Harper and Row, New York.
21. Davis, M. J. (1987). Phase space dynamics of bimolecular reactions and the breakdown of transition state theory. *J. Chem. Phys*, **86**, 3987.
22. Skodje, R. T. and Davis, M. J. (1988). A phase space analysis of the collinear I + HI reaction. *J. Chem. Phys.*, **88**, 2429.
23. Heller, E. J. (1984). Bound state eigenfunctions of classical chaotic Hamiltonian systems: scars of periodic orbits. *Phys. Rev. Lett.*, **53**, 1515.
24. Nicolaides, C. A., Clark, C. W., and Nayfeh, M. H. (1990). *Atoms in strong fields*, NATO ASI Series B212. Plenum, New York.
25. Wunner, G., Woelk, U., Zech, I., Zeller, G., Ertl, T., Geyer, P., Schweizer, W., and Ruder, H. (1986). Rydberg atoms in uniform magnetic fields: uncover-

ing the transition from regularity to irregularity in a quantum system. *Phys. Rev. Lett.*, **57**, 3261.

26. Delos, J. B., Knudson, S. K., and Noid, D. W. (1984). Trajectories of an atomic electron in a magnetic field. *Phys. Rev.* **A, 30**, 1208.
27. Wingten, D. and Freidrich, H. (1986). Matching the low field region and the high field region for the hydrogen atom in a uniform magnetic field. *J. Phys. B.*, **19**, 991.
28. Wingten, D. and Freidrich, H. (1986). Appropriate separability for the hydrogen atom in a uniform magnetic field. *J. Phys. B.*, **19**, 1261.
29. Delande, D. and Gay, J. C. (1986). Quantum chaos and the statistical properties of energy levels: numerical study of the hydrogen atom in a magnetic field. *Phys. Rev. Lett.*, **57**, 2006.
30. Lombardi, M., Labarti, P., Bordas, M. C., and Broyer, M. (1988). Molecular Rydberg states: classical chaos and its correspondence in quantum mechanics *J. Chem. Phys.*, **89**, 3479.
31. Noid, D. W., Kosykowski, M. L., and Marcus, R. A. (1979). Semiclassical calculations of bound states in multidimensional systems with Fermi resonance. *J. Chem. Phys.*, **71**, 2864.
32. Wunner, G. (1991). *Hydrogen atoms in strong magnetic fields*. Springer, Heidelberg.
33. Delande, D. and Gay, J. C. (1987). Scars of symmetry in quantum chaos. *Phys. Rev. Lett.*, **59**, 1809.

12
The quantum mechanics of chaotic systems or 'Can one hear the chaology of a drum?'

Jonathan Keating

This chapter has two titles. The first one is 'The quantum mechanics of chaotic systems', and this is indeed the general subject to be discussed. However, we mainly intend to focus on one particular aspect of what is becoming a very broad subject. This concerns the question raised in the subtitle, which is, with obvious influence from Kac [1], 'Can one hear the chaology of a drum?' Our aims are: first to explain what is meant by this question, second to give what appears to be an answer, and finally to describe how we are beginning to develop theories which explain to what extent this answer is, in fact, correct. On the way we shall also describe several interesting connections which have emerged between this relatively young area of physics and one of the longest-running pursuits of pure mathematics—the distribution of the prime numbers [2].

12.1 The quantum–classical relationship

Quantum mechanics is a wave theory of motion which contains, in its short-wavelength limit, classical mechanics. In this sense it is just like other wave theories, such as those of acoustics, optics, or the vibration of membranes (e.g. drums), which contain in their short-wavelength limits rays (straight line trajectories with specular reflection at boundaries). These rays are, of course, particularly obvious in optics.

Now the main title of this chapter mentions only quantum mechanics. However, let us emphasize right away that the physics to be described here does not represent some mysterious quantum phenomenon; it applies, in general, equally well to the other wave theories just mentioned. Thus in what follows, 'quantum mechanics' can usually be translated as 'any wave theory which has rays underlying its short-wavelength limit', and similarly 'classical mechanics' can be taken to mean 'any ray theory' [3]. In particular, results concerning the relationship between quantum and classical mechanics can be translated into the language of drums.

In quantum mechanics the short-wavelength limit has a special name; it is called the *semiclassical limit*. Mathematically this corresponds to allowing Planck's constant $\hbar$, here considered as a parameter in the appropriate quantum equation, to tend to zero. Of course Planck's constant is, as its name implies, really a constant of nature, so what does this mean? Well it actually involves allowing $\hbar$, when divided by some classical quantity with the same dimensions, to tend to zero. Physically this usually corresponds to probing the very highly excited states of a system. For drums, taking the short-wavelength limit means listening to the very high frequency modes of vibration.

'Quantization' is a process by which the quantum mechanical equations describing a given dynamical system may be obtained, starting from the classical mechanical description; that is, it gives a natural way to associate waves with a classical mechanical, or ray, structure. *Quantum chaology* is then essentially the study of the influence of the chaotic nature of the classical mechanics, or rays, on the associated quantum mechanics, or waves. Specifically it concerns itself with the semiclassical limit of quantum properties which themselves have no classical counterpart, that is, with essentially wave-like properties [4].

A natural question to ask in any work devoted to *chaos* is: given a system whose classical mechanics is chaotic, will the quantum dynamics satisfy any of the criteria for chaos as well? Curiously the answer appears to be *no*: quantum mechanical evolution is not persistently unstable. It can be unstable, but not for arbitrarily long times. This may be explained by noting that classical chaos can be regarded as exponential sensitivity to the structure of phase space on infinitely small scales. In quantum mechanics, however, the uncertainty principle generates a smoothing of this structure over areas which are of the order of Planck's constant; that is, infinitely complex structure on the finest scales no longer exists. This phenomenon has been called the *quantum suppression of classical chaos*, and it provides some of the most challenging problems in the study of the quantum mechanics of chaotic systems [5]. It is, however, not to be the subject of this Chapter. Instead we shall focus on the question which constitutes the subtitle. Our first task is to explain what it means.

The question that we would like to answer is: given a quantum system, can one decide, using only simple quantum properties, whether its underlying classical motion is chaotic or not? Alternatively, can one decide, solely by listening to it, whether the rays of elasticity underlying the vibrations of a given drum are bouncing around between its walls in a chaotic or a regular manner?

The answer lies in what some regard as the most remarkable and important result to have emerged from quantum chaology: it appears to be the case that any typical quantum system has energy levels which are distributed in one of a small number of universal ways, the particular distribution taken depending only upon whether the underlying classical mechanics is chaotic or not, and also on the presence of certain special symmetries. In terms of drums, what this means is that the frequencies of a given drum will have a particular distribution which typically depends only upon whether the underlying ray dynamics is chaotic or not; it will not depend upon the details of its shape. This universality emerges in the semiclassical limit, or, for drums, when we listen to the very high frequencies. It may be investigated by looking at the statistics of these eigenvalue distributions [6].

12.2 Universality and statistics

Consider a series of eigenvalues, scaled so as to have unit mean density. We shall concentrate upon two particular statistics of this sequence. The first is the distribution $P(s)$ of separations s between neighbouring scaled eigenvalues. The second, the *number variance* $\Sigma(L)$, is obtained by counting how many scaled eigenvalues lie in each of a series of ranges of length L. The average of the numbers found will, by construction, be L. The actual numbers will, however, fluctuate about this value and $\Sigma(L)$ is simply the variance of these fluctuations.

What universality then means is that, typically, all systems whose classical motion is integrable (that is, not chaotic) have the same statistics. By 'typically' we mean that this is the generic case. There will be a few exceptions, but this is what would be found with probability one for a system picked at random from a huge hat filled with all of the possible integrable systems. The universal statistics appropriate for classically integrable systems have the forms

$$P(s) = \exp(-s), \tag{12.1}$$

$$\Sigma(L) = L. \tag{12.2}$$

For systems which are chaotic, there is an analogous result. Typically, all systems whose classical motion is chaotic and time-reversal symmetric have

$$P(s) \approx \frac{\pi}{2} s \exp\left(-\frac{\pi s^2}{4}\right), \tag{12.3}$$

$$\Sigma(L) \approx \frac{2}{\pi^2} \ln L + k_1 \quad \text{if } L \gg 1, \tag{12.4}$$

where k_1 is a universal constant, and all systems which are classically chaotic but not time-reversal symmetric have

$$P(s) \approx \frac{32}{\pi^2} s^2 \exp\left(-\frac{4s^2}{\pi}\right) \tag{12.5}$$

$$\Sigma(L) \approx \frac{1}{\pi^2} \ln L + k_2 \quad \text{if } L \gg 1, \tag{12.6}$$

where k_2 is another universal constant. There is one more set of universal statistics, appropriate for chaotic systems containing an odd number of fermions, which will not be described here [7]. However, once this has been included there are then no more possibilities.

So 'universality' means that the statistics of a randomly chosen system will be one of the above possibilities. But that is not all. These universal statistics are very special in their own right. The statistics (12.1) and (12.2) are just those of random numbers; that is, they are Poisson statistics. The statistics (12.3) and (12.4) are just those that have been proved to hold for the eigenvalues of very large real symmetric matrices which have randomly picked entries; such matrices form what is called the *Gaussian Orthogonal Ensemble* (GOE) of random matrix theory [8]. Similarly, the statistics (12.5) and (12.6) are just those of the eigenvalues of very large complex Hermitian matrices which have randomly picked entries; such matrices form what is called the *Gaussian Unitary Ensemble* (GUE).

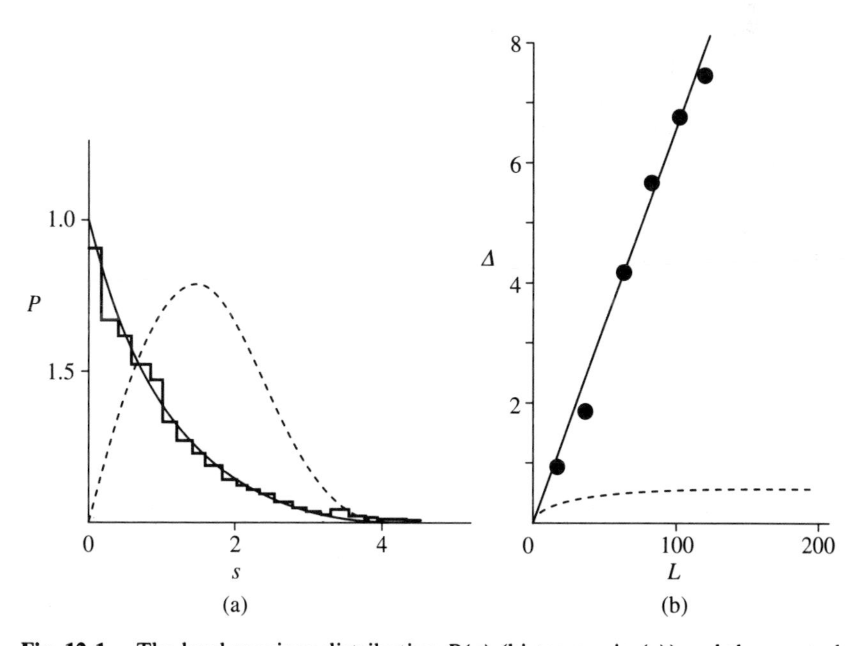

Fig. 12.1. The level spacings distribution $P(s)$ (histogram in (a)) and the spectral rigidity $\Delta(L)$ (filled circles in (b)) of the energy levels of the rectangular billiard. The full curves represent the statistics of eigenvalues which are Poisson distributed and the dashed curves represent the GOE random matrix forms. (Reproduced from [4], with permission.)

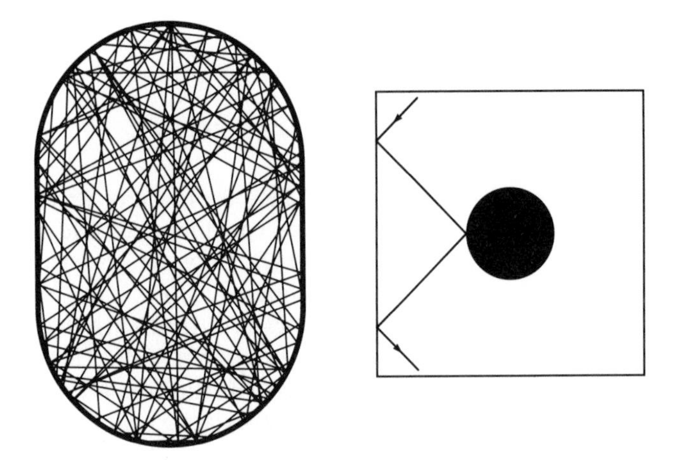

Fig. 12.2. The stadium of Bunimovich (left), with a chaotic orbit, and Sinai's billiard (right).

Hence, not only do we have semiclassical universality of eigenvalue statistics, but the statistical forms involved are highly characteristic and special. Let us present just some of the evidence for this remarkable phenomenon. The data to be shown were obtained by numerical computation of eigenvalues. However, beautiful experimental results have recently been reported [9] which are in excellent agreement with those obtained numerically. (Actually, instead of presenting results for $\Sigma(L)$, we will show those for a very closely related statistic: the *spectral rigidity* $\Delta(L)$ [10]. This behaves in a similar way to the number variance, but is computationally more convenient. The GOE and GUE forms for $\Delta(L)$ are the same

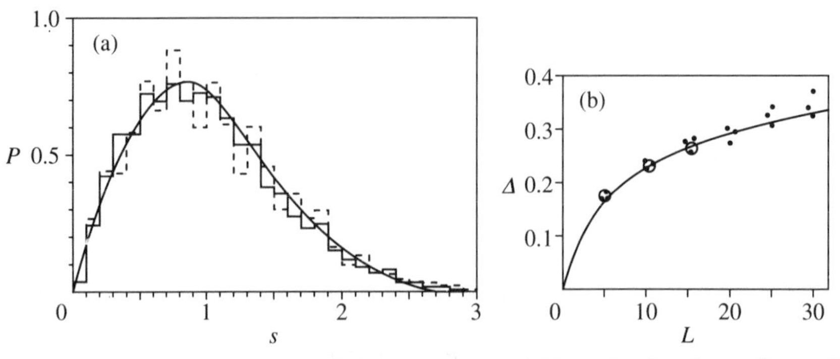

Fig. 12.3. The energy level statistics, (a) $P(s)$ and (b) $\Delta(L)$, for the stadium of Bunimovich (full line histogram and filled circles respectively—data from [11]) and for Sinai's billiard (dashed line histogram and open circles respectively—data from [12]). The GOE random matrix forms are represented by the smooth curves. (Reproduced from [4], with permission.)

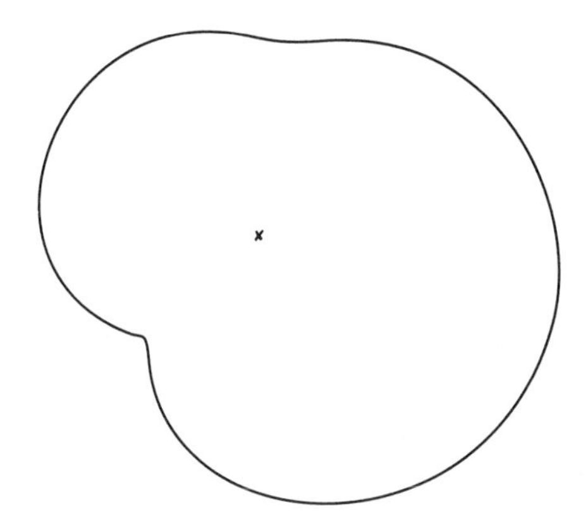

Fig. 12.4. The 'Africa' billiard. The cross marks the position of the magnetic flux line.

as (12.4) and (12.6), but with different universal constants. For Poisson-distributed numbers, $\Delta(L) = L/15$.)

Consider a particle bouncing around in a rectangular enclosure, that is, a pocketless billiard table. The classical motion of this particle is integrable. The statistics of the distribution of the energy levels of the corresponding quantum system are shown in Fig. 12.1 (These are, of course, just the frequencies of a drum of the same shape.) The agreement with Poisson statistics is obviously very good. Furthermore, these statistics may be seen to be completely different from those of the GOE.

Figure 12.2 shows two less conventionally shaped billiard tables. The first is known as the *stadium of Bunimovich*, and the second as *Sinai's billiard*. Billiards would be an entirely different game on these tables, because the classical mechanics of a particle bouncing around between their walls is chaotic, that is, typical neighbouring trajectories (or rays of the corresponding drums) separate exponentially quickly. The statistics of the energy levels are shown in Fig. 12.3. That these two very different systems appear to have the same eigenvalue statistics is an example of universality. These results are also in very good agreement with the GOE random matrix forms.

Finally, we have another completely chaotic system. This is a particle bouncing around in an enclosure which looks rather like the shape of Africa (see Fig. 12.4) [13]. In the middle of 'Africa' there is a single line of magnetic flux, which is sufficient to break time-reversal symmetry if the bouncing particle carries a charge. It is found that the eigenvalues of this system are distributed in a way which is obviously in good agreement with the GUE statistics, and obviously very different from those of the GOE (see Fig. 12.5).

So there is a universality of energy level statistics which has been observed in a wide range of systems, and it is this phenomenon which provides the answer to the question posed in the subtitle. In general you can indeed hear

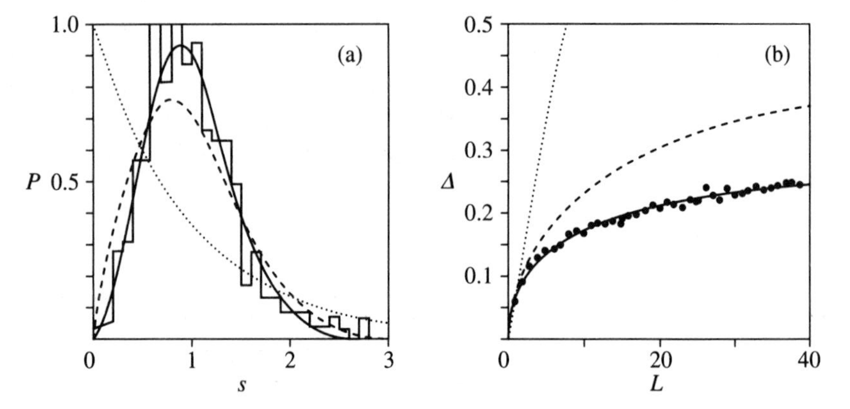

Fig. 12.5. The level spacings distribution $P(s)$ (histogram in (a)) and the spectral rigidity $\Delta(L)$ (filled circles in (b)) for the Africa billiard with magnetic flux line (data from [14]). The GUE and GOE random matrix forms are represented by the full and dashed curves respectively. The corresponding Poisson statistics are represented by the dotted curves. (Reproduced from [4], with permission.)

the chaology of a drum, provided that you have hypersensitive hearing. Specifically, you need to be able to hear all of the drum's frequencies. The statistics of the high frequencies will typically fall into one of the universality classes described above. Depending on which of the universal statistics are found, it is simple to decide whether the underlying rays are chaotic or not. This answer is based on extensive numerical and experimental investigations. To understand to what extent it is correct we require asymptotic theories for the statistics of high-lying eigenvalues [15], and the rest of the chapter will be devoted to describing our efforts to develop such theories in terms of the associated classical mechanics or ray structure.

12.3 Quantum energy levels and classical periodic orbits

Semiclassical theories of spectral statistics begin with the density of states $d(E)$, which is a series of delta functions, each centred on one of the energy levels of the system:

$$d(E) = \sum_n \delta(E - E_n). \tag{12.7}$$

This can also written in terms of the trace of the outgoing energy-dependent Green's function $G(\boldsymbol{q}_2, \boldsymbol{q}_1; E)$:

$$d(E) = -\frac{1}{\pi} \lim_{\varepsilon \to 0} \operatorname{Im} \int G(\boldsymbol{q}, \boldsymbol{q}; E + i\varepsilon) d\boldsymbol{q}. \tag{12.8}$$

Physically, this Green's function gives the amplitude at a point $\boldsymbol{q}_2$ of waves which are continuously emitted from the point q_1 with energy E. Semi-classically we expect these waves to collapse onto the underlying rays, or classical trajectories. Thus, in the semiclassical limit we expect contributions to this Green's function from all of the orbits of the classical motion which connect the two points q_1 and q_2; and this is indeed the case [16]. The actual contribution from a given classical path is proportional to $\exp[iS(q_2, q_1; E)/\hbar]$, where $S(q_2, q_1; E)$ is the classical action along that path, that is, the integral of the momentum along it:

$$S(q_2, q_1; E) = \int_{q_1}^{q_2} p(q, q_1; E)\mathrm{d}q. \tag{12.9}$$

Now what we need in order to take the trace of the Green's function are its diagonal elements and these will obviously be determined by closed orbits of the classical motion. In taking the trace we can evaluate the integral over the diagonal elements by the method of stationary phase. The stationary phase condition picks out from the set of closed orbits those which are self-retracing—the *periodic orbits*. Hence, there is a semiclassical relationship between the density of states, or the eigenvalues, of a quantum system, and the periodic orbits of the underlying classical motion.

Now there are two qualitatively different kinds of periodic orbits which contribute. Consider what happens as we take the diagonal elements of the Green's function: q_1 gets closer and closer to q_2. As the points approach each other, there are connecting trajectories whose lengths shrink to zero, and there are also trajectories whose lengths tend to nonzero values, that is, paths which loop away and come back again. This is shown in Fig. 12.6.

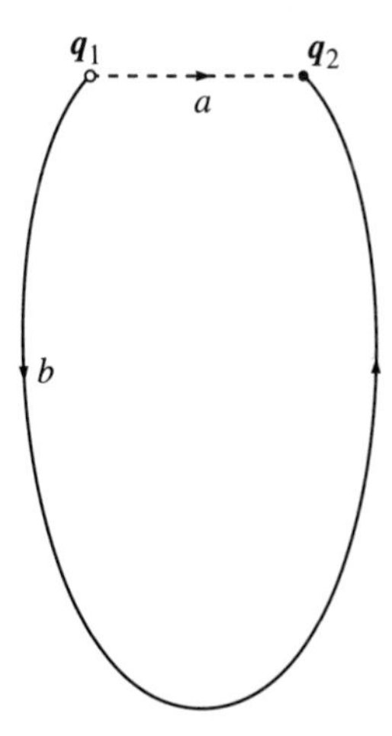

Fig. 12.6. Two possible classical trajectories connecting points q_1 and q_2. As the points approach each other, the length of path a tends to zero, but the length of path b tends to a nonzero value.

The orbits of zero length give rise to the average density of states $\langle d(E)\rangle$, which, as $\hbar \to 0$, takes the form [17]

$$\langle d(E)\rangle \sim \frac{1}{h^N}\frac{\mathrm{d}\Omega(E)}{\mathrm{d}E}, \qquad (12.10)$$

where N is the number of degrees of freedom and $\Omega(E)$ is the volume of the energy shell (i.e. a property of the classical mechanics determined by the geometry of the system). This result, along with the lower-order terms in the semiclassical approximation, means that one can indeed hear some important features of the shape of a drum, such as its area and perimeter [1, 18].

The contributions from the other periodic orbits, the ones of nonzero length, are described by the *Gutzwiller trace formula* [19]. This gives a representation of the semiclassical limit of the density of states as the sum of a smooth term—the average density of states—plus fluctuations which are related to the finite-length periodic orbits of the associated classical mechanics,

$$d(E) \approx \langle d(E)\rangle + \frac{2}{\hbar^{1+\eta}}\sum_p \sum_{k=1}^{\infty} A_{p,k}\cos\frac{k}{\hbar}(S_p + \mu_p), \qquad (12.11)$$

where $\eta = 0$ for systems which are classically chaotic, and $\eta = \frac{1}{2}(N-1)$ for systems which are classically integrable. It is this equation which forms the basis of our semiclassical theories of energy levels.

In (12.11), the periodic orbits have been divided into those orbits, labelled by p, which are not repetitions of any shorter orbit—the *primitive periodic orbits*—and their repetitions k. For chaotic systems, the contribution from a given orbit has an amplitude $A_{p,k}$, which is a classical quantity related to the period of the primitive orbit T_p, and the Poincaré map $\boldsymbol{M}_p$ linearized around it:

$$A_{p,k} = \frac{T_p}{2\pi[\det(\boldsymbol{M}_p^k - \boldsymbol{I})]^{1/2}}. \qquad (12.12)$$

Here the determinant depends only upon the Lyapunov exponents of the orbit and so this amplitude is directly related to its stability. For classically integrable systems an analogous formula for the amplitudes exists [20]. The phase of each periodic orbit contribution is the sum of its action S_p, defined as in (12.9) and its Maslov index μ_p, which will not be defined here because it plays no explicit role in the following story [21]. In fact, for simplicity we will henceforth incorporate μ_p into the symbol S_p.

Hence, the semiclassical limit of the density of states is the sum of its local average and a series of oscillations coming from all of the classical periodic orbits. Each such orbit gives a cosine wave, with an amplitude related to its stability and a wavelength given by Planck's constant divided by its period (i.e. it describes an oscillatory clustering of quantum energy levels on this scale). Now the mean separation between neighbouring levels is $1/\langle d(E)\rangle$, which is of the order of $\hbar^N$, where N is the number of degrees of freedom of the system. If N is greater than 1, then the fine structure of the spectrum is determined by orbits which have periods of the order of $\hbar^{1-N}$, which tends to infinity in the semiclassical limit. Thus the details of the individual levels are related to asymptotically long periodic orbits. The shortest orbits describe spectral clustering over a vast number of levels, indeed over a number which tends to infinity as $\hbar \to 0$. Thus, there are two important energy scales [17]: an *inner scale*, given by the mean level spacing and an *outer scale*, given by the longest scale of oscillatory spectral clustering, and hence related to the shortest periodic orbit.

In order to use the Gutzwiller trace formula in the investigation of spectral properties, we need information about the underlying classical mechanics. For integrable systems this is not too difficult to obtain, allowing the relationship between eigenvalues and periodic orbits to be studied in some detail [20]. As an example we will show the way in which this relationship works for an integrable system of the type considered earlier: a particle bouncing around in a square enclosure with, for simplicity, periodic boundary conditions. The quantum energy levels of this system are, when suitably scaled, simply sums of squares of integers:

$$E_{m,n} = m^2 + n^2. \tag{12.13}$$

Instead of considering the density of states, it is numerically more convenient to study the *spectral staircase*,

$$N(E) = \int_0^E d(x)\,\mathrm{d}x, \tag{12.14}$$

which gives the number of levels with energy less than E; that is, $N(E)$ has a step of unit height at each energy level. Substituting the trace formula (12.11) into (12.14) gives a representation of the spectral staircase in terms of the classical periodic orbits. Now, in the present example the sum over repetitions k can be performed analytically [22], leaving just the sum over the primitive periodic orbits. The result of using only the contributions from the three shortest primitive periodic orbits is shown in Fig. 12.7 for a range in which there are two energy levels, each eightfold degenerate ($40 = 6^2 + 2^2$ and $41 = 5^2 + 4^2$). It may be seen that these contributions are of a completely different form from the steps that we are seeking. They have discontinuities of slope, but of the wrong type; in fact, they are square-root singularities rather than steps. Not only that, they also lie at the wrong places. Consequently they are *false singularities*.

Hence, a given energy level is not associated with any one particular primitive periodic orbit. In fact, each level is determined by the whole set

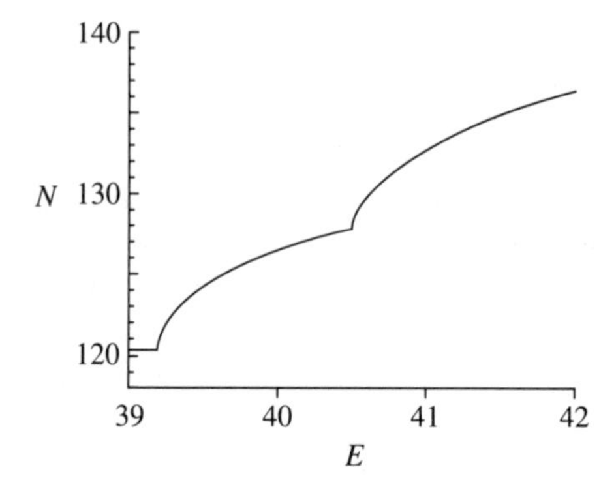

Fig. 12.7. The contribution from the three shortest primitive periodic orbits and all of their repetitions to the spectral staircase $N(E)$ of the square billiard, obtained by combining (11) and (14) [22].

of periodic orbits. In Fig. 12.8 it may be seen that adding in a further 247 orbit contributions gives a reasonable approximation to the correct staircase. Thus the steps of $N(E)$ are somehow born out of a seemingly miraculous conspiracy between the false singularities.

So for integrable systems the Gutzwiller formula can actually be used to obtain semiclassical approximations to the energy levels, and these turn out to be equivalent to the familiar Bohr–Sommerfeld quantization formula [21]. But for chaotic systems there are a number of fundamental difficulties

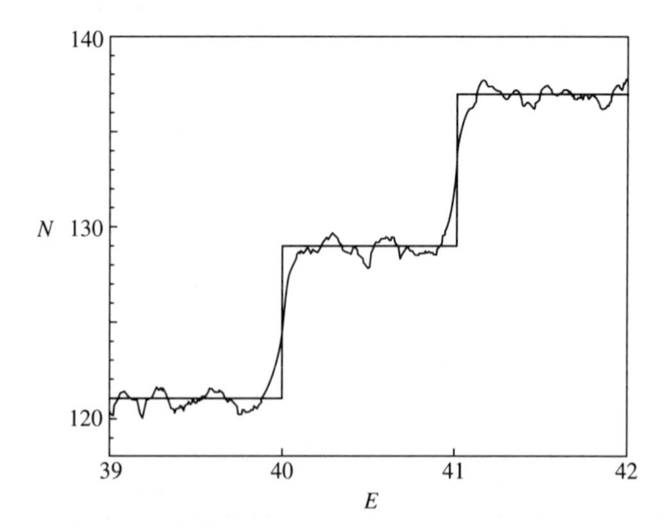

Fig. 12.8. The contribution from the 250 shortest primitive periodic orbits and all of their repetitions to the spectral staircase $N(E)$ of the square billiard, obtained by combining (11) and (14) [22]. The exact staircase is also shown.

which have so far hindered similar calculations. First, if the trace formula is to be used to calculate energy levels, then we need to know the actions, periods, and stabilities of all of the classical periodic orbits, especially the long ones which determine the fine structure of the spectrum. This presents a major difficulty, because in chaotic systems the number of these orbits proliferates exponentially with period, and they are particularly difficult to classify. However, this is not the most fundamental problem. Even if we did know all of the periodic orbits, even if we knew the action, period, and stability of every one of them, we still could not use the formula, because it does not converge; simply performing the sum directly would give $d(E) = \infty$. What is required for a sensible answer to be obtained is some clever method of *resummation*. Furthermore, even if there were such a resummation scheme available, it is not obvious what the δ-functions obtained would correspond to, because the trace formula is just a semi-classical approximation, and if $N > 1$ then it turns out that the error in making the approximation is at least of the same order as the mean separation between neighbouring levels.

So, at present, there are important difficulties which restrict the use of the trace formula in actually calculating energy levels. It turns out, however, that it can be used to calculate the form of their statistics and to show how universality arises. In fact, the universality in quantum statistics is directly related to universality in certain properties of long classical periodic orbits. For classically integrable systems this important idea has been used to show that generically the energy levels really are Poisson distributed in the semiclassical limit [23]. For classically chaotic systems no such general theory exists. However, a semiclassical theory for the number variance $\Sigma(L)$ has been developed by Berry [24], and we will now review the central ideas on which it is based.

12.4 Berry's semiclassical theory for $\Sigma(L)$

The number variance can be written in terms of the density of states

$$\Sigma(L) = \left\langle \left(\int_{E}^{E+L/\langle d \rangle} d(E')\,\mathrm{d}E' - L \right)^2 \right\rangle_E , \tag{12.15}$$

where the average is over an energy range which is much greater than the outer scale defined above. The semiclassical form of $\Sigma(L)$ (i.e. the number variance of high-lying energy levels) is then obtained by plugging in Gutzwiller's formula (12.5) for $d(E)$. This gives a relationship between the number variance and the classical periodic orbits. The result is that $\Sigma(L)$ may be written in terms of what is essentially the Fourier transform of the *spectral form factor* $K(\tau)$,

$$\Sigma(L) = \frac{2}{\pi^2}\int_0^\infty \frac{K(\tau)}{\tau^2}\sin^2\pi L\tau\,\mathrm{d}\tau, \tag{12.16}$$

where $K(\tau)$ is given in terms of a sum over pairs of classical periodic orbits,

$$K(\tau) \approx \frac{4\pi^2}{h\langle d\rangle\hbar^{2\eta}}\left\langle\sum_{i,j} A_i(E)A_j(E)\exp\left(\frac{\mathrm{i}}{\hbar}[S_i(E) - S_j(E)]\right)\right.$$
$$\left.\delta(T - \tfrac{1}{2}[T_i(E) + T_j(E)])\right\rangle_E, \tag{12.17}$$

with T related to the quantum scaled time τ via

$$\tau \equiv T/h\langle d\rangle. \tag{12.18}$$

The question is: how are we to evaluate the formula for $K(\tau)$?

The actions which appear in the exponent in (12.17) depend on E, which is being averaged over. Now we may expect that, as $\hbar \to 0$, the average will wash away all off-diagonal terms in the double sum, since these are then very rapidly oscillating cosine waves. Thus a good approximation to $K(\tau)$ should be given by simply taking the diagonal elements:

$$K(\tau) \sim K_{\mathrm{D}}(\tau) = \frac{4\pi^2}{h\langle d\rangle\hbar^{2\eta}}\sum_j A_j^2\delta(T - T_j). \tag{12.19}$$

Fortunately, we do have enough information about periodic orbits to evaluate this diagonal approximation. This is the classical information which allows us to understand eigenvalue statistics like $\Sigma(L)$. It takes the form of a remarkable and very general *classical sum rule*, discovered by Hannay and Ozorio de Almeida [25]. This result concerns the sum of the amplitudes of all of the periodic orbits with a given period T, in the limit as $T \to \infty$; that is, it concerns the asymptotically long orbits, which we know determine the important fine details of the spectrum. It turns out that these long orbits exhibit *classical universality*: typically, for all chaotic systems

$$\sum_j A_j^2\delta(T - T_j) \approx \frac{T}{4\pi^2}, \tag{12.20}$$

and for all classically integrable systems

$$\sum_j A_j^2\delta(T - T_j) \approx \frac{1}{(2\pi)^{1+N}}\frac{\mathrm{d}\Omega(E)}{\mathrm{d}E}, \tag{12.21}$$

as $T \to \infty$. This is, to say again, universal information about the long periodic orbits, and it is this property which ensures that the number variance of quantum levels also exhibit universality.

Where did this information come from? Well, only orbits with period T contribute to the sum. The sum rule follows from the fact that these orbits, when given the appropriate weighting, are uniformly dense in phase space when T is large; and the appropriate weighting is just the square of their amplitude A_j, earlier seen to be related to their stability. This uniform density in phase space is consequence of their *ergodicity* and this is all that is needed to obtain the universal classical behaviour described by (12.20) and (12:21).

The classical sum rule allows us to evaluate directly the diagonal approximation to $K(\tau)$. For simplicity, we will concentrate upon the results for systems which are either classically integrable, or classically chaotic and not time-reversal symmetric. These are that

$$K_{\rm D}(\tau) \approx 1, \tag{12.22}$$

and

$$K_{\rm D}(\tau) \approx \tau, \tag{12.23}$$

respectively. Is this then all that we need to understand $\Sigma(L)$? Does this alone explain the actual form of the statistic? Well no it doesn't, unfortunately. Simply substituting (12.20) into (12.16) gives a divergent integral. We need one more piece of information: the *semiclassical sum rule* [24].

The semiclassical sum rule is based on the following idea. Take the Dirac δ-functions which sit on the energy levels in the density of states and fatten them up a little, but not too much. In fact, the required fattening is so small that adjacent fattened δ-functions do not overlap. Semiclassically this is a very small fattening, because their mean separation is $1/\langle d(E)\rangle$, which tends to zero as $\hbar \to 0$. If it is sufficiently small, however, then the square of the fattened density of states is just the sum of the squares of similarly fattened δ-functions. Thus, if a Gaussian fattening with width $\varepsilon \ll 1/\langle d(E)\rangle$ is employed, then

$$\left[\sum_j \frac{1}{\varepsilon} \exp\left(-\frac{\pi}{\varepsilon^2}(E-E_j)^2\right)\right]^2 \approx \sum_j \frac{1}{\varepsilon^2} \exp\left(-\frac{2\pi}{\varepsilon^2}(E-E_j)^2\right) \tag{12.24}$$

This equation can be used to obtain a new formula for the average density of states, which we earlier saw was related to the periodic orbits of zero length, in terms of the periodic orbits of finite length:

$$\langle d(E)\rangle \approx \lim_{\varepsilon\to 0} \frac{\varepsilon\sqrt{2}}{\hbar^{2+2\eta}} \left\langle \sum_{i,j} A_i A_j \exp\left(\frac{\mathrm{i}}{\hbar}(S_i - S_j)\right) \exp\left(-\frac{\varepsilon^2}{4\pi\hbar^2}(T_i^2 + T_j^2)\right)\right\rangle_E . \tag{12.25}$$

Now, in this expression any number of short orbits can be neglected without affecting the result of taking the limit (because of the ε in front), and so it in fact provides a connection between the average density of states and the asymptotically long periodic orbits, the ones responsible for the fine details of the spectrum and hence for the individual δ-functions in $d(E)$.

Thus (12.25) represents a consistency condition on the long periodic orbits which is necessary to ensure that the δ-functions to which they give rise have the correct mean density.

That such a relationship must exist may be seen from Gutzwiller's trace formula (12.11). This represents the series of δ-functions as a smooth term plus contributions from all of the periodic orbits. The result of neglecting the smooth term and any finite number of short periodic orbit contributions will thus be a series of δ-functions with a smooth (and bounded) background. But from these δ-functions it would then be possible to calculate directly their mean density and their long-range correlation properties, which, as we have seen, are related to the neglected short periodic orbits. Thus somehow a knowledge of the short periodic orbits, including those of zero length, must be encoded in the long periodic orbits. This is called *bootstrapping* [24]. Late terms in the trace formula determine the contribution of the earlier terms, and (12.25) is an explicit example of this fact. This important concept will reappear later in the chapter.

It now follows from (12.17) and (12.25) that

$$K(\tau) \approx 1 \quad \text{for all } \tau \gg 1, \tag{12.26}$$

regardless of whether the classical trajectories are chaotic or not. But how does this tie in with the results of the diagonal approximation, (12.22) and (12.23)? The diagonal approximation must be good for small values of τ. For classically integrable systems, it follows from (12.26) that this approximation will also hold for $\tau \gg 1$. For classically chaotic systems, however, the diagonal approximation fails in this range, because the periodic orbits proliferate exponentially with period and so their actions will get so close that, for sufficiently large values of τ, some off-diagonal terms in (12.17) will not be washed away by the average. The effect of these nonnegligible off-diagonal terms is to cancel the diagonal contributions to $K(\tau)$ when $\tau \gg 1$, leaving just the remainder 1. Because of this cancellation, the integral (12.16) now converges.

So the classical sum rule determines the form of $K(\tau)$ when $\tau \ll 1$, and it is determined by the semiclassical sum rule when $\tau \gg 1$. The actual interpolation between these two regimes is not too important, but one very natural interpolation is that for classically integrable systems

$$K(\tau) \approx 1, \tag{12.27}$$

and for classically chaotic systems

$$K(\tau) \approx \begin{cases} \tau & \text{if } \tau \leqslant 1 \\ 1 & \text{if } \tau > 1. \end{cases} \tag{12.28}$$

Well, it is certainly natural in retrospect, because the form factors thus obtained are the exact form factors of Poisson statistics and of the GUE random matrix statistics, respectively. The corresponding results for $\Sigma(L)$ (i.e. (12.2) and (12.6)) can then be obtained by substituting these form factors into (12.16). Hence, not only does this theory explain how universality arises in the number variance, but it also gives the actual functional forms which are observed. Furthermore, by correctly taking into account the double degeneracy of orbits in chaotic systems which are time-reversal symmetric, the observed GOE form of the number variance can also be explained.

But that is not the end of the story. Real dynamical systems are not random matrices: they do have an underlying classical mechanics. Does this have any effect? Well, the number variance is closely related to the Fourier transform of the form factor $K(\tau)$ (see (12.16)), and so it effectively involves a smoothing over τ. As L is increased, the number variance becomes sensitive to the form of $K(\tau)$ only in the region where τ is small; the bigger L gets, the smaller is the range of values of τ near the origin which determines the form of $\Sigma(L)$. So let us magnify that range. The result for a typical classically chaotic system is shown schematically in Fig. 12.9 (a similar picture can be drawn for classically integrable systems). The form factor $K(\tau)$ is really a series of δ-functions, centred at values of τ which are related, via (12.8), to periods of periodic orbits. There is, of course, a shortest periodic orbit, and this gives rise to the first δ-function. The diagonal approximation gives the *average behaviour* of the form factor for $\tau \ll 1$. However, for sufficiently large values of L, $\Sigma(L)$ will be sensitive to such a small range of τ that the averaging provided by the transform

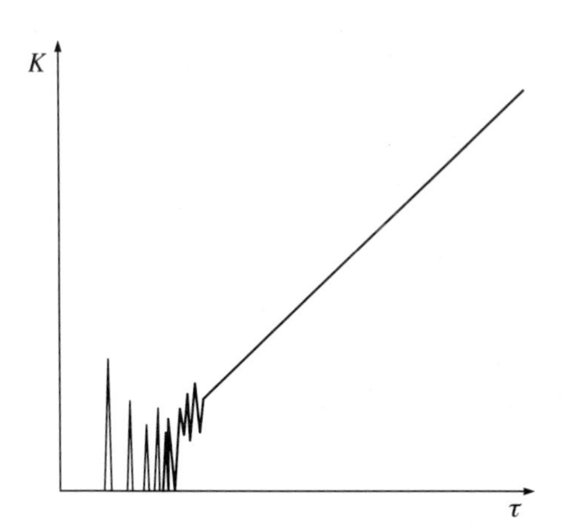

Fig. 12.9. The behaviour near $\tau = 0$ of the spectral form factor $K(\tau)$ for a typical classically chaotic system. The spikes are determined by the periodic orbits. These orbits proliferate exponentially with period, and so the spikes increase in density with τ. Their average contribution may be approximated by the straight line $K(\tau) = \tau$.

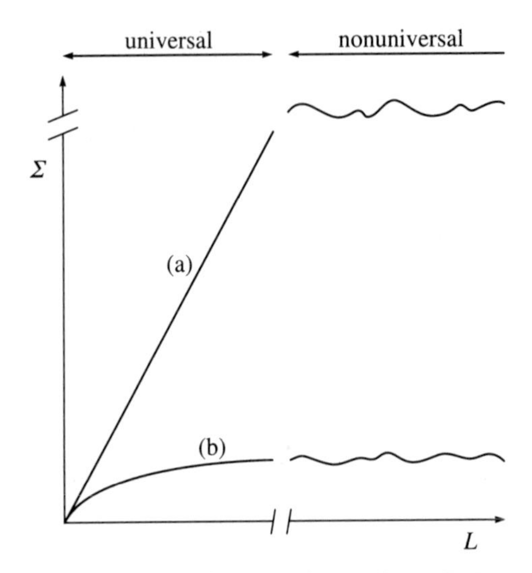

Fig. 12.10. The schematic form of the number variance $\Sigma(L)$) (or equivalently of the spectral rigidity $\Delta(L)$ for (a) a typical classically integrable system, and (b) a typical classically chaotic system. For small values of L the behaviour is universal, but over semiclassically long ranges the statistic exhibits system-dependent oscillations about some constant value.

will be over just the first few δ-functions nearest the origin. We would not then expect the diagonal approximation to give the correct form of $\Sigma(L)$. Instead, simply inserting the first few δ-functions into (12.16) gives a series of oscillations. As L is increased even further, there is no more structure to feel below the first δ-function. The number variance then saturates into oscillatory behaviour determined by the shortest periodic orbits. These short orbits are *not* universal: they depend upon the particular system in question. Thus, for large enough values of L universality breaks down, leaving nonuniversal system-dependent behaviour.

To summarize, the schematic form of the number variance (or equivalently, of the spectral rigidity) is shown in Fig. 12.10. It has a universal range, the same for all systems in a given class, which is related to the very long periodic orbits of the classical mechanics. Eventually over some semiclassically long range there will be a breakdown of this universality into nonuniversal behaviour which is related to the very short periodic orbits of the classical mechanics.

But is this actually observed? We have already seen that there is very good agreement between the energy level statistics of real systems and the universal random matrix forms. In Fig. 12.11 we return to the number variance of the rectangular billiard (or equivalently, the statistics of the eigenfrequencies of a rectangular drum). Earlier it was seen that this was in good agreement with the Poisson form. However, this new figure

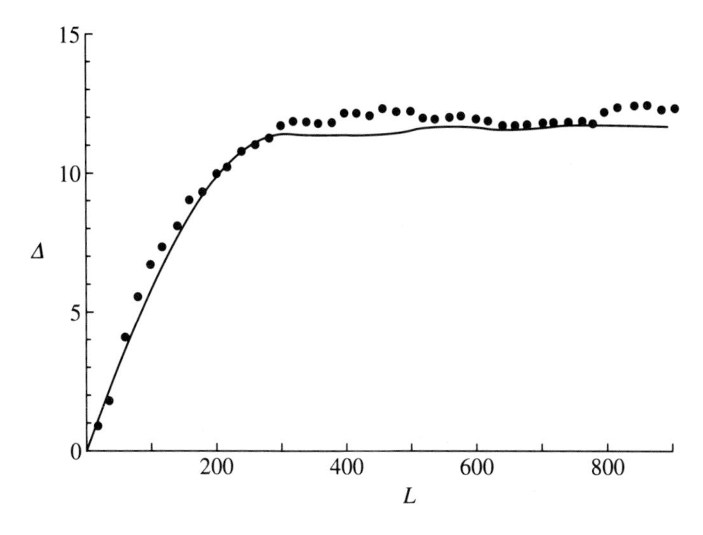

Fig. 12.11. The spectral rigidity $\Delta(L)$ of the energy levels of the rectangular billiard (full circles) as in Fig. 12.1, but over a larger range of values of L (data from [26]). The curve represents Berry's semiclassical theory. (Reproduced from [4], with permisson.)

contains data for a larger range of values of L. The universal form does indeed eventually degenerate into a series of quasi-periodic oscillations. The bold line represents the result of Berry's theory.

So we have seen how universality arises and also how it breaks down. However, Berry's theory is still incomplete. What we would really like to do would be to calculate all these things using only periodic orbits. Specifically, we need to understand the semiclassical sum rule and the bootstrapping relationships between the periodic orbits, using solely classical arguments. We also would like to be able to derive the correct interpolation between the limiting regimes in the form factor. And finally we would, of course, like to understand in greater detail the Gutzwilller formula which underlies semiclassical theory.

Well, in general, these are difficult problems and at present we have no clear idea of how to proceed. One approach is to study in detail particular 'model systems' which can be solved exactly. We will concentrate here on one family of such systems, which are knower as the *Arnol'd cat maps.*

12.5 A chaotic map of a cat

The cat maps are a family of linear maps on the 2-torus (i.e a regular doughnut); each map provides a rule for moving any point on the torus to another such point. The name comes from the fact that their chaotic nature is traditionally illustrated by showing the result of their action on the face of a cat [27]! They may be represented by 2 x 2 matrices with integer entries (this ensures that the mappings are continuous on the torus), unit determinant (this ensures that they are area-preserving), and a trace with a modulus greater than 2 (this ensures that the maps are completely chaotic—in fact, they are *Anosov* systems [27]). The coordinates on the torus can be chosen so as to have periodicity 1 in both directions. Furthermore they may also be considered as the position q and momentum p of the system; that is, the torus represents a *phase space*. One example of such a mapping is

$$\begin{bmatrix} q' \\ p' \end{bmatrix} = \begin{bmatrix} 2 & 1 \\ 3 & 2 \end{bmatrix} \begin{bmatrix} q \\ p \end{bmatrix} \quad (\text{mod } 1), \tag{12.29}$$

where 'mod 1' means that only the fractional parts of q' and p' are retained.

Although these systems are completely chaotic, we can actually get a great deal of information about both the classical mechanics and the quantum mechanics because of their strong number-theoretical character. Exact and explicit information about the classical periodic orbits can be obtained using a series of important results from the theory of quadratic number fields [28]. Asymptotic information about the long ones can then be derived by combining these number-theoretical results with methods from probability theory [29]. The properties of the long orbits thus obtained are just those which are needed in order to understand the relationship between the quantum mechanics of the maps and their classical dynamics.

For the quantized maps, Gutzwiller's formula is found to have three very important features [30]. It is exact; that is, it is not simply a semiclassical approximation, but is valid for all values of Planck's constant. It is also explicit, because we have simple formulae for the actions, periods, and stabilities of all of the periodic orbits. And finally, due to some special number-theoretical symmetries, it is a finite sum, so there are no problems with its convergence. As a result of these features, it can actually be used directly to calculate some important quantum properties. In fact, the cat maps are the only classically chaotic systems studied so far for which this is possible. Combining the results with the asymptotic information obtained for the long periodic orbits, it is possible to understand in detail the statistics of the eigenvalues, and in particular their number variance [30].

The result of the analysis is that $\Sigma(L)$ initially increases in a way which is the same for all cat maps. Its predicted form, obtained by combining classical information with the Gutzwiller formula, is found to be in good agreement with the results of numerical computation, as shown in Fig. 12.12. As L increases, we expect this smooth increase to degenerate into a series of quasi-periodic oscillations, and so it does. This is shown in Fig. 12.13.

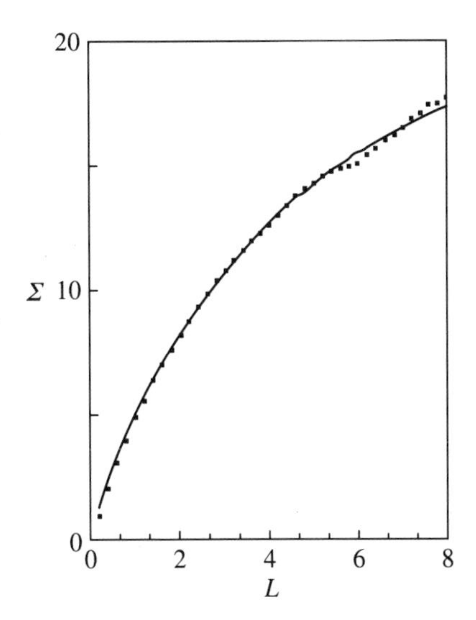

Fig. 12.12. The number variance $\Sigma(L)$ for the quantum analogue of the cat map (29) after averaging over a range of values of Planck's constant (filled squares). The curve represents the best fit to the theoretical form predicted by semiclassical theory [30].

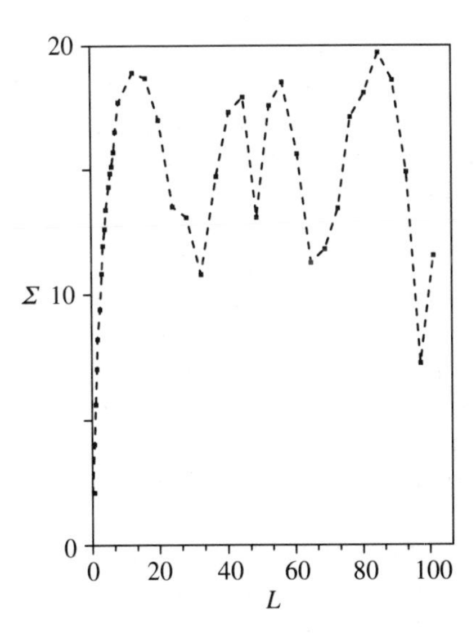

Fig. 12.13. The number variance $\Sigma(L)$ as in Fig. 12.12, but over a larger range of values of L. This shows the transition to the map-dependent oscillations determined by the shortest periodic orbits. Here the dashed line simply connects the points.

So have we solved our problems? Unfortunately not. The actual form of the number variance does not correspond to any of the universal forms listed earlier. The cat maps do not fall into any of the universality classes; they are very special systems, the reason being that they have such a strong number-theoretical character. This is, of course, the reason why we could do the calculations in the first place. It is the reason why the periodic orbit sum could be performed directly. But it means that the cat maps do not give any real clues as to how these sums may be dealt with in general.

Another set of completely chaotic systems which have been much investigated correspond to free motion on surfaces of constant negative curvature [31, 32]. (These surfaces are very difficult to visualize: they are the opposite of spheres, which are surfaces of constant positive curvature.) Here again, Gutzwiller's formula is exact (i.e. it is not just a semiclassical approximation), but it is not explicit (there is no formula for the periodic orbits, which have to be found on the computer) and it does diverge. But, here again, it has been found that the statistics of the energy levels are not typical, and this may also be linked to number-theoretical symmetries.

There is another way in which these systems are not typical: the connection between their quantum energy levels and classical periodic orbits does not go under the name 'Gutzwiller's formula'. It has a different name. It is called the *Selberg trace formula* [32], because it was first written down by the mathematician Atle Selberg more than ten years before physicists became interested in quantizing chaotic motion. Selberg was interested in this subject for an entirely different reason: he was trying to solve one of the most important problems in pure mathematics—the Riemann hypothesis—and this brings me to the last model system that we will discuss here, one which is intimately related to that hypothesis. Indeed, to call it a 'system' is really wishful thinking, because it can only exist if the hypothesis is correct!

12.6 Quantum chaology and the primes

Riemann's zeta-function is a function of a complex variable s [33]. It may be defined in terms of a sum over the positive integers,

$$\zeta(s) = \sum_{n=1}^{\infty} \frac{1}{n^s}, \tag{12.30}$$

which converges only if the real part of s is greater than 1. It can also be written as a product over all of the prime numbers p,

$$\zeta(s) = \prod_{p} \left(1 - \frac{1}{p^s}\right)^{-1}, \tag{12.31}$$

which again converges only if the real part of s is greater than 1. This second representation is the reason why the zeta-function is so important: its analytical properties are related to the distribution of the prime numbers. Specifically, the distribution of the primes is related to the distribution of the zeros of $\zeta(s)$.

How are these zeros distributed? Well, they are isolated points in the complex plane, and it is known that a particularly important set of them, the so-called *nontrivial zeros*, all have real parts between 0 and 1. The Riemann hypothesis is that they actually all lie on the line with real part $\frac{1}{2}$. That is, if E_n is a nontrivial solution of

$$\zeta(\tfrac{1}{2} + \mathrm{i}E) = 0, \tag{12.32}$$

then $\mathrm{Im}\, E_n = 0$. It was noted by Hilbert and Pólya in the early part of this century that the hypothesis would be proved if the E_n could be shown to be the eigenvalues of a Hermitian operator. The connection with quantum mechanics comes if we ask the question: could the E_n in fact be the energy levels of some quantum system which has a classical limit?

The density of states function $d(E)$ may be defined for the E_n exactly as in (12.7). Using (12.31) it may then be shown that $d(E)$ can also be written in the form [24, 34, 40]

$$d(E) = \langle d(E) \rangle - \frac{1}{\pi} \sum_p \sum_{k=1}^{\infty} \frac{\ln p}{p^{k/2}} \cos(Ek \ln p), \tag{12.33}$$

where $\langle d(E) \rangle$ is the average density of zeros, for which it is known that

$$\langle d(E) \rangle \approx \frac{1}{2\pi} \ln\left(\frac{E}{2\pi}\right) \quad \text{as } E \to \infty. \tag{12.34}$$

Now equation (12.33) is very similar in form to the trace formula (12.11) if the primes are taken to label primitive periodic orbits, the integers k their repetitions, and if we make the following identifications:

$$S_{p,k} = Ek \ln p, \tag{12.35}$$

$$T_{p,k} = \frac{\mathrm{d}S_{p,k}}{\mathrm{d}E} = k \ln p, \tag{12.36}$$

$$A_{p,k} = \frac{\ln p}{2\pi p^{k/2}}, \tag{12.37}$$

and $\hbar = 1$. In this case the density of states sum is exact, explicit (because we have formulae for the actions, periods, and stabilities of all of the periodic orbits), and, as in the general case, it does not converge. It diverges because in deriving it we used the product over primes (12.31) in a region where it is known not to converge.

So there is a very close analogy between the formula (12.33), which relates the zeros of the zeta-function to the primes, and Gutzwiller's trace formula, which relates quantum energy levels and classical periodic orbits. How seriously can it be taken? Can there really be a system whose quantum energy levels are the imaginary parts of the zeros of the zeta-function, and whose periodic orbits have periods which are the logarithms of the primes?

Well, if there really is such a system, we would expect it to satisfy the classical sum rule. Using (12.37) for the amplitudes, it is simple to show that the sum rule for classically chaotic systems (12.20) is obeyed, provided that the number of primes less than some value X increases asymptotically as $X/\ln X$. And indeed it does. This is just the prime number theorem, which was proved nearly one hundred years ago. So, if there is a classical system underlying the zeta-function, then it appears to be chaotic.

What about the statistics of the zeros: do they fall into any of the universality classes? Yes, they do. This was the result obtained from an important series of computations performed by Odlyzko [35]. He found a startling agreement between the statistics of the zeros and the analytical forms for the GUE. In Fig. 12.14 this is shown for the nearest-neighbour separation $P(s)$; however, the agreement holds not just for this statistic but for a whole range of others. This was a truly remarkable discovery.

So it really does look as if the nontrivial zeros could be the energy levels of a classical system which is chaotic and which does not have time-reversal symmetry (because the statistics were those of the GUE, and not the GOE). But we have already seen that for a real dynamical system (as opposed to

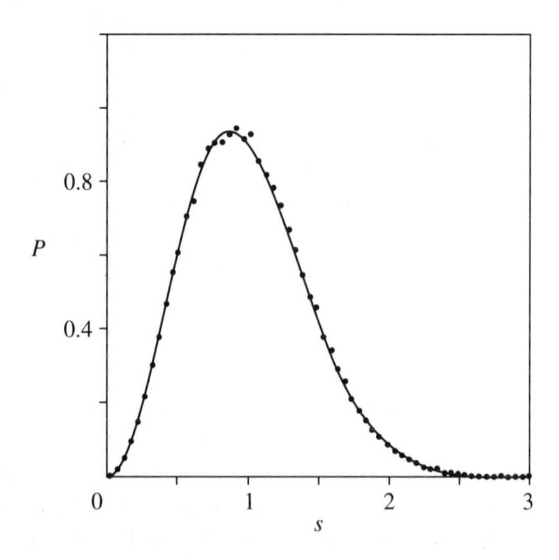

Fig. 12.14. The spacings distribution $P(s)$ calculated using the 10^5 zeros of the Riemann zeta-function nearest the 10^{12}th (filled circles—data from [35]). The curve represents the GUE random matrix form. (Reproduced from [4], with permission.)

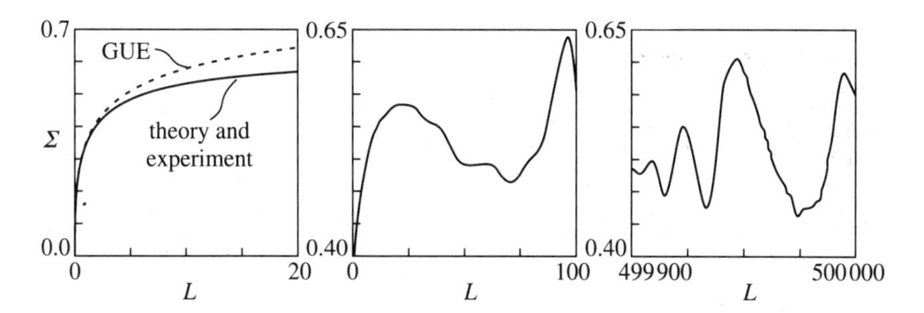

Fig. 12.15. The number variance $\Sigma(L)$ calculated using the 70 million zeros of the Riemann zeta-function nearest the 10^{20}th (data from [36]), with the GUE random matrix form (dashed curve) and the result of Berry's semiclassical theory (full curve). On this scale, differences between the data and the theoretical curve are unresolvable over the whole range. (Reproduced from [15], with permission.)

a randomly picked Hermitian matrix) the long-range statistics, such as the number variance, actually exhibit deviations from the universal forms. And remarkably Odlyzko found this for the zeta-function as well, in a second series of computations [36] which were epic in their proportions (he calculated the statistical properties of the 70 million closest neighbours of the 10^{20}th zero!). The number variance starts off in very close agreement with the GUE form, but eventually degenerates into a series of quasi-periodic oscillations. This is shown in Fig. 12.15. In fact, this figure contains two pieces of information. It shows Odlyzko's data along with Berry's semiclassical formula for the number variance of a classically chaotic system that is not time-reversal symmetric [37]. The two are completely indistinguishable on this scale. This agreement between the statistics of the zeta-function zeros and the semiclassical formula is one of the most exciting results of quantum chaology.

Hence there exists very strong evidence that there is in fact a real dynamical system lurking behind the zeta-function. Or at least there is a very strong analogy. So let's take it seriously and see what we can learn. It turns out that many of the important problems which plague our semiclassical theories can actually be solved for this particular system. Thus the semiclassical sum rule can be derived directly from properties of the periodic orbits—in fact, it is related to long-standing conjectures concerning the pairwise distribution of prime numbers [38] (there is bootstrapping among the primes!) It is also possible to derive the whole form factor, and not just the behaviour in the limiting regimes (i.e. when $\tau \ll 1$ or $\tau \gg 1$), directly from the periodic orbits. The result is that, even in the connecting region, $K(\tau)$ has the GUE form (12.28) [39, 40]. But, most importantly, this mathematical model gives a valuable clue as to how the divergence of the Gutzwiller sum can be tamed. This comes in the form of the Riemann–Siegel formula.

12.7 Bootstrapping and resummation

The sum over the periodic orbits in Gutzwiller's trace formula diverges. However, we know from bootstrapping that the long periodic orbits and the short periodic orbits are related. So maybe the divergent contributions from the long orbits could be resummed by expressing them in terms of the contributions from the short ones. This is the idea underlying the Riemann–Siegel formula [33], which gives an important asymptotic representation of $\zeta(\frac{1}{2} + \mathrm{i}E)$ as $E \to \infty$.

Formally we can write

$$\zeta(\tfrac{1}{2} + \mathrm{i}E) = \sum_{n=1}^{[(E/2\pi)^{1/2}]} \frac{1}{n^{1/2+\mathrm{i}E}} + \sum_{[(E/2\pi)^{1/2}]+1}^{\infty} \frac{1}{n^{1/2+\mathrm{i}E}}, \tag{12.38}$$

where [x] denotes the integer part of x. The second sum diverges. Now the Riemann–Siegel formula implies that

$$\zeta(\tfrac{1}{2} + \mathrm{i}E) \approx \sum_{n=1}^{[(E/2\pi)^{1/2}]} \frac{1}{n^{1/2+\mathrm{i}E}} + \exp\left(-2\pi\mathrm{i}\int_0^E \langle d(x)\rangle \,\mathrm{d}x\right) \sum_{n=1}^{[(E/2\pi)^{1/2}]} \frac{1}{n^{1/2-\mathrm{i}E}} \tag{12.39}$$

as $E \to \infty$. Hence the divergent second sum of (38) may be expressed in terms of the complex conjugate of the finite first sum, the contributions from integers $n > [(E/2\pi)^{1/2}]$ having been expressed in terms of those less than $([E/2\pi)^{1/2}]$.

So bootstrapping can be used as a basis for resummation in the hypothetical Riemann zeta system. But is there a Riemann–Siegel formula for all quantum systems which are classically chaotic [34]? Well, we can define a function $\Delta(E)$, called the *spectral determinant*, which is zero when E is equal to one of the energy levels of the system. This function may obviously be written in terms of the density of states $d(E)$, and hence, using Gutzwiller's formula, it can be expressed semiclassically as a sum of contributions from periodic orbits, the form of which can be shown to be closely analogous to the divergent sum (12.38). From this representation it is then possible to conjecture, on the grounds of a number of physical and mathematical arguments, a natural resummation which is exactly analogous to the Riemann–Siegel formula (12.39) [41], that is, a semiclassical representation in terms of a finite number of periodic orbits [42].

Recently, the conjectured formula underwent some preliminary tests using a system which corresponds to a billiard particle trapped between the x-axis, the y-axis, and the line $y = 1/x$. For this system a sufficient number of periodic orbits are known to allow $\Delta(E)$ to be calculated over a particular energy range which contains about 10 energy levels. The places where $\Delta(E) = 0$ were found to be in good agreement with the actual values of these levels, obtained by solving Schrödinger's equation [43]. This is a very good example of how, at present, it seems that number theory, and especially the Riemann zeta-function, is directing our research in chaology.

Well, what hopefully has been explained here is that typically you can indeed hear the chaology of a drum. All you have to do is to find the statistics of its eigenfrequencies, and these then tell you whether the underlying rays have a chaotic dynamics or not. What also has been shown is that in trying to understand why this is the case, we are led directly to some remarkably deep problems in classical mechanics, quantum mechanics, and also the theory of numbers. Who knows, perhaps one day some musician might even tap out a proof of the Riemann hypothesis!

12.8 Author's note

Subsequent to the lecture on which this chapter was based, there have been a number of significant advances concerning the use of Guzwiller's periodic orbit formula to obtain semiclassical approximations for the energy levels of classically chaotic systems. In particular, the conjectured generalization of the Riemann–Siegel formula discussed in the last section has since been derived [44, 40] and extended [45] to yield a complete asymptotic representation for the spectral determinant which uses only a finite number of periodic orbits. Furthermore, this method of derivation can, in turn, be applied to the zeta-function itself, giving a new asymptotic expansion which actually turns out to be more efficient than the original Riemann–Siegel formula [40, 45]. So, in sense, we have now begun to repay the important inspiration which the zeta-function provided for quantum chaology.

It now also turns out that some of these results can be obtained from entirely different approaches [46, 47]. Many of them have been tested numerically and found to yield significant improvements in the efficiency of the periodic orbit formula [48]. However, in a number of cases these developments actually lead to more questions than they answer and it seems that we are still very far from a full understanding of the semiclassical properties of quantum energy levels.

Notes and References

1. Kac, M. (1966). Can one hear the shape of a drum? *Am. Math. Monthly*, **73**(4), part II, 1–23.
2. For a more general description of the connection between the quantum mechanics of chaotic systems and the theory of numbers, see Keating, J. P. (1990).Physics and the queen of mathematics. *Physics World*, **3**(4), 46–50.
3. Technically 'ray theory' means 'Hamiltonian dynamical system'. The general relationship between waves and rays is dealt with in Synge, J. L. (1954).

Geometrical mechanics and de Broglie waves, Cambridge Monographs on Mechanics and Applied Mathematics. Cambridge University Press.
4. Berry, M. V. (1987). Quantum chaology. *Proc. R. Soc. Lond.* A, **413**, 183–98.
5. For a general review of the chaotic properties of time-dependent quantum systems, see the lectures given by B. Chirikov at the Les Houches school on Chaos and Quntum Physics, session no. 52, August 1989. Chirikov, B. (1990). In *Les Houches Lecture Series 52* (eds M. J. Giannoni, A. Voros, and J. Zinn-Justin), pp. 443–545. North-Holland, Amsterdam.
6. For a general review of energy-level statistics, see Bohigas, O., Giannoni, M. J., and Schmit, C. (1986). Spectral fluctuations of classically chaotic systems. In: *Quantum chaos and statistical nuclear physics*, lecture notes in physics 263 (ed. T. H. Seligman and H. Nishioka), pp. 18–40. Springer, Berlin.
7. For an example of a system exhibiting these statistics, see Scharf, R., Dietz, B., Kús, M., Haake, F., and Berry, M. V. (1988). Kramers' degeneracy and quartic level repulsion. *Europhys. Lett.*, **5**, 383–9.
8. Porter, C. E. (1965). *Statistical theories of spectra: Fluctuations*. Academic Press, New York. Mehta, M. L. (1991). *Random matrices*, 2nd edn. Academic Press, New York.
9. Stockmann, H. J. and Stein, J. (1990). 'Quantum' chaos in billiards studied by microwave absorption. *Phys. Rev. Lett.*, **64**, 2215–18.
10. $\Delta(L)$ is, like $\Sigma(L)$, a simple bilinear quadratic statistic. If $N(x)$ is the number of scaled eigenvalues less than the scaled energy x, then $\Delta(L)$ is the mean-squared deviation of $N(x)$ from the straight line which best fits it over a length L:

$$\Delta(L) = \left\langle \min_{A,B} \frac{1}{L} \int_{-L/2}^{L/2} [N(x+X) - Ax - B]^2 \, dx \right\rangle_x .$$

11. Bohigas, O., Giannoni, M. J., and Schmit, C. (1984). Spectral properties of the laplacian and random matrix theories. *J. Phys. Lett.*, **45**, L1015–22.
12. Bohigas, O., Giannoni, M. J., and Schmit, C. (1984). Characterization of chaotic quantum spectra and universality of level fluctuation laws. *Phys. Rev. Lett.*, **52**, 1–4.
13. The boundary of the 'Africa' billiard is the conformal transformation of the unit circle generated by $w(z) = z + 0.2z^2 + 0.2e^{i\pi/3}z^3$.
14. Berry, M. V. and Robnik, M. (1986). Statistics of energy levels without time reversal symmetry: Aharonov–Bohm chaotic billiards. *J. Phys.* A, **19**, 649–68.
15. The importance of the role played by asymptotics in these investigations is discussed in Berry, M. V. (1990). Some quantum-to-classical asymptotics. In *Les Houches Lecture Series 52* (eds M. J. Giannoni, A. Voros, and J. Zinn-Justin), pp. 251–303. North-Holland, Amsterdam.
16. Van Vleck, J. H. (1928). The correspondence principle in the statistical interpretation of quantum mechanics. *Proc. Natl. Acad. Sci. USA*, **14**, 178–88.
17. Berry, M. V. (1983). Semiclassical mechanics of regular and irregular motion. In *Chaotic behaviour of deterministic systems*, Les Houches Lectures 36 (ed. G. Iooss, R. G. H. Helleman, and R. Stora), pp. 171–271. North-Holland, Amsterdam.
18. Berry, M. V. (1987). Improved eigenvalue sums for inferring quantum billiard geometry. *J. Phys.* A, **20**, 2389–2403 (1987).

19. Gutzwiller, M. C. (1971). Periodic orbits and classical quantization conditions. *J. Math. Phys.*, **12**, 343–58. For a review of trace formulae in general, see Littlejohn, R. G. (1990). Semiclassical structure of trace formulas. Preprint from the Theoretical Physics Institute, University of Minnesota. See also Gutzwiller, M. C. (1991). The semi-classical quantization of chaotic Hamiltonian systems. In *Les Houches Lecture Series 52* (eds. M. J. Giannoni, A. Voros, and J. Zinn-Justin), pp. 201–49. North-Holland, Amsterdam.
20. Berry, M. V. and Tabor, M. (1976). Closed orbits and the regular bound spectrum. *Proc. R. Soc.* A, **349**, 101–23.
21. Maslov, V. P. and Fedoriuk, M. V. (1981). *Semiclassical approximation in quantum mechanics*. Reidel, Dordrecht. For recent results concerning Maslov indices, see Robbins, J. M. (1991). Maslov indices in the Gutzwiller trace formula. *Nonlinearity*, **4**, 343–63.
22. Keating, J. P. and Berry, M. V. (1987). False singularities in partial sums over closed orbits. *J. Phys.* A, **20**, L1139–41.
23. Berry, M. V. and Tabor, M. (1977). Level clustering in the regular spectrum. *Proc. R. Soc. Lond.* A, **356**, 375–94.
24. Berry, M. V. (1985). Semiclassical theory of spectral rigidity. *Proc. R. Soc. Lond.* A, **400**, 229–51.
25. Hannay, J. H. and Ozorio de Almeida, A. M. (1984). Periodic orbits and a correlation function for the density of states. *J. Phys.* A, **17**, 3429–40.
26. Casati, G., Chirikov, B. V., and Guarneri, I. (1985). Energy level statistics of integrable quantum systems. *Phys. Rev. Lett.*, **54**, 1350–53.
27. Arnol'd, V. I. and Avez, A. (1968). *Ergodic problems of classical mechanics*. Benjamin, New York.
28. Percival, I. C. and Vivaldi, F. (1987). Arithmetical properties of strongly chaotic motions. *Physica* D, **25**, 105–30.
29. Keating, J. P. (1991). Asymptotic properties of the periodic orbits of the cat maps. *Nonlinearity*, **4**, 277–307.
30. Keating, J. P. (1991). The cat maps: Quantum mechanics and classical motion. *Nonlinearity*, **4**, 309–41.
31. Aurich, R., Sieber, M., and Steiner, F. (1988). Quantum chaos of the Hadamard–Gutzwiller model. *Phys. Rev. Lett.*, **61**, 483–7.
32. Balazs, N. L. and Voros, A. (1986). Chaos on the pseudosphere. *Phys. Rep.*, **143**, 109–240.
33. Properties of the Riemann zeta-function are reviewed in, for example, Edwards, H. M. (1974). *Riemann's zeta function*. Academic Press, New York. Titchmarsh, E. C. and Heath-Brown, D. R. (1986). *The theory of the Riemann zeta-function*. Oxford University Press.
34. Berry, M. V. (1986). Riemann's zeta function: A model for quantum chaos? In: *Quantum chaos and statistical nuclear physics*, Lecture Notes in Physics 263 (ed. T. H. Seligman and H. Nishioka), pp. 1–17. Springer, Berlin.
35. Odlyzko, A. M. (1987). On the distribution of spacings between zeros and the zeta function. *Math. Comp.*, **48**(177), 273–308.
36. Odlyzko, A. M. (1989). The 10^{20}-th zero of the Riemann zeta function and 70 million of its neighbours. Preprint from AT&T Bell Labs.
37. Berry, M. V. (1988). Semiclassical formula for the number variance of the Riemann zeros. *Nonlinearity*, **1**, 399–407.
38. Keating, J. P. (1991). The semiclassical sum rule and Riemann's zeta function.

In *quantum chaos* (eds H. Cerdeira and R. Ramaswamy). World Scientific, Singapore.

39. Montgomery, H. L. (1973). The pair correlation of the zeros of the zeta function. *Proc. Symp. Pure Math.*, **24**, 181–93.
40. Keating, J. P. (1990). The Riemann zeta-function and quantum chaology, lectures given at the International School of Physics 'Enrico Fermi', course CXIX (Quantum Chaos), summer 1991. Proceedings edited by G. Casati, I. Guarneri, and U. Smilansky. (In press.)
41. Berry, M. V. and Keating, J. P. (1990). A rule for quantizing chaos? *J. Phys.* A, **23**, 4839–49.
42. Different resummation techniques from the one described here have been proposed for systems which admit a complete symbolic-dynamical description. See, for example, Gutzwiller, M. C. (1982). Quantization of a classically ergodic system. *Physica* D, **5**, 183–207. Cvitanovic, P. and Eckhardt, B. (1989). Periodic-orbit quantization of chaotic systems. *Phys. Rev. Lett.*, **63**, 823–6.
43. Sieber, M. (1990). Personal communication.
44. Keating, J. P. (1992). Periodic orbit resummation and the quantization of chaos. *Proc. Roy. Soc. Lond. A*, **436**, 99–108.
45. Berry, M. V. and Keating, J. P. (1992). A new asymptotic representation for $\zeta(\frac{1}{2} + it)$ and quantum spectral determinants. *Proc. Roy. Soc. Lond. A*, **437**, 151–73.
46. Bogomolny, E. B. (1992). Semiclassical quantization of multidimensional systems. *Nonlinearity*, (in press).
47. Doron, E. and Smilansky, U. (1992). Semiclassical quantization of chaotic billiards—a scattering theory approach. *Nonlinearity*, (in press).
48. For a general review of recent developments, see the focus issue on periodic orbit theory: *Chaos*, **2**(1), (1992).

Index